T0231880

Biostatistics

A Computing Approach

Chapman & Hall/CRC Biostatistics Series

Chapman & Hall/CRC Biostatistics Series

Chapman & Hall/CRC Biostatistics Series

Biostatistics
A Computing Approach

Stewart Anderson, PhD, MA

Professor, Department of Biostatistics
University of Pittsburgh Graduate School of Public Health
Pennsylvania, USA

CRC Press
Taylor & Francis Group
Boca Raton London New York

CRC Press is an imprint of the
Taylor & Francis Group, an **informa** business

A CHAPMAN & HALL BOOK

CRC Press
Taylor & Francis Group
6000 Broken Sound Parkway NW, Suite 300
Boca Raton, FL 33487-2742

© 2012 by Taylor & Francis Group, LLC
CRC Press is an imprint of Taylor & Francis Group, an Informa business

No claim to original U.S. Government works

Version Date: 20111101

International Standard Book Number: 978-1-58488-834-5 (Hardback)

Visit the Taylor & Francis Web site at
http://www.taylorandfrancis.com

and the CRC Press Web site at
http://www.crcpress.com

Contents

Preface

For Deb, Joey and Naomi with all my love

Introduction

The nature of statistical science has been altered drastically in the last few years due to the advent of easy access to high speed personal computers and the wide availability of powerful and well–designed software. In the past, much of the focus of statistical science was understanding both the very large sample (asymptotic) behavior and the very small sample behavior of test statistics. Today, we have the computational tools to investigate how statistical modeling and testing work when the sample sizes are in a "moderate range." Many real-world applications involve these moderately sized datasets.

At the other end of the spectrum, we now have datasets that are extremely large but the complexity is such that large sample behavior may be difficult to describe mathematically. Modern computating and graphical tools facilitate the identification of patterns in these complex datasets that can be helpful in the initiation of understanding their mathematical behavior. Consequently, the emergence of high speed computing has facilitated the development of many exciting statistical and mathematical methods in the last 25 years. These methods have greatly broadened the landscape of available tools in statistical investigations of complex data.

The purpose of this book is to introduce the reader who has had a little background in statistics to both classical and modern methods currently being used in statistics. Many of the classical methods are typically taught in a basic course in statistics and are still widely used in practice. However, with the amount of data available and the wide availability of powerful software, newer methods are now being employed more frequently for applications in medicine and other fields. The focus here will be on visualization and computational approaches associated with both classical and modern techniques. Furthermore, since the computer often serves as the laboratory for understanding statistical and mathematical concepts, we will focus on computing as a tool for doing both analyses and simulations which, hopefully, can facilitate such understanding. As a practical matter, programs in R (R core team, 2010) [117],

S-plus [143], and SAS [128] are presented throughout the text. (SAS and all other SAS Institute Inc. product or service names are registered trademarks or trademarks of SAS Institute Inc. in the USA and other countries. ® indicates USA registration.) In addition to these programs, appendices describing the basic use of SAS and R are provided. The appendices are admittedly somewhat superficial but do facilitate a rapid introduction of these packages to the new user. That being said, it is not my intention to provide a comprehensive overview of these packages. Luckily, there is a wealth of SAS, R and S-plus tutorials online. In particular, SAS provides extensive online support (see, for example, http://support.sas.com/documentation/onlinedoc/ code.samples.html). Also, there are a number of excellent books available on the use of SAS (Delwiche and Slaughter [27], Der and Everitt [29], Wilkin [158]); R (Dalgaard [26], Jones, Maillardet, and Robinson [73], Maindonald and Braun [89], and Verzani [154]); and S or S-plus (Chambers and Hastie [15], Everitt [38], and Venables and Ripley [153]). There are also books that give overviews of both SAS and R, see Kleinman and Horton [80] and Muenchen [103].

Many practical applications and examples will be presented. The applications will be largely in the medical area although other application areas will also be explored. This book will differ somewhat from other books currently available in that it will emphasize the importance of simulation and numerical exploration in a modern day statistical investigation. However, I will also provide many "hand calculations" that can be implemented with a simple pocket calculator. Many times, exercises that employ such calculations give an individual a feel for how methods really work.

This book is best suited for students who have had some exposure to the basics of probability and statistics. However, short "reviews" of basic topics should enable someone who is reasonably new to the field to become more familiar with most concepts. In cases where basic methods are introduced, broader issues of applicability will be explored through discussion of relevant literature or directly by numerical simulation. Furthermore, I will sometimes refer the reader to a number of excellent elementary statistics books that have been written by a host of different authors, e.g., Altman [4], Armitage and Berry [8], Fisher and van Belle [46], Fleiss [49], Milton [100], Pagano and Gauvreau [110], Rao [120], Rosner [124], Selvin [137], Snedecor and Cochran [140], Shork and Remington [138], Sokal and Rohlf [141], and Zar [161].

The level of exposition of this book will be suitable for students who have an interest in the application of statistical methods but do not necessarily intend to become statisticians. It will also be valuable to a beginning statistics student who has not yet been exposed to a broad range of methods applied in practice. The student will not need to have had a lot of mathematics but the exposition will not shy away from the use of mathematics. It will be assumed that all readers are familiar with concepts in basic algebra. Calculus is not a prerequisite for understanding this material but would be useful in conceptualizing many ideas in probability and statistics. Simple linear algebra will be reviewed in the text for the purpose of the development of regression and other linear models.

Modern-day high level statistical packages are extremely powerful. I would argue that sometimes they are too powerful. Often, they give a dazzling array of results, many of which are not relevant to the question being asked. However, in the right hands, these modern tools allow investigators to tailor analyses to almost any situation they encounter. Furthermore, through both visualization and analysis they can facilitate discovery of novel features about data.

In this book, I have chosen to illustrate statistical concepts using SAS and R. One can argue that these two packages represent the opposite ends of the spectrum of statistical packages. SAS, which is the standard in the biopharmaceutical industry, was established in the 1976 by Dr. James Goodnight. The software was first used on mainframe computers and evolved over the years to become available on mini-computers, Unix workstations, IBM compatible PCs and Macintoshs. A brief history of SAS can be found at http://www.sas.com/presscenter/bgndr_history.html. S was originally developed by Rick Becker, John Chambers, Doug Dunn, Paul Tukey, and Graham Wilkinson in the late 1970s and early 1980s. It became a popular program because of its cutting-edge procedures and its ability to easily produce innovative graphics. It was also developed as an open-source package that could be obtained directly from the developers. Early versions of the program were available on a Unix platform. In 1987, Douglas Martin founded Statistical Sciences, Inc. (StatSci) which developed a commercialized version called S-plus. This company later became a division of MathSoft. In 2001, the company re-emerged as the independent company, Insightful. Over the years, they developed a professional support group for users and made the software available on many different platforms including Windows, Linux platform and later, Macintosh. R was developed originally developed by Ross Ihaka and Robert Gentleman from 1991–1993 at the University of Auckland. It had an "S like" syntax but with a number of notable differences. Like its predecessor, S, R can be freely accessed. It can presently be downloaded at the website, http://www.cran.r-project.org/. Brief histories of S, S-plus and R can be found at the following web-sites: http://cm.bell-labs.com/cm/ms/departments/sia/S/history.html, http://www.stat.berkeley.edu/ rodwong/Stat131a/History_R.pdf and http://cran.r-project.org/doc/html/interface98-paper/paper.html.

SAS and R (or S-plus) are very general programs and provide great flexibility to the statistical user when (s)he is either interested in doing quick and dirty analyses or in developing very complex systems for graphics and analyses.

Outline of the Rest of the Book

Chapter 1: This chapter will review some basics of probability including the definition of random variables, conditional distributions, expectation, and variance. Several probability distributions will be reviewed including the uniform, normal, binomial, Poisson, t, chi-squared (χ^2), F, hypergeometric, and exponential distributions.

Chapter 2: This chapter will introduce the concept of using simulations in statistical investigations. Some background about the generation of (pseudo)random numbers will be covered. To test the validity of sequences of random numbers, simple time series models are reviewed. The time series concepts will also be used in Chapter 8 when repeated measures analyses are introduced.

Chapter 3: In this chapter, the strong law of large numbers and the central limit theorem will be introduced. Simulated examples are given that help to visualize these two important concepts.

Chapter 4: This chapter will begin with a review of correlation and simple linear regression. The next section will emphasize the role of visualization of raw data, fitted models and residuals in the process of regression modeling. A general approach to multiple regression using a general linear models approach will then be introduced. This introduction is followed by a discussion of regression diagnostics, partial and multiple correlation. The integration of these techniques into the complete model building process will be emphasized. Next, smoothing techniques are briefly introduced and illustrative examples are given. Finally, a brief appendix reviewing some simple matrix algebra and how simple computations are performed in R and SAS.

Chapter 5: In this chapter, the one-way analysis of variance (ANOVA) model will be reviewed and will be motivated according to the general linear models approach introduced in Chapter 4. Special topics in one-way ANOVA will include the formulation of linear contrasts, testing trends in outcomes across treatments that involve increasing doses of a drug or compound, and an application of Dunnett's test for testing a number of experimental groups against a control will be explored. Multiple comparisons procedures will be examined with examples in medical data. Some examples of the analysis of multi-way ANOVA designs will also be introduced. Specific examples of factorial and randomized block designs will be given.

Chapter 6: This chapter starts with a short review of several methods in the analysis of contingency tables including odds ratios, relative risk, and the Mantel-Haenszel procedure. Additionally, logistic regression will be introduced and an example will be given to facilitate the interpretation of these models. Finally, two methods, McNemar's test, and Cohen's Kappa, for analyzing paired count data are introduced.

Chapter 7: A general discussion of the analysis of multivariate outcomes is introduced in this chapter. Analogs to univariate t-tests, ANOVA and regression models are introduced. Particular topics include Hotelling's T^2, MANOVA and multivariate regression methods. In addition, several methods of classifying data into different groups are introduced. The classical method of discriminant analysis will be compared to logistic regression methods of classifying data into separate populations. An example of a newer method involving the use of classification trees is given. Some of the multivariate techniques will be used in later chapters in the discussion of repeated measures analyses.

Chapter 8: This chapter introduces methods used for analyzing data that are repeatedly measured on each subject or experimental unit. The chapter will focus on the analysis of repeated measures data where the measurements are continuous. Comparisons of standard ANOVA methods, multivariate methods, and random effects models are made. Classical growth curve analysis will be introduced. Finally, modern approaches using both fixed and random effects, that is *mixed* effects models will be introduced to enable one to address a broad array of potential questions regarding repeated measures data.

Chapter 9: In Chapter 9, nonparametric techniques for analyzing both paired and unpaired data are reviewed. For paired data, the sign test and the signed rank test will be detailed. For unpaired data, the rank-sum test and the Kruskal–Wallis test and their relationships to parametric procedures will be explored. The Spearman ρ test for correlation will also be discussed. Finally, the bootstrap will be introduced as a powerful computer intensive method for investigating complex statistical questions.

Chapter 10: In Chapter 10, methods of analysis of time to event data are discussed. First, a general discussion of the analysis of incidence density is given. Next, a general discussion of survival data is given, followed by simple examples of life-table and Kaplan–Meier methods for summarizing survival data. The log-rank procedure and its relationship to the Mantel–Haenszel procedure is introduced. An example of the implementation and interpretation of Cox's proportional hazards model is given. Cumulative incidence curves for displaying event probabilities over time in the presence of other competing risks will also be explored.

Chapter 11: In the last chapter, issues regarding sample size and power are addressed. A general discussion of the construction of power curves is given. Specific formulas are given for the determination of sample size and power for unpaired and paired continuous data, comparisons of proportions, repeated measures and survival data. In each case, a practical example is presented to illustrate how a particular formula is used.

Chapter 12 (Appendix A): A brief tutorial on SAS is given. This will focus on the DATA step, some simple summary procedures, and a couple of the graphics features. My discussion here will focus more on procedures and less on graphics.

Chapter 13 (Appendix B): A brief tutorial on R is given. Most of the R commands can be also implemented in S-plus. A few comments will be given on the similarities and differences between the two packages. The general focus of this chapter will be on reading in data, and describing a few graphical and statistical functions.

Acknowledgements

This book is roughly based on a course, Introduction to Biostatistics II, which I taught 13 times at the University of Pittsburgh. Over the years, the students challenged me to update the course during a time when the availability of computing tools for statistical analysis was rapidly changing. One reason I loved teaching the course is that, through countless interactions with very talented students, I learned as much as I taught.

Several individuals, all of who had extremely strong computing skills, influenced my presentation of computational methods within this book. Some of these individuals include the late John Bryant, James Dignam, Greg Yothers, Andriy Bandos, Hanna Bandos, Maria Mor, Meredith Lotz Wallace, Nick Christian, and Chi Song. I thank Jong Hyeon-Jeong for his encouragement many years ago to learn R. I thank Drs. Joseph Costantino, Samuel Wieand, John Bryant and Carol Redmond for their support of my efforts within the biostatistical center of the National Surgical Adjuvant Breast and Bowel Project (NSABP). I apologize to others who have influenced me but who have been, inadvertently, left out.

I must also thank the countless contributers to R, SAS and other packages who have made statistical tools easily accessible. Such contributions have allowed statistical methodology to be easily accessible to the general statistical and scientific community for addressing the many problems encountered in analyzing increasingly complex datasets.

I wish to also thank the reviewers of the manuscript for their comments. Each review provided me with suggestions that were incorporated into the manuscript. I owe particular thanks to Dr. Dean Nelson of the University of Pittsburgh at Greensburg, for his thorough reading of the manuscript and insightful commments which helped me to greatly improve the flow of the book and make useful additions to a couple of chapters in the book. The proofer of the text was also instrumental in removing annoying errors and ensuring consistency throughout the book. Jim McGovern also deserves much credit for his careful corrections. The responsibility for any remaining errors that still exist within the manuscript lies solely with me.

I thank my editors at Chapman Hall, Rob Calver and Sarah Morris, for their encouragement and remarkable patience over time. I also appreciate the technical support given by Shashi Kumar.

Three colleagues, Drs. William C. Troy, Howard Rockette and Sati Mazumdar, have given invaluable advice and encouragement to me over the years. Other mentors, namely, my father, John F. Anderson, and Drs. Hyman Zuckerman, Bernard Fisher, Richard H. Jones, Gary O. Zerbe and George D. Swanson have inspired me to try to be a lifelong scholar. Most of all, I owe thanks to my wife, Deb, for her love, encouragement and support over the last 27 years.

Stewart J. Anderson
October 2011

## CHAPTER 1

Review of Topics in Probability and Statistics

In this chapter, a brief review of probability is given. We will briefly summarize the basic principles of random variables, probability density functions, cumulative density functions, expected value and variance. Several distributions will be reviewed including the uniform, the normal, the binomial, the Poisson, the chi-squared, the t, the F, the hypergeometric, and the exponential distributions.

1.1 Introduction to Probability

Much of the formalization of probability began in the 17th century with the French mathematicians Blaise Pascal (1623–1662) and Pierre Fermat (1601–1665). This was a time when gambling games, e.g., roulette, dice, etc., were popular among nobility across Europe.

1.1.1 Schools of Thought

In the last 100 years, two primary schools of thought have emerged regarding the philosophical interpretation of probability: the *frequentist* approach and the subjective or *Bayesian* approach. These two philosophies are briefly summarized below.

1. Frequentist: interprets probability as the relative frequency of an outcome that would occur if an experiment were repeated a large number of times.
2. Subjective (Bayesian): each person has a subjective notion of what the probability of an outcome occurring is and experience can either substantiate or refute this notion.

In most modern statistical applications, the frequentist approach is employed. However, that is slowly changing with the wealth of methods that are currently being developed which employ Bayesian approaches and are tailored to situations where the quantification of an expert's (or set of experts') prior notions can facilitate complex decision making in scientific and economic applications. In this book, we will focus primarily on the frequentist approaches but will also touch on some Bayesian approaches as well.

1.1.2 Basic Definitions and Assumptions

An <u>event</u> is an outcome or a set of outcomes of interest. Examples of events are a "tails" on a coin flip, the Pittsburgh Pirates winning the World Series, a person contacting AIDS, a systolic blood pressure being 125 mm Hg, or a systolic blood pressure being ≥ 125 mm Hg. Events can be *simple*, that is, consisting of a finite (or countable such as the integers) set of outcomes or *composite*, that is, consisting of an uncountable number of outcomes such as the real numbers (or a subset thereof). Among the five examples of events given above, the first four are simple whereas the last event is composite. The *sample space*, S, is the set of all possible outcomes in a given experience.

It is often useful to characterize the relationship of different events using *sets*. Sets are merely collections of things such as events, outcomes, measurements, people, things, etc. Some set notation and useful properties of sets are summarized below.

<u>Set Notation and properties</u>

1. $A \equiv \{x : x \in A\}$ ("the set of all x such that x is in A")
2. $A^c \equiv \overline{A} \equiv \{x : x \notin A\}$ ("the set of all x such that x is *not* in A")
3. $A \cup B \equiv \{x : x \in A \cup B\}$ ("the set of all x that are in *either* A or B [or both]")
4. $A \cap B \equiv \{x : x \in A \cap B\}$ ("the set of all x that are in *both* A and B")
5. $A \setminus B \equiv A \cap \overline{B} \equiv \{x : x \in A \cap \overline{B}\}$ ("the set of all x that are in A *but not* B")
6. $\{\} \equiv \phi$ (the empty or "null" set): The empty set has the property that $A \cap \phi = \phi$ and $A \cup \phi = A$
7. $S \equiv$ the sample space
8. $A \subset B$ (read "A contained in B" or "A is a subset of B") $\Longrightarrow$ if $x \in A$ then $x \in B$
9. $A = B \iff A \subset B$ and $B \subset A$

A *probability* is a number, usually expressed as a value between 0 and 1 which is assigned to the chance of the occurence of one or more events. Typically, the basic principles of probability are governed by the following three rules or axioms:

The Axioms of Probability

1. $\Pr(A) \geq 0$ for all sets A
2. $\Pr(S) = 1$
3. $\Pr(A_1 \cup A_2 \cup \ldots) = \Pr(A_1) + \Pr(A_1) + \ldots$ if $A_i \cap A_j = \phi \; \forall \, i \neq j$

Item 3 is called the *property of countable additivity*, i.e., if $A_1, A_2, \ldots$ form a set of countable *mutually exclusive* events, then the probability of their union is the sum of the individual probabilities.

Properties derived from the axioms of probability

1. $\Pr(\phi) = 0$
   Proof: $\Pr(S) = \Pr(S \cup \phi) = \Pr(S) + \Pr(\phi)$. But, $\Pr(S) = 1 \implies \Pr(\phi) = 0$ because $\Pr(A) \geq 0$ for all sets A.
2. $\Pr(\overline{A}) = 1 - \Pr(A)$
   Proof: $1 = \Pr(S) = \Pr(A \cup \overline{A}) = \Pr(A) + \Pr(\overline{A}) \implies \Pr(\overline{A}) = 1 - \Pr(A)$.
3. $\Pr(A \cup B) = \Pr(A) + \Pr(B) - \Pr(A \cap B)$
   Proof: First, we note that $A \cup B$ consists of three mutually exclusive and exhaustive sets, $(A \cup B) = (A \cap \overline{B}) \cup (\overline{A} \cap B) \cup (A \cap B)$. Using axiom 3, we know that $\Pr(A \cup B) = \Pr(A \cap \overline{B}) + \Pr(\overline{A} \cap B) + \Pr(A \cap B)$. But,

$$\Pr(A) = \Pr(A \cap \overline{B}) + \Pr(A \cap B)$$

$$\Pr(B) = \Pr(\overline{A} \cap B) + \Pr(A \cap B)$$

$$\implies \Pr(A) + \Pr(B) = \Pr(A \cap \overline{B}) + \Pr(\overline{A} \cap B) + 2\Pr(A \cap B).$$

Thus, $\Pr(A) + \Pr(B) - \Pr(A \cap B) = \Pr(A \cap \overline{B}) + \Pr(\overline{A} \cap B) + \Pr(A \cap B) = \Pr(A \cup B)$.

A generalization of property 3, which can be derived by a simple mathematical induction argument can be written as

$$Pr(A_1 \cup A_2 \cup \ldots \cup A_k)$$

$$= \sum_{i=1}^{k} Pr(A_i) - \sum_{i=1}^{k} \sum_{j \neq i} Pr(A_i \cap A_j) + \ldots + (-1)^k Pr(A_1 \cap A_2 \cap \ldots \cap A_k) \quad (1.1)$$

Example 1.1. Of 100 newborn babies, roughly 49 are girls and 51 are boys. In a family with three children, what is the probability that at least one is a boy? For this example, *we'll assume that the sexes of successive children are independent events.*

If we let $B_1 = $ the event that the first child is a boy, $B_2 = $ the event that the second child is a boy, and $B_3 = $ the event that the third child is a boy, then the probability of interest can be written as

$$Pr(B_1 \cup B_2 \cup B_3)$$

$$= Pr(B_1) + Pr(B_2) + Pr(B_3) - Pr(B_1 \cap B_2) - Pr(B_1 \cap B_3) - Pr(B_2 \cap B_3)$$

$$+ Pr(B_1 \cap B_2 \cap B_3)$$

or *more simply*

$$Pr(B_1 \cup B_2 \cup B_3) = 1 - Pr(\overline{B}_1 \cap \overline{B}_2 \cap \overline{B}_3) = 1 - (0.49)^3 \approx 1 - 0.118 = 0.882 \ .$$

1.2 Conditional Probability

In many cases where the probability of a set of events is being calculated, we have information that reduces a given sample space to a much smaller set of outcomes. Thus, by *conditioning* on the information available, we're able to obtain the *conditional probability* of a given event.

<u>Definition</u> Given that A and B are events such that $Pr(B) > 0$, then $Pr(A|B) = \frac{Pr(A \cap B)}{Pr(B)}$. $Pr(A|B)$ is read as "the probability of A *given* (or conditional on) B."

<u>If</u> events are *independent*, then

$$Pr(A|B) = \frac{Pr(A \cap B)}{Pr(B)} = \frac{Pr(A) \cdot Pr(B)}{Pr(B)} = Pr(A) \ \ \text{or}$$

$$Pr(B|A) = \frac{Pr(B \cap A)}{Pr(A)} = \frac{Pr(B) \cdot Pr(A)}{Pr(A)} = Pr(B).$$

Thus, <u>if</u> A and B are independent, B *provides no information about the probability of A* or vice versa.

Example 1.2. In Example 1.1, what is the probability that *exactly* two of the three children are boys *given that* at least one child is a boy? Let $A = B_1 \cup B_2 \cup B_3$ and $B = (B_1 \cap B_2 \cap \overline{B}_3) \cup (B_1 \cap \overline{B}_2 \cap B_3) \cup (\overline{B}_1 \cap B_2 \cap B_3)$ so that the desired probability can be expressed as

$$Pr(B|A) = \frac{Pr(B \cap A)}{Pr(A)} = \frac{Pr(B)}{Pr(A)}$$

$$= \frac{Pr(B_1 \cap B_2 \cap \overline{B}_3) + Pr(B_1 \cap \overline{B}_2 \cap B_3) + Pr(\overline{B}_1 \cup B_2 \cap B_3)}{Pr(B_1 \cup B_2 \cup B_3)}$$

$$= \frac{3[(0.51)^2 0.49]}{1 - 0.49^3} \approx \frac{0.383}{0.882} \approx 0.433$$

<u>Bayes' Rule</u> One consequence resulting from the definition of conditional probability is known as *Bayes' Rule.* Assuming that both $Pr(A) > 0$ and $Pr(B) > 0$, we can rewrite the definition of conditional probability as $Pr(A \cap B) = Pr(A|B)Pr(B)$ or equivalently as $Pr(A \cap B) = Pr(B|A)Pr(A)$. Setting the right-hand sides of the above equations equal to each other we have $Pr(B|A)Pr(A) = Pr(A|B)Pr(B)$, which yields the formula originally discovered by Bayes:

$$Pr(B|A) = \frac{Pr(A|B)Pr(B)}{Pr(A)} . \qquad (1.2)$$

This formula allows one to switch the order of conditioning of two events and as we will see, is very useful in medical applications.

1.2.1 Sensitivity, Specificity, and Predictive Value

One area where conditional probability is widely applicable is in fields where one wishes to evaluate a test or instrument. For example, in medicine, one may want to test a new, potentially better, and/or cheaper diagnostic tool against a known standard. Examples in the past are nuclear magnetic resonance (NMR) assay versus the standard bioassay in determining components of blood, or the alpha-fetoprotein test versus amniocentesis for the detection of Down's Syndrome. Once an easier diagnostic tool has been developed, its usefulness is established by comparing its results to the results of some definitive (but more time-consuming and cumbersome) test of the disease status. For our discussion, let $T^+ = \{$test or symptom is positive$\}$, $T^- = \{$test or symptom is negative$\}$, $D^+ = \{$disease is positive$\}$, and $D^- = \{$disease is negative$\}$.

Test (Symptom) Result	Gold No. Disease +	Raw Counts Standard No. Disease −	Total No.
+	a True positive	b False positive	$a + b$
−	c False negative	d True negative	$c + d$
	$a + c$	$b + d$	$N = a + b + c + d$

<u>Important definitions for the evaluation of diagnostic tests</u>

i) Sensitivity: $Pr(T^+|D^+) = \frac{a/N}{(a+c)/N} = \frac{a}{a+c}$

ii) Specificity: $Pr(T^-|D^-) = \frac{d/N}{(b+d)/N} = \frac{d}{b+d}$

iii) Predictive value positive (PV^+): $Pr(D^+|T^+) = \frac{a/N}{(a+b)/N} = \frac{a}{a+b}$

iv) Predictive value negative (PV^-): $Pr(D^-|T^-) = \frac{d/N}{(c+d)/N} = \frac{d}{c+d}$

v) Prevalence is the probability of currently having a disease, that is, $Pr(D^+)$

Bayes' rule (as applied to Sensitivity, Specificity, PV^+, and PV^-)

Test Result	Gold Standard Disease +	Gold Standard Disease −	Total
Test pos. (T^+)	$Pr(T^+ \cap D^+)$	$Pr(T^+ \cap D^-)$	$Pr(T^+)$
Test neg. (T^-)	$Pr(T^- \cap D^+)$	$Pr(T^- \cap D^-)$	$Pr(T^-)$
	$Pr(D^+)$	$Pr(D^-)$	1

Figure 1.1 *Temporal Relationships Between Sensitivity, Specificity and Predictive Value (PV)*

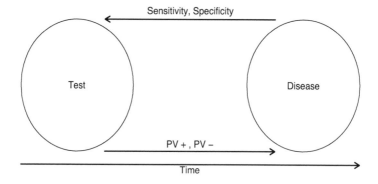

The diagram in Figure 1.1 demonstrates the typical temporal relationship among sensitivity, specificity, predictive value positive (PV^+), and negative (PV^-). Usually, the experimental test (or diagnosis) is one that can be employed long before the definitive measure or test of disease status can be assessed. Sensitivity (specificity) often provides a retrospective assessment of how well an experimental test performed, that is, it provides a measure of what proportion of the truly positive (negative) cases were correctly identified by the experimental test as being positive (negative). In contrast, PV^+ (PV^-) usually provides a prospective assessment, that is, given that the experimental test was positive (negative), what proportion of the cases ultimately turned out to be positive (negative). Sensitivity and specificity can be related to predictive value via Bayes' rule. The accuracy of the relationships depends primarily on how well the disease *prevalence*, $Pr(D^+)$, is estimated. Accordingly, a meaningful relationship between sensitivity, specificity, and predictive value can be established in prospective studies where a cohort is followed for a period of time and the proportion with the disease outcome is recorded. In retrospective studies, however, where the proportion of outcomes has been established and the test results are examined, the relationship between the entities is often not meaningful unless the study is designed so that the disease prevalence is similar to that of the general population.

Example 1.3. Suppose that a screening test for a blood disease yields the following results: $Pr(T^+ \cap D^+) = 0.05$, $Pr(T^+ \cap D^-) = 0.14$, $Pr(T^- \cap D^+) = 0.01$ and $Pr(T^- \cap D^-) = 0.80$. The results could be arranged as follows:

Joint Probabilities of Blood Tests and Disease Status

Test	Disease Status		
	Positive	Negative	Total
Positive	0.05	0.14	0.19
Negative	0.01	0.80	0.81
Total	0.06	0.94	1.00

Then, for this test, Sensitivity $= 0.05/0.06 \approx 0.833$, Specificity $= 0.80/0.94 \approx 0.851$, $PV^+ = 0.05/0.19 \approx 0.263$, and $PV^- = 0.8/0.81 \approx 0.988$.

1.2.2 Total Probability Rule and the Generalized Bayes' Rule

Bayes' rule can be generalized in cases where we're interested in obtaining conditional probabilities for more than two events. For example, if there are more than two events, we can set up a table so that the entries of the table represent the *joint* probabilities of *mutually exclusive and exhaustive* events. Consider two variables, A and B, which have r and k possible categories, respectively. The *joint* probabilities of the two sets of events can be displayed as in the table below.

Table of Joint Probabilities Useful
for the Total Probability Rule and for Generalized Bayes' Rule

	1	2	$\ldots$	m	Total
1	$Pr(A_1 \cap B_1)$	$Pr(A_1 \cap B_1)$	$\ldots$	$Pr(A_1 \cap B_k)$	$Pr(A_1)$
2	$Pr(A_2 \cap B_1)$	$Pr(A_2 \cap B_2)$	$\ldots$	$Pr(A_2 \cap B_k)$	$Pr(A_2)$
$\vdots$	$\vdots$	$\vdots$	$\ldots$	$\vdots$	$\vdots$
k	$Pr(A_r \cap B_1)$	$Pr(A_r \cap B_2)$	$\ldots$	$Pr(A_r \cap B_k)$	$Pr(A_r)$
Total	$Pr(B_1)$	$Pr(B_2)$	$\ldots$	$Pr(B_k)$	1

To find the probability, of say, A_2, we can use the <u>total probability rule</u>, for example,

$$Pr(A_2) = Pr(A_2 \cap B_1) + Pr(A_2 \cap B_2) + \ldots + Pr(A_2 \cap B_k)$$
$$= Pr(A_2|B_1)Pr(B_1) + Pr(A_2|B_2)Pr(B_2) + \ldots + Pr(A_2|B_k)Pr(B_k)$$

<u>Theorem</u> Let $B_1, B_2, \ldots, B_k$ be a set of mutually exclusive and exhaustive events, that is, one event *and only one event can* occur. Let A represent another type of event. Then

$$Pr(B_i|A) = \frac{Pr(A|B_i)Pr(B_i)}{\sum_{j=1}^{k} Pr(A|B_j)Pr(B_j)} \tag{1.3}$$

Equation (1.3) is known as *Generalized* Bayes' Rule.

Two examples of the Generalized Bayes' Rule are formulae that relate predictive value positive and negative to sensitivity and specificity.

$$PV^+ = Pr(D^+|T^+) = \frac{Pr(T^+|D^+)Pr(D^+)}{\left[Pr(T^+|D^+)Pr(D^+) + Pr(T^+|D^-)Pr(D^-)\right]}$$

or, equivalently,

$$PV^+ = \frac{\text{sensitivity} \cdot \text{prevalence}}{\left[\text{sensitivity} \cdot \text{prevalence} + (1 - \text{specificity}) \cdot (1 - \text{prevalence})\right]} \tag{1.4}$$

Similarly, for predictive value negative

$$PV^- = Pr(D^-|T^-) = \frac{Pr(T^-|D^-)Pr(D^-)}{\left[Pr(T^-|D^-)Pr(D^-) + Pr(T^-|D^+)Pr(D^+)\right]}$$

or, equivalently,

$$PV^- = \frac{\text{specificity} \cdot (1 - \text{prevalence})}{\left[\text{specificity} \cdot (1 - \text{prevalence}) + (1 - \text{sensitivity}) \cdot \text{prevalence}\right]} \tag{1.5}$$

Example 1.4. Consider a disease with a prevalence of 2% in the population. A test for the disease is determined to have a sensitivity of 99% and a specificity of 91%. Thus, the predictive values positive and negative, respectively, are

$$PV^+ = \frac{(.99)(.02)}{(.99)(.02) + (.09)(.98)} = \frac{.0198}{.108} \approx 0.1833$$

$$\text{and } PV^- = \frac{(.91)(.98)}{(.91)(.98) + (.01)(.02)} = \frac{.8918}{.8920} \approx 0.9998 \, .$$

Hence, in this example, because the disease is fairly rare in the population, the ability of the test to identify positive cases given that the test is positive is fairly poor *even though the sensitivity and specificity are very high*. However, if the test is negative for the disease in an individual, then chances are extremely high that the person is really negative for the disease.

1.3 Random Variables

Sometimes we're interested in calculating the probability of a *number* of events occuring rather than that of an isolated event. To do this we use *random variables*.

Definition (Random Variable)

A variable whose value changes according to a probabilistic model is called a *random variable*.

When we consider the probability of each value of the random variable a function of the value itself, we can derive a (discrete) *probability density function* (p.d.f.) denoted $f(k) = Pr(X = k)$. This is also sometimes referred to as a probability *mass* function. The probability density (mass) function follows the same rules as the axioms of probability. By considering the cumulative probability, $F(k) = Pr(X \leq k) = \sum_{i=1}^{k} f(i)$ as a function of k, we can derive the *cumulative distribution function* (c.d.f.). If the values of a random variable fall on a continuum then it is referred as *continuous*. The cumulative density function, $Pr(X \leq x)$ is given by $F(x) = \int_{-\infty}^{x} f(u)du$. In some cases, the probability density function is only defined on a finite interval and is set equal to 0 outside that interval.

Example 1.5. A simple example of the above principles can be illustrated by formulating the probability density function and cumulative density function of the outcomes that can be obtained when rolling a fair die.

If we let $X = \#$ of spots on a fair die, then the probability density function and cumulative density function can be written as follows:

r	1	2	3	4	5	6
$Pr(X = r)$	1/6	1/6	1/6	1/6	1/6	1/6
$Pr(X \leq r)$	1/6	1/3	1/2	2/3	5/6	1

Note that $Pr(X > r) = 1 - Pr(X \leq r) = 1 - F(r)$. Thus, $Pr(X > 4) = 1 - Pr(X \leq 4) = 1 - 2/3 = 1/3 = Pr(X \geq 5)$. Plots of both the probability density function and cumulative density function are given in Figure 1.2.

An important property of a random variable, X, is its *expected value*, denoted $E(X)$ or by the Greek letter, μ. The expected value is a *weighted average* that weights each value of X times the probability of that value occurring. Those products are then summed over all of the values of X. The expected value can also be thought of as the "balancing point" of the distribution. Mathematically, it is written as $\sum_{j=1}^{n} i f(i)$ if

Figure 1.2 *Example of the Probability Density and Cumulative Distribution Functions Associated with the Tossing of a Fair Six-Sided Die*

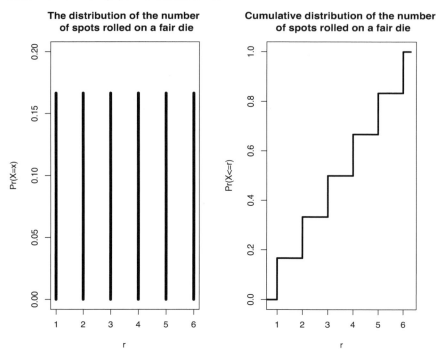

the probability density function has n discrete values and $\int_{-\infty}^{\infty} x f(x) dx$ if the probability density function is continuous. Another important property of a distribution of a population is its *variance*, which is often denoted as σ^2. The variance, which quantifies the *expected* squared variation about the mean, is defined to be

$$\sigma^2 \equiv E((X - E(X))^2) = E((X - \mu)^2) = E(X^2) - ((E(X))^2 = E(X^2) - \mu^2 \ .$$

Example 1.6. In Example 1.5,

$$\mu = E(X) = \sum_{k=1}^{6} \{k \cdot Pr(X = k)\} = 1(1/6) + 2(1/6) + 3(1/6) + 4(1/6) + 5(1/6) + 6(1/6)$$

$$= \frac{21}{6} \cdot \left(\sum_{i=1}^{6} i\right) = \frac{21}{6} = 3.5$$

To calculate σ^2, we can first calculate

$$E(X^2) = \sum_{k=0}^{2} \{k^2 \cdot Pr(X = k)\} = 1^2(1/6) + 2^2(1/6) + 3^2(1/6) + 4^2(1/6) + 5^2(1/6) + 6^2(1/6)$$

$$= \frac{1}{6} \cdot \left(\sum_{i=1}^{6} i^2 \right) = \frac{91}{6}$$

and then subtracting off $\mu^2 = (7/2)^2$. Thus,

$$\sigma^2 = E(X^2) - \mu^2 = 182/12 - 147/12 = 35/12 \approx 2.91667$$

Therefore, this population has an expected value $\mu = 3.5$ and a variance $\sigma^2 = 2.9$.

Finally, we can derive the *cumulative distribution function* (c.d.f.) of X by calculating $F(k) = Pr(X \le k)$ for $k = 1, 2, 3, 4, 5$ and 6. Thus,

$$F(x) = \begin{cases} 0, & \text{if } x < 1; \\ 1/6, & \text{if } 1 \le x < 2; \\ 2/6, & \text{if } 2 \le x < 3; \\ 3/6, & \text{if } 3 \le x < 4; \\ 4/6, & \text{if } 4 \le x < 5; \\ 5/6, & \text{if } 5 \le x < 6; \\ 1, & \text{if } x \ge 6. \end{cases} \qquad (1.6)$$

1.4 The Uniform Distribution

The distribution depicted in Example 1.5 is a special case of a *discrete uniform* distribution. The general form of the probability density function for a discrete uniform distribution is

$$f(i) = \frac{1}{b-a}, \text{ for } i = a+1, a+2, \ldots, b-a .$$

In Example 1.5, $a = 0$ and $b = 6$.

The probability density function of a *continuous* uniform distribution can be written as

$$f(x) = \frac{1}{b-a}, a \le x \le b; \ 0, \text{otherwise} .$$

The shape of this distribution is flat with a height of $1/(b-a)$ and 0 outside of the boundaries of $a \le x \le b$. If $X \sim U(a, b)$, then $E(X) = \frac{a+b}{2}$ and $\text{Var}(X) = \frac{(b-a)^2}{12}$. If $a = 0$ and $b = 1$, then the distribution is said to have a *standard* (continuous) uniform distribution. As we will see in later chapters of the book, this distribution is key for the generation of random numbers.

1.5 The Normal Distribution

1.5.1 Introduction

The normal distribution, which is sometimes referred to as the "bell-shaped curve" or the *Gaussian distribution*, as we will see, plays a central role in the development of statistical methods. As a practical matter, the values of a large number of physical measurements made on people, animals, plants, and objects are well approximated by the distribution. Theoretically, the normal family of distributions can take on any values from $-\infty$ to ∞ so that, strictly speaking, the normal distribution is an inexact approximation to characterizing real data. However, the probabilities of a particular random variable taking on very small or large values are so infinitesimally small that they are essentially 0. Consequently, the normal approximation is a very good approximation of the distributions of many types of real-world data.

The probability density function of the normal distribution is

$$f(x) = \frac{1}{\sqrt{2\pi}\sigma} e^{-(x-\mu)^2/2\sigma^2}, \quad -\infty < x < \infty$$

A random variable, X, which is normally distributed and has mean μ and variance σ^2 is denoted as $X \sim N(\mu, \sigma^2)$.

To obtain a probability associated with a normal random variable, say, $Pr(X < x)$, we must find the area under a normal curve from $-\infty$ to x. In general, to find the area under a curve for a continuous function, one must use *definite integration*. Hence, the aforementioned probability, $Pr(X < x)$ would be calculated as

$$F(x) = Pr(X < x) = Pr(X \le x) = \int_{-\infty}^{x} \frac{1}{\sqrt{2\pi}\sigma} e^{-(t-\mu)^2/2\sigma^2} dt.$$

Typically, we don't have to perform this integration but use tables or programs that give values for $Pr(Z \le z)$ where Z is a *standard normal distribution*, that is, $Z \sim N(0, 1)$. The standardization of a random variable, $X \sim N(\mu, \sigma^2)$, is accomplished by <u>shifting</u> X by μ units to yield a 0 mean and then to <u>scale</u> $X - \mu$ by a factor of σ yielding a variance of 1. Figure 1.3 shows the effects of shifting and scaling a random variable that is distributed as $N(4, 4)$ into a standard normal distribution (dotted curves to solid curves).

Example 1.7. (from *National Vital Statistics Report*, Vol. 54, No. 2, September 8, 2005, p.19 [104]) The mean birthweight for infants delivered in single deliveries in 2003 was 3325 grams (g) with a standard deviation of 571 g. A low birthweight is less than $2,500$ g. If birthweight in the general population is assumed to be normally distributed, then we might be interested in determining the predicted proportion of infants with low birthweights.

Letting X = the weight in grams of of infants, we assume that $X \sim N(3325, 571^2) = N(3325, 326041)$. We're asked to find $Pr(X < 2500)$ (see Figure 1.4 for a graphical representation). Thus, $Pr(X < 2500) = Pr(Z < \frac{2500-3325}{571}) = Pr(Z < \frac{-825}{571}) \approx Pr(Z <$

Figure 1.3 *Effects of Shifting and Scaling Observations on a Normal Distribution*

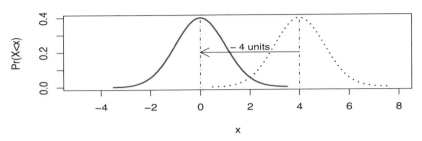

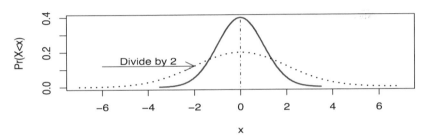

$-1.4448) = Pr(Z > 1.4448) \approx 0.074$. The probability can either be obtained from normal probability tables or from programs such as R with either the statement,

```
pnorm(2500,3325,571) or pnorm((2500-3325)/571).
```

1.5.2 How to Create Your Own Normal Tables

Traditionally, when one was interested in calculating probabilities derived from a normal distribution, (s)he would consult a table of cumulative probabilities of a standard normal distribution, that is, $\Phi(z) = Pr(Z < z)$. These "look-up" tables were useful for a wide range of applications but would have to be accessed from a textbook or other source. One immediate advantage of having powerful computing at one's fingertips is the fact that one can easily obtain standard probabilities without having to consult tables. In fact, one can easily produce standard statistical tables using most modern statistical packages. For example, suppose we wish to create a standard normal table. Thus, we want to produce a random variable, Z such that $Z \sim N(0,1)$. To use this distribution for statistical inference, we need to calculate the cumulative density function. We might also want to calculate quantities like $Pr(|Z| < z)$ and $Pr(Z > z)$, which are useful for calculating p-values. The code

Figure 1.4 *Theoretical Population Distribution of Single Infant Birthweights*

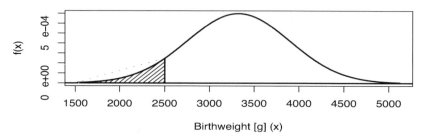

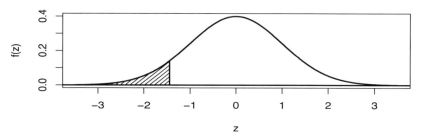

below (shown in an interactive session), which will run in either R or S-plus, creates such a table.

```
> # Create a normal table
> z <- 0:300/100
> pr.z <- pnorm(z)
> pr.abs.z <- 2*(pnorm(z)-1/2)
> pr.Z.gt.z <- 1-pnorm(z)
> Z.tbl<-cbind(z,round(pr.z,4),round(pr.abs.z,4),round(pr.Z.gt.z ,4))
> dimnames(Z.tbl)<-list(c(rep("",length(z))),c("z ","Pr(Z<z)",
+   "Pr(|Z|<z)","Pr(Z>z)"))
> Z.tbl
```

The above code would arrange the output in four columns with the headings, z Pr(Z<z) Pr(|Z|<z) Pr(Z>z). That would create a 301×4 matrix that could be printed out on 5 or 6 pages of paper. A rearranged version of this table is given below (Table 1.1).

Table 1.1. Standard Normal Table

z	Pr(Z<z)	Pr(|Z|<z)	Pr(Z>z)	z	Pr(Z<z)	Pr(|Z|<z)	Pr(Z>z)	z	Pr(Z<z)	Pr(|Z|<z)	Pr(Z>z)
0.00	0.5000	0.0000	0.5000	0.34	0.6331	0.2661	0.3669	0.68	0.7517	0.5035	0.2483
0.01	0.5040	0.0080	0.4960	0.35	0.6368	0.2737	0.3632	0.69	0.7549	0.5098	0.2451
0.02	0.5080	0.0160	0.4920	0.36	0.6406	0.2812	0.3594	0.70	0.7580	0.5161	0.2420
0.03	0.5120	0.0239	0.4880	0.37	0.6443	0.2886	0.3557	0.71	0.7611	0.5223	0.2389
0.04	0.5160	0.0319	0.4840	0.38	0.6480	0.2961	0.3520	0.72	0.7642	0.5285	0.2358
0.05	0.5199	0.0399	0.4801	0.39	0.6517	0.3035	0.3483	0.73	0.7673	0.5346	0.2327

z	Pr(Z<z)	Pr(|Z|<z)	Pr(Z>z)	z	Pr(Z<z)	Pr(|Z|<z)	Pr(Z>z)	z	Pr(Z<z)	Pr(|Z|<z)	Pr(Z>z)
0.06	0.5239	0.0478	0.4761	0.40	0.6554	0.3108	0.3446	0.74	0.7704	0.5407	0.2296
0.07	0.5279	0.0558	0.4721	0.41	0.6591	0.3182	0.3409	0.75	0.7734	0.5467	0.2266
0.08	0.5319	0.0638	0.4681	0.42	0.6628	0.3255	0.3372	0.76	0.7764	0.5527	0.2236
0.09	0.5359	0.0717	0.4641	0.43	0.6664	0.3328	0.3336	0.77	0.7794	0.5587	0.2206
0.10	0.5398	0.0797	0.4602	0.44	0.6700	0.3401	0.3300	0.78	0.7823	0.5646	0.2177
0.11	0.5438	0.0876	0.4562	0.45	0.6736	0.3473	0.3264	0.79	0.7852	0.5705	0.2148
0.12	0.5478	0.0955	0.4522	0.46	0.6772	0.3545	0.3228	0.80	0.7881	0.5763	0.2119
0.13	0.5517	0.1034	0.4483	0.47	0.6808	0.3616	0.3192	0.81	0.7910	0.5821	0.2090
0.14	0.5557	0.1113	0.4443	0.48	0.6844	0.3688	0.3156	0.82	0.7939	0.5878	0.2061
0.15	0.5596	0.1192	0.4404	0.49	0.6879	0.3759	0.3121	0.83	0.7967	0.5935	0.2033
0.16	0.5636	0.1271	0.4364	0.50	0.6915	0.3829	0.3085	0.84	0.7995	0.5991	0.2005
0.17	0.5675	0.1350	0.4325	0.51	0.6950	0.3899	0.3050	0.85	0.8023	0.6047	0.1977
0.18	0.5714	0.1428	0.4286	0.52	0.6985	0.3969	0.3015	0.86	0.8051	0.6102	0.1949
0.19	0.5753	0.1507	0.4247	0.53	0.7019	0.4039	0.2981	0.87	0.8078	0.6157	0.1922
0.20	0.5793	0.1585	0.4207	0.54	0.7054	0.4108	0.2946	0.88	0.8106	0.6211	0.1894
0.21	0.5832	0.1663	0.4168	0.55	0.7088	0.4177	0.2912	0.89	0.8133	0.6265	0.1867
0.22	0.5871	0.1741	0.4129	0.56	0.7123	0.4245	0.2877	0.90	0.8159	0.6319	0.1841
0.23	0.5910	0.1819	0.4090	0.57	0.7157	0.4313	0.2843	0.91	0.8186	0.6372	0.1814
0.24	0.5948	0.1897	0.4052	0.58	0.7190	0.4381	0.2810	0.92	0.8212	0.6424	0.1788
0.25	0.5987	0.1974	0.4013	0.59	0.7224	0.4448	0.2776	0.93	0.8238	0.6476	0.1762
0.26	0.6026	0.2051	0.3974	0.60	0.7257	0.4515	0.2743	0.94	0.8264	0.6528	0.1736
0.27	0.6064	0.2128	0.3936	0.61	0.7291	0.4581	0.2709	0.95	0.8289	0.6579	0.1711
0.28	0.6103	0.2205	0.3897	0.62	0.7324	0.4647	0.2676	0.96	0.8315	0.6629	0.1685
0.29	0.6141	0.2282	0.3859	0.63	0.7357	0.4713	0.2643	0.97	0.8340	0.6680	0.1660
0.30	0.6179	0.2358	0.3821	0.64	0.7389	0.4778	0.2611	0.98	0.8365	0.6729	0.1635
0.31	0.6217	0.2434	0.3783	0.65	0.7422	0.4843	0.2578	0.99	0.8389	0.6778	0.1611
0.32	0.6255	0.2510	0.3745	0.66	0.7454	0.4907	0.2546	1.00	0.8413	0.6827	0.1587
0.33	0.6293	0.2586	0.3707	0.67	0.7486	0.4971	0.2514	1.01	0.8438	0.6875	0.1562

z	Pr(Z<z)	Pr(|Z|<z)	Pr(Z>z)	z	Pr(Z<z)	Pr(|Z|<z)	Pr(Z>z)	z	Pr(Z<z)	Pr(|Z|<z)	Pr(Z>z)
1.02	0.8461	0.6923	0.1539	1.36	0.9131	0.8262	0.0869	1.70	0.9554	0.9109	0.0446
1.03	0.8485	0.6970	0.1515	1.37	0.9147	0.8293	0.0853	1.71	0.9564	0.9127	0.0436
1.04	0.8508	0.7017	0.1492	1.38	0.9162	0.8324	0.0838	1.72	0.9573	0.9146	0.0427
1.05	0.8531	0.7063	0.1469	1.39	0.9177	0.8355	0.0823	1.73	0.9582	0.9164	0.0418
1.06	0.8554	0.7109	0.1446	1.40	0.9192	0.8385	0.0808	1.74	0.9591	0.9181	0.0409
1.07	0.8577	0.7154	0.1423	1.41	0.9207	0.8415	0.0793	1.75	0.9599	0.9199	0.0401
1.08	0.8599	0.7199	0.1401	1.42	0.9222	0.8444	0.0778	1.76	0.9608	0.9216	0.0392
1.09	0.8621	0.7243	0.1379	1.43	0.9236	0.8473	0.0764	1.77	0.9616	0.9233	0.0384
1.10	0.8643	0.7287	0.1357	1.44	0.9251	0.8501	0.0749	1.78	0.9625	0.9249	0.0375
1.11	0.8665	0.7330	0.1335	1.45	0.9265	0.8529	0.0735	1.79	0.9633	0.9265	0.0367
1.12	0.8686	0.7373	0.1314	1.46	0.9279	0.8557	0.0721	1.80	0.9641	0.9281	0.0359
1.13	0.8708	0.7415	0.1292	1.47	0.9292	0.8584	0.0708	1.81	0.9649	0.9297	0.0351
1.14	0.8729	0.7457	0.1271	1.48	0.9306	0.8611	0.0694	1.82	0.9656	0.9312	0.0344
1.15	0.8749	0.7499	0.1251	1.49	0.9319	0.8638	0.0681	1.83	0.9664	0.9328	0.0336
1.16	0.8770	0.7540	0.1230	1.50	0.9332	0.8664	0.0668	1.84	0.9671	0.9342	0.0329
1.17	0.8790	0.7580	0.1210	1.51	0.9345	0.8690	0.0655	1.85	0.9678	0.9357	0.0322
1.18	0.8810	0.7620	0.1190	1.52	0.9357	0.8715	0.0643	1.86	0.9686	0.9371	0.0314
1.19	0.8830	0.7660	0.1170	1.53	0.9370	0.8740	0.0630	1.87	0.9693	0.9385	0.0307
1.20	0.8849	0.7699	0.1151	1.54	0.9382	0.8764	0.0618	1.88	0.9699	0.9399	0.0301
1.21	0.8869	0.7737	0.1131	1.55	0.9394	0.8789	0.0606	1.89	0.9706	0.9412	0.0294
1.22	0.8888	0.7775	0.1112	1.56	0.9406	0.8812	0.0594	1.90	0.9713	0.9426	0.0287
1.23	0.8907	0.7813	0.1093	1.57	0.9418	0.8836	0.0582	1.91	0.9719	0.9439	0.0281
1.24	0.8925	0.7850	0.1075	1.58	0.9429	0.8859	0.0571	1.92	0.9726	0.9451	0.0274
1.25	0.8944	0.7887	0.1056	1.59	0.9441	0.8882	0.0559	1.93	0.9732	0.9464	0.0268
1.26	0.8962	0.7923	0.1038	1.60	0.9452	0.8904	0.0548	1.94	0.9738	0.9476	0.0262
1.27	0.8980	0.7959	0.1020	1.61	0.9463	0.8926	0.0537	1.95	0.9744	0.9488	0.0256
1.28	0.8997	0.7995	0.1003	1.62	0.9474	0.8948	0.0526	1.96	0.9750	0.9500	0.0250
1.29	0.9015	0.8029	0.0985	1.63	0.9484	0.8969	0.0516	1.97	0.9756	0.9512	0.0244
1.30	0.9032	0.8064	0.0968	1.64	0.9495	0.8990	0.0505	1.98	0.9761	0.9523	0.0239
1.31	0.9049	0.8098	0.0951	1.65	0.9505	0.9011	0.0495	1.99	0.9767	0.9534	0.0233
1.32	0.9066	0.8132	0.0934	1.66	0.9515	0.9031	0.0485	2.00	0.9772	0.9545	0.0228
1.33	0.9082	0.8165	0.0918	1.67	0.9525	0.9051	0.0475	2.01	0.9778	0.9556	0.0222
1.34	0.9099	0.8198	0.0901	1.68	0.9535	0.9070	0.0465	2.02	0.9783	0.9566	0.0217
1.35	0.9115	0.8230	0.0885	1.69	0.9545	0.9090	0.0455	2.03	0.9788	0.9576	0.0212

z	Pr(Z<z)	Pr(|Z|<z)	Pr(Z>z)	z	Pr(Z<z)	Pr(|Z|<z)	Pr(Z>z)	z	Pr(Z<z)	Pr(|Z|<z)	Pr(Z>z)
2.04	0.9793	0.9586	0.0207	2.38	0.9913	0.9827	0.0087	2.72	0.9967	0.9935	0.0033
2.05	0.9798	0.9596	0.0202	2.39	0.9916	0.9832	0.0084	2.73	0.9968	0.9937	0.0032
2.06	0.9803	0.9606	0.0197	2.40	0.9918	0.9836	0.0082	2.74	0.9969	0.9939	0.0031
2.07	0.9808	0.9615	0.0192	2.41	0.9920	0.9840	0.0080	2.75	0.9970	0.9940	0.0030
2.08	0.9812	0.9625	0.0188	2.42	0.9922	0.9845	0.0078	2.76	0.9971	0.9942	0.0029
2.09	0.9817	0.9634	0.0183	2.43	0.9925	0.9849	0.0075	2.77	0.9972	0.9944	0.0028
2.10	0.9821	0.9643	0.0179	2.44	0.9927	0.9853	0.0073	2.78	0.9973	0.9946	0.0027
2.11	0.9826	0.9651	0.0174	2.45	0.9929	0.9857	0.0071	2.79	0.9974	0.9947	0.0026
2.12	0.9830	0.9660	0.0170	2.46	0.9931	0.9861	0.0069	2.80	0.9974	0.9949	0.0026
2.13	0.9834	0.9668	0.0166	2.47	0.9932	0.9865	0.0068	2.81	0.9975	0.9950	0.0025
2.14	0.9838	0.9676	0.0162	2.48	0.9934	0.9869	0.0066	2.82	0.9976	0.9952	0.0024
2.15	0.9842	0.9684	0.0158	2.49	0.9936	0.9872	0.0064	2.83	0.9977	0.9953	0.0023
2.16	0.9846	0.9692	0.0154	2.50	0.9938	0.9876	0.0062	2.84	0.9977	0.9955	0.0023
2.17	0.9850	0.9700	0.0150	2.51	0.9940	0.9879	0.0060	2.85	0.9978	0.9956	0.0022
2.18	0.9854	0.9707	0.0146	2.52	0.9941	0.9883	0.0059	2.86	0.9979	0.9958	0.0021
2.19	0.9857	0.9715	0.0143	2.53	0.9943	0.9886	0.0057	2.87	0.9979	0.9959	0.0021
2.20	0.9861	0.9722	0.0139	2.54	0.9945	0.9889	0.0055	2.88	0.9980	0.9960	0.0020
2.21	0.9864	0.9729	0.0136	2.55	0.9946	0.9892	0.0054	2.89	0.9981	0.9961	0.0019
2.22	0.9868	0.9736	0.0132	2.56	0.9948	0.9895	0.0052	2.90	0.9981	0.9963	0.0019
2.23	0.9871	0.9743	0.0129	2.57	0.9949	0.9898	0.0051	2.91	0.9982	0.9964	0.0018
2.24	0.9875	0.9749	0.0125	2.58	0.9951	0.9901	0.0049	2.92	0.9982	0.9965	0.0018
2.25	0.9878	0.9756	0.0122	2.59	0.9952	0.9904	0.0048	2.93	0.9983	0.9966	0.0017
2.26	0.9881	0.9762	0.0119	2.60	0.9953	0.9907	0.0047	2.94	0.9984	0.9967	0.0016

2.27	0.9884	0.9768	0.0116	2.61	0.9955	0.9909	0.0045	2.95	0.9984	0.9968	0.0016
2.28	0.9887	0.9774	0.0113	2.62	0.9956	0.9912	0.0044	2.96	0.9985	0.9969	0.0015
2.29	0.9890	0.9780	0.0110	2.63	0.9957	0.9915	0.0043	2.97	0.9985	0.9970	0.0015
2.30	0.9893	0.9786	0.0107	2.64	0.9959	0.9917	0.0041	2.98	0.9986	0.9971	0.0014
2.31	0.9896	0.9791	0.0104	2.65	0.9960	0.9920	0.0040	2.99	0.9986	0.9972	0.0014
2.32	0.9898	0.9797	0.0102	2.66	0.9961	0.9922	0.0039	3.00	0.9987	0.9973	0.0013
2.33	0.9901	0.9802	0.0099	2.67	0.9962	0.9924	0.0038				
2.34	0.9904	0.9807	0.0096	2.68	0.9963	0.9926	0.0037				
2.35	0.9906	0.9812	0.0094	2.69	0.9964	0.9929	0.0036				
2.36	0.9909	0.9817	0.0091	2.70	0.9965	0.9931	0.0035				
2.37	0.9911	0.9822	0.0089	2.71	0.9966	0.9933	0.0034				

1.6 The Binomial Distribution

The *binomial distribution* is the distribution associated with observing exactly r "successes" in n trials. This is equivalent to $n - r$ "failures" in n trials.

1.6.1 Assumptions Associated with the Binomial Distribution

A *binomially distributed* random variable, X, denoted as $X \sim b(n, p)$, is used when the following conditions are true:

1. The number of trials, n, is *fixed*.
2. The trials are independent events. They are sometimes called *Bernoulli trials*.
3. There are only two outcomes for each trial: a success or a failure.
4. Each trial has a *fixed* probability, p, of success and conversely, a fixed probability, $1 - p$ of failure.

For a given Bernoulli random variable, Y_i, the probability density function is given as

$$f(y_i) = \begin{cases} 1 - p, & \text{if } y_i = 0; \\ p, & \text{if } y_i = 1; \\ 0, & \text{otherwise.} \end{cases} \qquad (1.7)$$

Now, if we sum n such independent Bernoulli random variables, i.e., $X = \sum_{i=1}^{n} Y_i$, then the resulting random variable is distributed as a *binomial* random variable, X, with probability density function

$$f(x) = \begin{cases} \binom{n}{x} p^x (1 - p)^{n-x}, & \text{for } x = 0, 1, 2, \ldots, n \\ 0, & \text{otherwise} \end{cases}.$$

The cumulative density function for the binomial is as follows:

$$F(k) = Pr(X \leq k) = \sum_{x=0}^{k} \binom{n}{x} p^x (1 - p)^{n-x} \qquad (1.8)$$

The cumulative density function for the binomial distribution is usually very easy to calculate. In cases *where both n and k are reasonably large*, say $n \geq 10$, and $k \geq 4$, it may be useful to use the following recursive formula:

$$Pr(X = k+1) = \frac{n-k}{k+1} \cdot \frac{p}{1-p} \cdot Pr(X = k) \tag{1.9}$$

The above formula is <u>not</u> particularly useful when either n or k are small. It is easier in those cases just to calculate the probability directly from either equation (1.7) or equation (1.8).

Example 1.8. In Example 1.1, let $B = \#$ of boys born in a family having three children. There are four possibilities: $B = 0$, $B = 1$, $B = 2$, or $B = 3$. The probability associated with each of the values of B can be calculated as

$$Pr(B = 0) = \binom{3}{0} 0.51^0 \, 0.49^3 = 1 \times 1 \times 0.1176 \approx 0.1176,$$

$$Pr(B = 1) = \binom{3}{1} 0.51^1 \, 0.49^2 = 3 \times 0.51 \times 0.2401 \approx 0.3674,$$

$$Pr(B = 2) = \binom{3}{2} 0.51^2 \, 0.49^1 = 3 \times 0.2601 \times 0.49 \approx 0.3823,$$

$$\text{and } Pr(B = 3) = \binom{3}{3} 0.51^3 \, 0.49^0 = 1 \times 0.1327 \times 1 \approx 0.1327;$$

and hence, the complete probability density function can be specified. In this particular instance, it is not useful to employ the recursion given above.

1.6.2 Other Properties of the Binomial Distribution

If $X \sim b(n,p)$, then $\mu = E(X) = np$ and $\sigma^2 = Var(X) = np(1-p)$. Thus, in the above example, since $X \sim b(3, 0.51)$, then $\mu = 3(0.51) = 1.53$ and $\sigma^2 = 3(0.51)(0.49) = 0.7497$. Assuming that this model is a good one for the general population, we would expect an average of about 1.53 boys in families with three children with a standard devation of $\sqrt{.7497} \approx 0.8659$.

Example 1.9. Graphed below are eight different binomial probability density functions, with different values of n; four have $p = 0.3$ and four have $p = 0.1$. Notice that when, say, $n = 20$, if we ignore the right-hand tail (values greater than, say, 12) of the distribution, which has probabilities that are very close to 0, then the distribution looks much like a normal probability density function with mean equal to $np = 20 \times 0.3 = 6$ and $\sigma = \sqrt{20 \times 0.7 \times 0.3} \approx 2.05$.

Figure 1.5 *Probability Density Functions of a Binomial Distribution with p = .3 and p = .1*

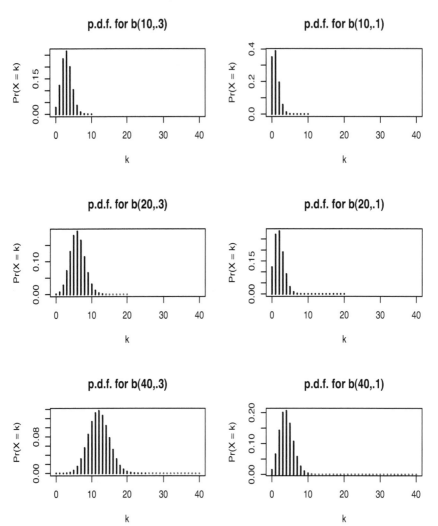

1.6.3 Estimation of Proportions in One Population

In most applications, we are not interested in estimating the number of successes but rather the *proportion*, p, of successes that occur. The population binomial proportion is estimated from a random sample by simply dividing the number of events or successes (that is, X) by the total number of trials (that is, n). Thus, $\widehat{p} = \frac{X}{n}$.

To make inference about the proportion of successes, we typically use the fact that when n is large enough, the probability density function is approximately normally distributed (see Figure 1.5). Many authors use the rule of thumb that the variance of X should be greater or equal to 5 in order to employ this approximation. Thus, if $\hat{\sigma}^2 = \widehat{\text{Var}}(X) = n\hat{p}(1 - \hat{p}) \geq 5$, $X \stackrel{.}{\sim} N(np, np(1 - p))$. Furthermore, if $X \stackrel{.}{\sim} N(np, np(1 - p))$, then $\hat{p} = \frac{X}{n} \stackrel{.}{\sim} N(p, p(1 - p)/n)$. From this relationship, we can derive two-sided confidence intervals for p as being

$$\left(\hat{p} - Z_{1-\alpha/2}\sqrt{\frac{\hat{p}(1 - \hat{p})}{n}} \ , \ \hat{p} + Z_{1-\alpha/2}\sqrt{\frac{\hat{p}(1 - \hat{p})}{n}} \right) , \quad (1.10)$$

where $Z_{1-\alpha/2}$ is the $\left(\frac{\alpha}{2}\right)^{\text{th}}$ percentile of the normal distribution. Typical values that are used are when $\alpha = 0.05$, $Z_{1-\alpha/2} = Z_{.975} = 1.96$ and when $\alpha = .01$, $Z_{1-\alpha/2} = Z_{.995} = 2.576$. The previous two values of α would be associated with two-sided 95% and 99% confidence intervals, respectively.

One can also use the relationship of the binomial and normal distributions to employ hypothesis testing for p. One way to do this is to first form the test statistic,

$$Z = \frac{\hat{p} - p_0}{\sqrt{p_0(1 - p_0)/n}} , \quad (1.11)$$

where p_0 is the proportion that is assumed if the null hypothesis were true. Then, we compare this test statistic to some prespecified number that depends on the α-level of the test and whether or not the test is one-sided or two-sided. For example, if one wishes to test say, $H_0 : p = p_0$ versus $H_1 : p > p_0$, then Z would be compared to $Z_{1-\alpha}$. If $Z \geq Z_{1-\alpha}$ then one would reject H_0; otherwise, one would not reject H_0. If the alternative hypothesis were two-sided, that is, if one were testing $H_0 : p = p_0$ versus $H_1 : p \neq p_0$, then one would reject the null hypothesis if $|Z| \geq Z_{1-\alpha/2}$; otherwise, one would not reject H_0.

Example 1.10. From the Framingham study, the probability of a 40-year old woman having coronary heart disease in her lifetime is calculated to about 31.7% (Lloyd-Jones et al. [88]). Suppose that in a sample of 58 of diabetic women, 28 have at least one coronary heart event in their lifetime. Is this result consistent with that expected in the general population?

The question here is consistent with a two-sided alternative hypothesis. (Had the question been, "Is this rate *higher than* the general population?," then the test of interest would be one-sided.) We will arbitrarily choose the α-level to be .01. Thus, the estimated p in our sample is $\hat{p} = 28/58 \approx 0.483$. The estimated variance is $n\hat{p}(1 - \hat{p}) = 58 \times (28/58) \times (30/58) = 14.483 > 5 \implies$ a normal approximation appears to be very reasonable and we, therefore, form the test statistic,

$$Z = \frac{0.483 - 0.317}{\sqrt{\frac{(.317)(.683)}{58}}} \approx \frac{.166}{0.0611} \approx 2.713 > 2.576 = Z_{.995}$$

$\implies$ we would reject H_0 at the $\alpha = .01$ level and conclude that women with diabetes have a higher rate of coronary heart disease. Assuming that this (hypothetical) sample of women is randomly chosen from a diabetic population, we would conclude that the smallest 99%

confidence interval for the lifetime probability, p, of coronary heart disease in the diabetic population of women would be given by

$$\left(0.483 - 2.576 \cdot 0.0611, \; 0.483 + 2.576 \cdot 0.0611\right) \approx \left(0.483 - 0.157, \; 0.483 + 0.157\right)$$

$$\approx \left(0.325, \; 0.640\right).$$

Even though the confidence interval is calculated without any knowledge of a hypothesis being tested, notice that it excludes 0.317, which is consistent with rejecting the test of $H_0 : p = 0.317$ at the $\alpha = .01$ level.

1.7 The Poisson Distribution

When one is interested in calculating the probability that a number of *rare* events occur among a *large* number of independent trials then the *Poisson distribution* is employed.

Assumptions associated with the Poisson Distribution

1. For each Δt_i, $Pr(1 \text{ success}) \approx \lambda \Delta t_i$, $Pr(0 \text{ successes}) \approx 1 - \lambda \Delta t_i$,

2. λ, the number of successes per unit time is <u>constant</u> over time,

3. $\Pr\left([\text{success in } \Delta t_i] \cap [\text{success in } \Delta t_j]\right)$
$= \Pr\left(\text{success in } \Delta t_i\right) \times \Pr\left(\text{success in } \Delta t_j\right)$ for all $i \neq j$. Thus, the probabilities of success in different time periods *are assumed to be independent*. Using a bit of calculus, one can derive the p.d.f.:

$$f(y) = \begin{cases} e^{-\lambda t} \cdot \frac{(\lambda t)^x}{x!}, & \text{for } x = 0, 1, 2, \dots; \\ 0, & \text{otherwise.} \end{cases} \tag{1.12}$$

The expected value of the Poisson distribution is $\mu = E(X) = \lambda t$ and the variance is $\sigma^2 = \text{Var}(X) = \lambda t$. Thus, *for a Poisson distribution, the mean and the variance are equal*.

1.7.1 A Recursion for Calculating Poisson Probabilities

Probabilities for the Poisson distribution are easier to calculate than for the binomial distribution. Like the binomial distribution, it may be useful to use a recursive formula in cases where a number of tedious calculations are required, that is,

$$Pr(X = j + 1) = \frac{\mu}{j+1} \cdot Pr(X = j) \tag{1.13}$$

You must start this recursion by first calculating $Pr(X = 0) = e^{-\lambda t}$. It may be noted again that the above formula is *not* particularly useful when j is small (Rosner [124]).

Example 1.11. In a paper by Ye et al. [160], investigators found that the daily number of hospital emergency room transports over the summer months in Tokyo for males > 65 of age for hypertension was approximately a Poisson distrubtion with $\lambda \approx 1$. This condition as others, such as asthma, may be related both to temperature and air pollution. Suppose in a particularly hot three-day period, six individuals are transported into the emergency room with hypertension. From a public health perspective one would ask "Is this cluster of events unusual and thus, a cause for concern or is it likely to be due to random variation?"

Let $X = \#$ of males > 65 years old transported daily to emergency rooms in Tokyo for hypertension. $X \sim \text{Poisson}(\lambda = 1)$. We're asked to calculate $Pr(X \geq 6)$ over a three-day period. We must first calculate the mean number of hypertension transports over the three-day period, that is, $\mu = \lambda t = 1(3) = 3$. Next, observe that $Pr(X \geq 6) = 1 - Pr(X < 6) = 1 - Pr(X \leq 5) = 1 - \sum_{j=0}^{5} Pr(X = j)$.

A nonrecursive way to solve this problem would be by directly calculating of $Pr(X \leq 6)$ and then subtracting the result from one. That is,

$$Pr(X \geq 6) = 1 - \sum_{j=0}^{5} Pr(X = j) = 1 - e^{-3}\left(\frac{3^0}{0!} + \frac{3^1}{1!} + \frac{3^2}{2!} + \frac{3^3}{3!} + \frac{3^4}{4!} + \frac{3^5}{5!}\right)$$

$$\approx 1 - 0.04979\left(1 + 3 + \frac{9}{2} + \frac{27}{6} + \frac{81}{24} + \frac{243}{120}\right) \approx 1 - 0.91608 = 0.08392 .$$

In this example, the recursive method would yield the following results:

Step	x	$Pr(X = x)$
(Initialization)	0	$e^{-3}\frac{3^0}{0!} = e^{-3} \approx 0.04979$
$j+1 = 1$	1	$\frac{3}{1} \cdot Pr(X = 0) \approx 0.14936$
$j+1 = 2$	2	$\frac{3}{2} \cdot Pr(X = 1) \approx 0.22404$
$j+1 = 3$	3	$\frac{3}{3} \cdot Pr(X = 2) \approx 0.22404$
$j+1 = 4$	4	$\frac{3}{4} \cdot Pr(X = 3) \approx 0.16803$
$j+1 = 5$	5	$\frac{3}{5} \cdot Pr(X = 4) \approx 0.10081$
		$\sum_{j=0}^{5} Pr(X = j) \approx \mathbf{.91608}$

So, $Pr(X \geq 6) = 1 - Pr(X \leq 5) = 1 - 0.91608 = .08392 \approx .083$. The conclusion here is that six or more emergency transports for hypertension over a three-day period is reasonably common since this would occur by chance about 8.3% of the time.

1.7.2 Poisson Approximation to the Binomial Distribution

If $X \sim b(n, p)$ where n is large and p is near 0 ($\Longrightarrow 1 - p = q \approx 1$), then the distribution of X can be approximated by the Poisson distribution with $\lambda = np$, i.e., $X \stackrel{.}{\sim} \text{Pois}(\lambda = np)$. Note also that if $q \approx 1$, then $npq \approx np$ yielding a distribution whose mean and variance are approximately equal such as in the Poisson distribution.

Example 1.12. Suppose that we use a Poisson distribution with $\mu = 5$ to approximate $X_1 \sim b(100, 0.05)$ and a Poisson distribution with $\mu = 1$ to approximate $X_2 \sim b(100, 0.01)$, respectively. The first eight values ($j = 0, 1, \ldots, 7$) of those two approximations along with the exact values and the differences are given in the Table 1.2. As can be seen, the absolute values of all of the differences in the first approximation are less than 0.005 and are less than 0.002 in the second approximation. Most authors recommend the use of the Poisson approximation when $p \leq 0.05$ and when $n \geq 100$ although some such as Rosner [124] even recommend that p be < 0.01 and $n \geq 100$.

Table 1.2. Poisson Approximations to the Binomial for $n = 100$ and $p = 0.01, \ 0.05$

j	Approximate $\Pr(X_1 = j)$	Exact $\Pr(X_1 = j)$	Difference	Approximate $\Pr(X_2 = j)$	Exact $\Pr(X_2 = j)$	Difference
0	0.0067	0.0059	0.0008	0.3679	0.3660	0.0018
1	0.0337	0.0312	0.0025	0.3679	0.3697	−0.0019
2	0.0842	0.0812	0.0030	0.1839	0.1849	−0.0009
3	0.1404	0.1396	0.0008	0.0613	0.0610	0.0003
4	0.1755	0.1781	−0.0027	0.0153	0.0149	0.0004
5	0.1755	0.1800	−0.0046	0.0031	0.0029	0.0002
6	0.1462	0.1500	−0.0038	0.0005	0.0005	0.0000
7	0.1044	0.1060	−0.0016	0.0001	0.0001	0.0000

1.8 The Chi-Squared Distribution

If $Z \sim N(0, 1)$, then $Z^2 \sim \chi_1^2$. The sum of the squares of independent standard normal random variables is a chi-squared distribution, that is, if $Z_1, Z_2, \ldots, Z_r$ are independently and identically distributed as $N(0, 1)$ then $S^2 = Z_1^2 + Z_2^2 + \ldots Z_r^2 \sim \chi_r^2$ where r is called the number of *degrees of freedom*. For the χ_r^2 distribution, $E(S^2) = r$ and $Var(S^2) = 2r$. Unlike the normal distribution, the χ^2 family of distributions is <u>not</u> symmetric for any r degrees of freedom (although it becomes more and more symmetric as the degrees of freedom increase). The probability density functions for three χ^2 distributions are shown in Figure 1.6.

The χ^2 distribution is widely employed in statistical applications. Two of the most common uses of the χ^2 involve making inferences about the variance of a normally distributed random variable and comparing differences in proportions in a standard contingency table analysis.

Figure 1.6 *Probability Density Functions for χ^2 Distributions*

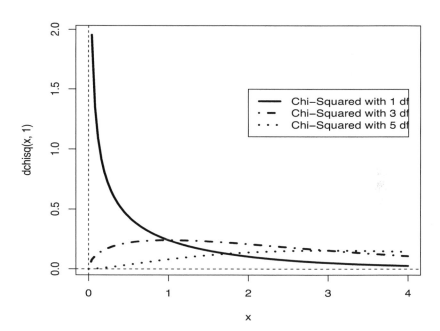

1.8.1 Using the Chi-Squared Distribution to Compare Sample Variances

For a set of random variables, $X_1, X_2, \ldots, X_n$ that are normally distributed as $N(\mu, \sigma^2)$, the usual *point estimates* of μ and σ^2 are $\widehat{\mu} = \overline{x} = \sum_{j=1}^{n} x_i/n$ and $\widehat{\sigma}^2 = s^2 = \sum_{j=1}^{n}(x_i^2 - \overline{x})^2/(n-1) = (\sum_{j=1}^{n} x_i^2 - n\overline{x}^2)/(n-1)$, respectively. Even if these estimates, based on the data at hand, of the population parameters are from a random sample, they are associated with uncertainty because of sampling error. It turns out that

$$\frac{(n-1)S^2}{\sigma^2} \sim \chi_{n-1}^2, \text{ where } S^2 = \frac{\sum_{j=1}^{n}(X_i - \overline{X})^2}{n-1} \tag{1.14}$$

So, if a sample of data comes from a normal distribution, we can test whether its variance, σ^2 is equal to some value by the use of the χ^2 distribution. Formally, we test $H_0 : \sigma^2 = \sigma_0^2$ versus some alternative, for example, $H_1 : \sigma^2 \neq \sigma_0^2$. To do this, we form a test statistic,

$$X^2 = \frac{(n-1)S^2}{\sigma_0^2} \text{ and compare it to } \begin{cases} \chi_{n-1,1-\frac{\alpha}{2}}^2 \text{ or } \chi_{n-1,\frac{\alpha}{2}}^2 & \text{if } H_1 \text{ is two-sided} \\ \chi_{n-1,1-\alpha}^2 \text{ or } \chi_{n-1,\alpha}^2 & \text{if } H_1 \text{ is one-sided.} \end{cases} \tag{1.15}$$

We reject H_0 if X^2 is greater than a prespecified critical value and do not reject otherwise.

Example 1.13. In a study by Pearson and Lee (1903) [111] of the physical characteristics of men, a group of 140 men was found to have a mean forearm length of 18.8021 inches with a standard deviation of 1.12047 inches. If we want to test the hypothesis of whether or not the variance is equal to 1, then we can formally test $H_0 : \sigma^2 = 1$ versus $H_1 : \sigma^2 \neq 1$. The sample size is $n = 140$ and the sample *variance* is $s^2 = 1.2555$. To perform the test, we form

$$X^2 = \frac{139(1.12047)^2}{1^2} \approx 174.509.$$

Since $\chi^2_{139,.025} \approx 108.25$ and $\chi^2_{139,.975} \approx 173.53$, we have that $X^2 > \chi^2_{139,.975}$ which implies that we reject H_0 at the two-sided $\alpha = .05$ level. But, we also know that $99.809 \approx \chi^2_{139,.005} < X^2 < \chi^2_{139,.995} \approx 185.69$, which means we would *not* reject the two-sided hypothesis if $\alpha = .01$. A more precise p-value obtained from the `probchi` function in SAS or the command, `2*(1-pchisq(174.509,139))` in R or S-plus, is $p \approx 0.044$.

We can form confidence intervals for σ^2 by using the following relationships:

$$Pr\left(\frac{(n-1)s^2}{\sigma} \leq \chi^2_{n-1,1-\frac{\alpha}{2}}\right) = Pr\left(\frac{(n-1)s^2}{\chi^2_{n-1,1-\frac{\alpha}{2}}} \leq \sigma\right) = 1 - \frac{\alpha}{2} \text{ and}$$

$$Pr\left(\frac{(n-1)s^2}{\sigma} \leq \chi^2_{n-1,\frac{\alpha}{2}}\right) = Pr\left(\frac{(n-1)s^2}{\chi^2_{n-1,\frac{\alpha}{2}}} \leq \sigma\right) = \frac{\alpha}{2}$$

Therefore, given a random sample from a normal population, a $(1-\alpha)\%$ confidence interval for σ is given by

$$\left(\frac{(n-1)s^2}{\chi^2_{n-1,1-\frac{\alpha}{2}}} , \frac{(n-1)s^2}{\chi^2_{n-1,\frac{\alpha}{2}}}\right) \tag{1.16}$$

Example 1.14. In the previous example, $s \approx 1.1205 \implies s^2 \approx 1.2555$, $\chi^2_{139,.025} \approx 108.254$, $\chi^2_{139,.975} \approx 173.53$, $\implies$ 95% confidence interval for σ^2 is approximately $\left(\frac{139(1.2555)}{173.53}\right.$, $\left.\frac{139(1.2555)}{108.254}\right) \approx \left(1.006 , 1.612\right)$. These boundaries are associated with standard deviation boundaries of $\left(1.003 , 1.270\right)$. Also, $\chi^2_{139,.005} \approx 99.809, \chi^2_{139,.995} \approx 185.693$, $\implies$ 99% confidence interval for σ is approximately $\left(0.940 , 1.748\right)$ and is associated with standard deviation boundaries of $\left(0.970 , 1.322\right)$. Notice that the 99% confidence intervals *are wider than* the 95% confidence intervals.

1.8.2 Comparing Proportions in Two or More Populations

To compare proportions in two or more populations with reasonably large sample sizes, we use the fact that, as was demonstrated earlier, estimated proportions are well approximated by a normal distribution that can be standardized so that its mean is 0 and its variance is 1. We can then exploit the fact that if a random variable is distributed as a standard normal, then, as was stated in the previous section, its square is distributed as a χ_1^2 distribution.

Example 1.15. The following example was originally presented in Rosner [122, 124]. Suppose we are performing an epidemiologic investigation of persons entering a clinic treating venereal diseases (VD). We find that 160 of 200 patients who are diagnosed as having gonorrhea and 50 of 105 patients as having nongonoccocal urethritis (NGU) have had previous episodes of urethritis. *Is there any association between the present diagnosis and having prior episodes of urethritis?*

For this problem, let $Gon^+ = \{$diagnosis of gonorrhea$\}$, $NGU = \{$diagnosis of nongonoc-cocal urethritis$\}$, $Ureth^+ = \{$prior episode of urethritis$\}$, and $Ureth^- = \{$no prior episode of urethritis$\}$.

Answering this question is equivalent to answering the question of whether or not the probability of having gonorrhea *conditional on having urethritis* is the same as that of having gonorrhea *conditional on not having urethritis*, that is,

$$Pr(Gon^+|Ureth^+) = Pr(Gon^+|Ureth^-)$$

is true. A statistical formulation of this hypothesis is as follows:

$$H_0 : Pr(Gon^+|Ureth^+) = Pr(Gon^+|Ureth^-) \text{ or } p_1 = p_2 \text{ versus}$$
$$H_1 : Pr(Gon^+|Ureth^+) \neq Pr(Gon^+|Ureth^-) \text{ or } p_1 \neq p_2 \,.$$

Under H_0: no difference in proportions of patients with gonorrhea (in those with urethritis versus those without), one would expect the proportion of cases of gonorrhea to be the same in both groups, that is, $\hat{p} = 200/305 \approx 0.6557$, which is the total proportion of patients having gonorrhea. Consequently, the expected number of cases of patients having gonorrhea in each group (according to urethritis status) would be proportional to the number of patients in that group.

To start the formal analysis, the data are arranged into a convenient form called a 2×2 *contingency table*:

Contingency Table of VD Data

	Observed Table				Expected Table under H_0	
	Gon^+	Gon^-			Gon^+	Gon^-
$Ureth^+$	$a = 160$	$b = 50$	210	$Ureth^+$	137.7	72.3
$Ureth^-$	$c = 40$	$d = 55$	95	$Ureth^-$	62.3	32.7
	200	105	305			

In Figure 1.7, three ways of visualizing the association between gonorrhea and urethritis are depicted. In panel A, an *association plot* is displayed. Cells with observed values greater than expected values under H_0 are displayed as having positive associations whereas those with observed values less than expected under H_0 are displayed as having negative associations. Hence, the cells where the urethritis and gonorrhea levels are either both positive or both negative have positive associations whereas the other two cells have negative associations. A similar idea used in the fourfold plot displayed in panel B. In that plot, the cells where observed values > expected values are displayed as sections of larger circles. In panel C, a *mosaic* plot is displayed. Like the other two plots, the mosaic plot is arranged so that the cells with observed values > expected values are associated with the larger rectangles. Also, the widths of the rectangles reflect the fact that the *marginal* value of the urethritis positive patients is larger than that of the urethritis negative patients (210 versus 95). Likewise, the heights of the rectangles reflect the fact that the *marginal* value of the gonorrhea positive patients is larger than that of the gonorrhea negative patients (200 versus 105).

Figure 1.7 *Three Visualizations of the Association Between Gonorrhea and Urethritis*

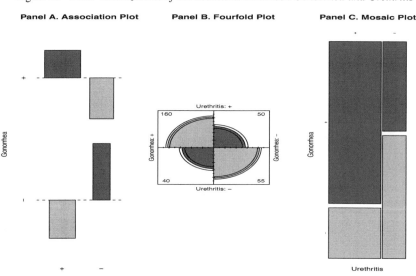

The R code to produce the plots in Figure 1.7 is given below. (Note that the `col` statements below will only work in cases where the plots are printed in color.)

```
Urethritis<-c(rep("+",210),rep("-",95))
Gonorrhea<-c(rep("+",160),rep("-",50),rep("+",40),rep("-",55))
z<-table(Urethritis,Gonorrhea)[2:1,2:1]

par(mfrow=c(1,3)) # create a 1 x 3 panel of plots
assocplot(z,col=c("red","green"), main="Panel A. Association Plot")
fourfoldplot(z,col = c("red","green"))
title("Panel B. Fourfold Plot",cex=.7)
mosaicplot(z,col=c("red","green"),main="Panel C. Mosaic Plot")
par(mfrow=c(1,1)) # return plot
```

We will formally analyse the data using two <u>equivalent</u> methods: using a normal analysis and using a χ^2 analysis. Typically, however, one would first determine whether or not the sample is large enough to warrant such approximations. For a 2×2 table, some authors, for example, Rosner [124], conservatively require that the expected number of counts in each of the four cells be ≥ 5. Other authors, for example, Cochran [18], use a less stringent criterion, that is allow expected values to be, say, ≥ 1 but < 5 to employ the approximation. We'll use the conservative criterion here to ensure accurate approximations. Therefore, we must check to see that $n_1 \hat{p}\hat{q} > 5$ and $n_2 \hat{p}\hat{q} > 5$. In our case,

$$n_1 = 210, \ \hat{p}_1 = \frac{160}{210}, \ \hat{q}_1 = \frac{50}{210}, \quad n_2 = 95, \ \hat{p}_2 = \frac{40}{95}, \ \hat{q}_2 = \frac{55}{95}$$

$$\Longrightarrow \hat{p} = \frac{n_1 \hat{p}_1 + n_2 \hat{p}_2}{(n_1 + n_2)} = \frac{200}{305}, \quad \hat{q} = 1 - \hat{p} = \frac{105}{305}$$

$$n_1 \hat{p}\hat{q} \approx 47.4 \gg 5 \quad \text{and} \quad n_2 \hat{p}\hat{q} \approx 21.4 \gg 5.$$

The first type of analysis uses a normal approximation to test a binomial proportion. This analysis can be done with or without what is known as a *continuity correction* (Yates [159]). The continuity correction adjusts for the fact that we use a continuous distribution to approximate a statistic associated with discrete quantities, that is, the number of cases of gonorrhea in each group. The continuity corrected statistic is as follows:

$$Z_C = \frac{\hat{p}_1 - \hat{p}_2 - \frac{1}{2}\left(\frac{1}{n_1} + \frac{1}{n_2}\right)}{\sqrt{\hat{p}\hat{q}\left(\frac{1}{n_1} + \frac{1}{n_2}\right)}}, \tag{1.17}$$

where $\hat{p}$ and $\hat{q} = 1 - \hat{p}$ are the *pooled* estimates of p and q, respectively. In our example, $Z_C = \frac{.333}{.059} = 5.672$. The *uncorrected* normal statistic is:

$$Z_U = \frac{\hat{p}_1 - \hat{p}_2}{\sqrt{\hat{p}\hat{q}\left(\frac{1}{n_1} + \frac{1}{n_2}\right)}}. \tag{1.18}$$

In this problem, $Z_U = 5.802$. Notice that the uncorrected test statistic is slightly larger than the corrected version and yields a more significant p-value. Some authors, such as Mantel and Greenhouse [93], Mantel [92], and Miettinen [98], argue that the corrected version is a more accurate estimate of the exact distribution underlying the counts in a 2×2 contingency table whereas others, such as Grizzle [60] and Conover [20], argue that the uncorrected version gives a more accurate estimate of the χ^2 distribution. Practically speaking, however, unless the associated p-value is near the designated α-level or $N = n_1 + n_2$ is small, then either method is acceptable. The uncorrected test gives a more extreme p-value ($p \approx 6.55 \times 10^{-9}$) than the corrected version ($p \approx 1.44 \times 10^{-8}$) but in both methods, $p < 0.0001$, a highly significant result.

The χ^2 approach also requires that use the *expected* table given above, that is, the values in each cell of the 2×2 contingency table we would expect if H_0 is true. Under $H_0 : p_1 = p_2$, our expected values are $(a+b)\hat{p} = 137.7$, $(a+b)\hat{q} = 72.3$, $(c+d)\hat{p} = 32.7$, and $(c+d)\hat{q} = 62.3$. If these values were computed by a hand on a calculator or in a computer program, one should *store* the intermediate calculations to the maximum number of digits allowed. Intermediate values should <u>not</u> be rounded off and reentered. For example, 137.7 is actually stored in the calculator or in the memory of a computer as 137.7049180 but is *reported as* 137.7. Notice also that the numbers in the cells of the expected tables add up to the numbers in the margins.

This is a good check to make sure that your intermediate calculations are correct. To do the *continuity corrected* χ^2 analysis, one calculates the following statistic:

$$X_C^2 = \sum_{i=1}^{4} \frac{(|O_i - E_i| - \frac{1}{2})^2}{E_i} = \frac{N(|ad - bc| - \frac{N}{2})^2}{(a+b)(c+d)(a+c)(b+d)}, \tag{1.19}$$

where the O_i and the E_i are the observed and expected values, respectively, in the 4 cells and a, b, c, d, and N are as in the contingency table above. The uncorrected χ^2 statistic is:

$$X_U^2 = \sum_{i=1}^{4} \frac{(O_i - E_i)^2}{E_i} = \frac{N(ad - bc)^2}{(a+b)(c+d)(a+c)(b+d)}. \tag{1.20}$$

In our example, $X_C^2 = 305[(160(55) - 40(50)) - 152.5]^2/(210)(95)(200)(105) \approx 32.170$ and $X_U^2 = 305[160(55) - 40(50)]^2/(210)(95)(200)(105) \approx 33.663$.

Both X_C^2 and X_U^2 are compared to a $\chi_{1,1-\alpha}^2$ value located in a table (or calculated by pocket calculator or computer). Notice that $X_C^2 = Z_C^2$, i.e., $(5.672)^2 = 32.17$ and that $X_U^2 = Z_U^2$. These relationships hold because if $Z \sim N(0,1)$ then $Z^2 \sim \chi_1^2$. The SAS program and output below it produce the χ^2 analyses given above. The numbers below the cell counts represent the row and column percentages associated with each cell count.

```
* SAS program to analyze 2 x 2 table with cell counts as input;
options nocenter nonumber nodate ls=60;
data twobytwo;
input ureth gon cnt @@;
cards;
1 1 160  1 2 50 2 1 40 2 2 55
;
proc format;
value pn 1='       +' 2='       -';
proc freq;
table ureth*gon/chisq nocum nopercent;
format ureth gon pn.; weight cnt;
title 'Urethritis data from Rosner';
run;
```

```
*------------------------ PARTIAL OUTPUT FROM SAS -------------------------
                    Urethritis data from Rosner


                        The FREQ Procedure

                    Table of ureth by gon


            ureth       gon

            Frequency
            Row Pct  |
            Col Pct  |     +  |     -  |  Total
            -------------------------------
                +    |   160  |    50  |   210
                     | 76.19  | 23.81  |
                     | 80.00  | 47.62  |
            -------------------------------
                -    |    40  |    55  |    95
                     | 42.11  | 57.89  |
                     | 20.00  | 52.38  |
            -------------------------------
            Total        200      105      305

            Statistics for Table of ureth by gon
```

```
Statistic                              DF      Value       Prob
------------------------------------------------------------
Chi-Square                              1     33.6632     <.0001
Continuity Adj. Chi-Square              1     32.1702     <.0001
```

1.9 Student's *t*-Distribution

If each member of a set of random variables is normally distributed as $N(\mu, \sigma^2)$ and the variance is estimated from the sample (that is, *not* assumed to known a priori), then the ratio of the sample mean minus μ over the sample standard error of the mean is distributed as a "Student's" *t*-distribution. This distribution was discovered by Gosset (1908) [146], an employee of the Guiness Brewery, who used the pseudonym "Student" when he published his work. Formally, $\frac{\overline{X}-\mu}{s/\sqrt{n}} \sim t_{n-1}$ where s is the *sample* standard deviation and t_{n-1} denotes a *central Student's t-distribution* with $\nu = n - 1$ degrees of freedom [df]. This distribution has an expected value of 0 and a variance equal to $\frac{\nu}{\nu-2} = \frac{n-1}{n-3}$. The degrees of freedom can be thought of as pieces of information. The estimate of the sample standard deviation is given by $s = \sqrt{\sum_{j=1}^{n}(x_i - \overline{x})^2/(n - 1)}$ so that once the mean, $\overline{x}$, is calculated then only $n - 1$ pieces of information are needed to calculate s.

Each of the members of the family of central *t*-distributions look very similar to the standard normal distribution except for the fact that the *t*-distributions have wider tails than does the standard normal distribution. It can be shown that as $n \to \infty$, $t_{n-1} \to N(0, 1)$. What this says is that as your sample size, n, gets large, then any analysis where a *t*-distribution is used could be well approximated by the standard normal distribution. Figure 1.8 depicts three different *t*-distributions: one with 5 df, one with 30 df and one with ∞ df, that is, the normal distribution.

The tail values of a *t*-distribution and its corresponding Z value for various values of the sample size, n, are summarized in Table 1.3.

Table 1.3. Ratios of t-Statistics to Z-Statistics for Different Sample Sizes (n)

n	5	10	15	20	25	30	100	∞
df $= n - 1$	4	9	14	19	24	29	99	∞
$t_{n-1,.975}$	2.776	2.262	2.145	2.093	2.064	2.045	1.984	1.960
$Z_{.975}$	1.960	1.960	1.960	1.960	1.960	1.960	1.960	1.960
t/Z	1.417	1.154	1.094	1.068	1.053	1.044	1.012	1.000

Figure 1.8 *Probability Density Functions for Three Central t-Distributions*

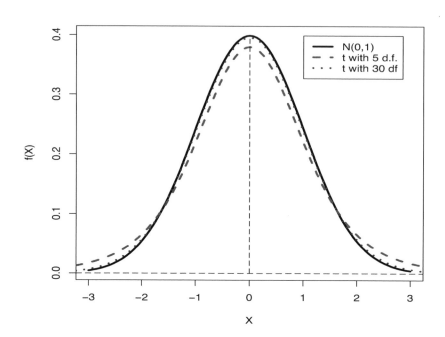

To intuitively see why the t-distribution has wider tails than the standard normal distribution, consider a situation where $X_1, X_2, \ldots, X_n$ are independently and identically distributed $N(\mu, \sigma^2)$. Then

$$\frac{\overline{X} - \mu}{\sigma/\sqrt{n}} \sim N(0,1) \quad \text{and} \quad \frac{\overline{X} - \mu}{s/\sqrt{n}} \sim t_{n-1}. \tag{1.21}$$

In the first expression of equation (1.21), the population variance is assumed to be known. In the second expression, the population variance is *estimated* from the sample at hand. Thus, there is an added uncertainty due to the fact that we do not know the value of σ but, rather, we have to use an estimate, s, of σ, which can vary from sample to sample. Hence, the added uncertainty in s causes the t-distribution to have wider tails. The formal relationship between the normal, t- and χ^2-distributions can easily be derived mathematically.

From equation (1.21), we can derive the *shortest* $(1 - \alpha)\%$ confidence interval for μ, which is given by $\overline{x} \mp t_{n-1, 1-\frac{\alpha}{2}} s/\sqrt{n}$ or

$$\overline{x} - t_{n-1, 1-\frac{\alpha}{2}} s/\sqrt{n} < \mu < \overline{x} + t_{n-1, 1-\frac{\alpha}{2}} s/\sqrt{n} \tag{1.22}$$

Equation (1.22) is very important because in most applications, σ is not assumed to be known.

Example 1.16. In Example 1.13, the shortest 95% confidence for the mean forearm length is given by

$$\left(18.8021 - t_{139,0.975} \frac{1.12047}{\sqrt{140}}, \; 18.8021 + t_{139,0.975} \frac{1.12047}{\sqrt{140}} \right)$$

$$\approx \left(18.8021 - 1.9773\,(1.12047), 18.8021 + 1.9773\,(1.12047) \right) \approx \left(16.587, \; 21.018 \right).$$

The t-distribution is also used for comparing the means of two samples of normally distributed data. In studies involving two samples, two basic experimental designs can be employed:

Design #1: Match or pair individuals (experimental units) so that all relevant traits *except* the one of interest are closely related. One way to do this is to pair different individuals with similar traits so that two treatments or interventions can be compared in individuals with very similar characteristics. In this way, one can test the effect of a treatment or trait on the observation of interest while adjusting for other traits that can confound the hypothesized relationship between the treatments and the observations. For example, a common confounder for treatment of high blood pressure is age. Thus, in a paired study, pairs of individuals of similar ages might be matched to test the efficacy of two blood pressure treatments. Another way to adjust for confounding is to use the individual as "his/her/its own control." When the study design calls for two or more observations to be made over time for each individual, then the design is said to *longitudinal* in nature. Longitudinal studies often lead to paired analyses. For this type of design, one analyzes the differences, Δ, of the n pairs using a one-sample t-test with $n-1$ degrees of freedom. Usually, the null hypothesis is $H_0: \; \overline{\Delta} = 0$, which is tested against either a one-sided or two-sided alternative hypothesis. However, more general tests of the mean of the differences can also be employed where H_0 is $H_0: \; \overline{\Delta} = \delta \neq 0$.

Design #2: Randomly assign individuals to different groups according to the different treatments or characteristics of interest. The number of individuals in one group does not have to equal the number of individuals in the other group(s). Also, in this design, *no attempt is made* to pair observations in one group to observations in another. Therefore, we assume that the observations are *independent* across groups as well as within each group. This type of study design obviously leads to an unpaired analysis. It is sometimes referred to as a *parallel groups* design.

Example 1.17. (Sources: Hand et al., *Small Data Sets*, Chapman–Hall, 1996 [61], p. 37 and Sternberg et al. [144]) The activity level of dopamine, a substance in the central nervous system, has been found to be associated with psychotic behavior in schizophrenic patients. In a study of 25 hospitalized schizophrenic patients, an antipsychotic drug was administered and the patients were later classified as nonpsychotic ($n = 15$) or psychotic ($n = 10$). In each patient, a sample of cerebrospinal fluid was removed and the concentration levels of dopamine b-hydroxylase [DBP] were measured in nmol/(ml)(h)/(mg) of protein. The data are given as follows:

Non-psychotic	.0104	.0105	.0112	.0116	.0130	.0145	.0154	.0156
	.0170	.0180	.0200	.0200	.0210	.0230	.0252	
Psychotic	.0150	.0204	.0208	.0222	.0226	.0245	.0270	.0275
	.0306	.0320						

In these two samples, $n_1 = 15$, $\bar{x}_1 = 0.016427$, $s_1 = 0.004695$ and $n_2 = 10, \bar{x}_2 = 0.02426$, $s_2 = 0.00514$. These data are *unpaired* and we can perform a t-test by forming the statistic, $T = (\bar{x}_1 - \bar{x}_2)/\{SE (\bar{x}_1 - \bar{x}_2)\}$ where "SE" denotes the standard error of the difference of the two means. We assume here that the individual patient observations are statistically independent and so that $\text{Var}(\bar{X}_1 - \bar{X}_2) = \text{Var}(\bar{X}_1) + \text{Var}(\bar{X}_1) = \text{Var}(X_1)/n_1 + \text{Var}(X_2)/n_2$. The sum of variances is associated with $(n_1 - 1) + (n_2 - 1) = n_1 + n_2 - 2$ degrees of freedom (d.f.). One assumption of a standard t-test is that the variances of the two underlying populations are equal and hence, t-tests are performed by *pooling* the variances, that is, $s^2 = ((n_1 - 1)s_1^2 + (n_2 - 1)s_2^2)/(n_1 + n_2 - 2)$. From that relationship, one can calculate the *pooled* SE of the differences of the means as $SE (\bar{x}_1 - \bar{x}_2) = s\sqrt{\frac{1}{n_1} + \frac{1}{n_2}}$. Ultimately, under $H_0 : \mu_1 = \mu_2$,

$$T = \frac{\bar{x}_1 - \bar{x}_2}{s\sqrt{\frac{1}{n_1} + \frac{1}{n_2}}} \sim t_{n_1 + n_2 - 2} . \tag{1.23}$$

In our example,

$$s = \sqrt{\frac{14(0.004695^2) + 9(0.00514^2)}{23}} \approx \sqrt{\frac{0.0005464}{23}} \approx 0.004874$$

$$\Rightarrow T = \frac{0.016427 - 0.02426}{0.004874\sqrt{\frac{1}{15} + \frac{1}{10}}} \approx \frac{-0.00783}{0.00198} \approx -3.9364 .$$

Since $|T| > 2.807 = t_{23, .995}$, we would reject the null hypothesis at the $\alpha = 0.01$ level. Alternatively, the p-value could be calculated as $2 \times \Pr(T < -3.9364) \approx 2(0.00033) = 0.00066$.

1.10 The F-Distribution

One tacit assumption that was made in the preceding example was that the two underlying populations each shared the same variance. Therefore, we *pooled* the sample variances to get a more precise estimate of the underlying variance, σ^2. What if the underlying variances were not equal? This question lead researchers in the 1930s to a classical statistical problem known as the **Behrens-Fisher** problem.

Because of the Behrens–Fisher problem, we should first test for the equality of variances before we can make inference on population means. Thus, we may test, for example, $H_0 : \sigma_1^2 = \sigma_2^2$ versus $H_1 : \sigma_1^2 \neq \sigma_2^2$. One possibility for constructing a test statistic for comparing equality of variances is to use the difference of the sample variances. However, it turns out that the statistic formed by the *ratio of sample variances* has better statistical properties for testing the above hypothesis. Therefore, we will form the statistic,

$$F = \frac{s_1^2}{s_2^2} \ \left(\text{or } F = \frac{s_2^2}{s_1^2} \right) \tag{1.24}$$

to test $H_0 : \sigma_1^2 = \sigma_2^2$ versus some H_1. In general, F is formed by dividing the larger s_i^2 by the smaller one. If H_0 is true (and if $F = \frac{s_1^2}{s_2^2}$), then F is distributed as an "F" distribution with $n_1 - 1$ and $n_2 - 1$ degrees of freedom (d.f.), respectively (or $n_2 - 1$ and $n_1 - 1$ degrees of freedom, respectively). The degrees of freedom indicate the number of pieces of information needed to estimate s_1^2 and s_2^2 given $\overline{x}_1$. $\overline{x}_2$., respectively. The F-distribution was discovered by George Snedecor and was named after the famous statistician, R.A. Fisher. The family of F-distributions like the χ^2 family of distributions is a family of skewed distributions. In Figure 1.9, plots of F-distributions with 4 numerator d.f. and 4, 10 and 40 denominator d.f. are displayed. The F-distribution also has the peculiar property that $F_{d_1,d_2,p} = 1/F_{d_2,d_1,1-p}$. Thus, for example, $F_{2,5,.975} = 8.43 \Longrightarrow F_{5,2,.025} = 1/8.43 \approx 0.1186$.

Figure 1.9 *Plots of a Family of F-Distributions with 4 D.F. in the Numerator*

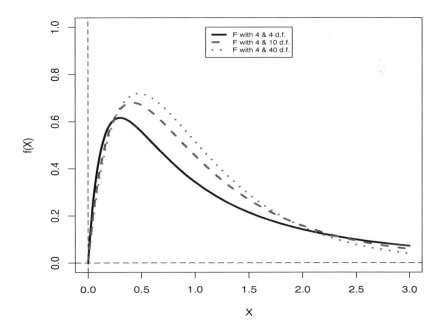

Example 1.18. In doing the test of equality of means earlier, we really should have first tested $H_0 : \sigma_1^2 = \sigma_2^2$ versus $H_1 : \sigma_1^2 \neq \sigma_2^2$, that is, whether or not there was any significant difference in the population variances between the two groups.

To perform this test, we form

$$F = \frac{s_2^2}{s_1^2} = \frac{(0.00514)^2}{(0.004695)^2} = \frac{0.00002642}{0.00002205} \approx 1.199 \sim F_{9,14} \text{ under } H_0 .$$

Notice that the test statistic, that is, F, in the example was formed by *dividing the larger sample sum of squares* (in this case, s_2^2) *by the smaller sample sum of squares* (s_1^2). This is typically done because published F tables have values that are all ≥ 1. This convention saves paper!! If this test were done at the 5% α-level, then we would conclude that there was not enough evidence to declare a difference in the variances of the two samples. Thus, we did the right thing by pooling variances in the earlier analysis.

1.10.1 The t-Test Revisited

In the data example, everything with regard to the "nuisance" parameters, σ_1^2 and σ_2^2 worked out well, that is, we did not reject $H_0 : \sigma_1^2 = \sigma_2^2$ and so, concluded that it was reasonable to pool the variances when performing the t-test. But, what if we rejected $H_0 : \sigma_1^2 = \sigma_2^2$? That is, *how do we test the equality of means when the variances between two groups are significantly different?* One way to do this is to use a method first proposed by Satterthwaite [130]. First,

$$\text{form} \quad T_s^* = \frac{\overline{x}_1 - \overline{x}_2}{\sqrt{\frac{s_1^2}{n_1} + \frac{s_2^2}{n_2}}} \tag{1.25}$$

Under H_0, T_s^* is *not* distributed as a central t-distribution because we cannot pool the variances and, thus, cannot get the correct degrees of freedom. However, T_s^* is close to being a t-distribution under H_0, if we modify the degrees of freedom a little. The most widely used modification of the degrees of freedom in this situation was provided by Satterthwaite [130]. He proposed to construct

$$d' = \frac{(s_1^2/n_1 + s_2^2/n_2)^2}{\frac{(s_1^2/n_1)^2}{(n_1-1)} + \frac{(s_2^2/n_2)^2}{(n_2-1)}} .$$

Then, the d' were *truncated* to the nearest integer, d''. (For example, if $d' = 10.9$, then $d'' = 10$.) So, under $H_0 : \mu_1 = \mu_2$, $T_u^* \sim t_{d''}$. We then refer to the familiar t-table (or a computer program) to find the appropriate critical value and proceed as we did in the other unpaired t-test.

Example 1.19. Suppose that two groups of subjects, one a group with borderline obese individuals and one with normal weight individuals, are measured for systolic blood pressure. Suppose also that there are 13 individuals in the first group and 25 in the second group, respectively. The raw data is given in the table below.

Normal	131	121	118	121	115	124	120	122	115
	130	120	121	128	110	120	119	121	126
	131	119	126	123	119	125	113		
Nearly obsese	137	135	149	139	111	144	130	128	140
	146	134	150	125					

Testing $H_0 : \sigma_1^2 = \sigma_2^2$ yields $F = s_2^2/s_1^2 = (10.824)^2/(5.3395)^2 \approx 4.11 \sim F_{12,24}$. Since $F_{12,24,.995} \approx 3.42$, we reject H_0 at the $\alpha = 0.01$ level. An analysis using PROC TTEST in SAS is presented below. The results given below summarize both the pooled and unpooled (Satterthwaite) versions of the t-test as well as the test of equality of the variances from the two samples.

```
data tt1;
input y @@;
cards;
131 121 118 121 115 124 120 122 115 130 120 121 128 110 120 119 121 126
131 119 126 123 119 125 113
;
data tt2;
input y @@;
cards;
137 135 149 139 111 144 130 128 140 146 134 150 125
;
run;

data tt; set tt1(in=d1) tt2(in=d2);
if d1 then group = 1;
else group = 2;
run;

options ls=50 pageno=1 nodate;
proc ttest;
class group;
var y;
title 'Example of a two-sample t-test with unequal variances';
run;
```

OUTPUT

```
-----------------------------------------------------------------------
       Example of a two-sample t-test with unequal variances        1

                    The TTEST Procedure

                         Statistics

                   Lower CL           Upper CL  Lower CL
Variable  group       N      Mean    Mean      Mean    Std Dev  Std Dev
y                    25    119.32  121.52    123.72    4.1692   5.3395
             1
y                    13    129.46     136    142.54    7.762    10.824
             2
y         Diff (1-2)       -19.76  -14.48    -9.196    6.1965   7.6199

                         Statistics

                   Upper CL
       Variable  group    Std Dev   Std Err   Minimum    Maximum
       y                    7.428    1.0679       110        131
                 1
```

y		17.868	3.0021	111	150
y	2 Diff (1-2)	9.8979	2.6055		

T-Tests

Variable	Method	Variances	DF	t Value	Pr > |t|
y	Pooled	Equal	36	-5.56	<.0001
y	Satterthwaite	Unequal	15.1	-4.54	0.0004

Equality of Variances

Variable	Method	Num DF	Den DF	F Value	Pr > F
y	Folded F	12	24	4.11	0.0031

Since the test of equality of variances is highly significant, one would use the results from Satterthwaite's procedure. Notice that the degrees of freedom are much smaller when we adjust for the unequal variances as compared to when we pool the variances (standard deviations). Nevertheless, the overall inference regarding the mean level of blood pressure is unchanged either way as we can conclude that it is higher in individuals who are slightly obese.

1.11 The Hypergeometric Distribution

Another probability distribution that is used for some statistical applications is the *hypergeometric* distribution. Suppose that a group of N subjects in a study are divided into two groups of size $n_1 = a + c$ and $n_2 = N - n_1 = b + d$ and suppose also that a failures are recorded in group 1 and c failures are experienced in group 2. A 2×2 table summarizing this scenario is shown below.

	Group 1	Group 2	
Fail	a	b	$a + b$
$\overline{\text{Fail}}$	c	d	$c + d$
	$a + c$	$b + d$	N

Assuming that we know that $a + b$ subjects fail out of $N = a + b + c + d$ patients, we can calculate the probability of a out of $a + c$ patients failing in population #1 and b out of $b + d$ subjects failing in population #2 as

$$\frac{\binom{a+c}{a}\binom{b+d}{b}}{\binom{N}{a+b}} = \frac{(a+c)! \, (b+d)! \, (a+b)! \, (c+d)!}{N! \, a! \, b! \, c! \, d!}.$$

This distribution has expected value $\frac{(a+b)(a+c)}{N}$ and variance $\frac{(a+b)(a+c)(c+d)(b+d)}{N^2(N-1)}$.

The hypergeometric distribution is used in statistical applications to calculate what is known as *Fisher's exact test.* Fisher's exact test is often used when the *expected* cell counts under a null hypothesis, H_0, of equal failure proportions are < 5 in one or more cells of a 2×2 table. To reduce the amount of calculation time, one can develop a recursion to calculate the probability density function of a hypergeometric distribution. This is accomplished by first arranging the 2×2 table so that the first row has the smallest value. Then start with the case none of the failures are in group 1 leaving all of the failures in group 2. Then arrange the next table so that its margins remain the same as the previous table but the entry in the upper left-hand corner of the new table is incremented by one (see Tables "a" and "$a + 1$" below). By examining successive tables, one can easily see that the probability of Table $a + 1$ is obtained by multiplying the probability of table a times the product of the off-diagonal elements of table a over the diagonal elements of table $a + 1$. A more thorough discussion of this recursion is outlined in Rosner [124].

Table a

a	b	$a + b$
c	d	$c + d$

| $a + c$ | $b + d$ | N |

Table a + 1

$a + 1$	$b - 1$	$a + b$
$c - 1$	$d + 1$	$c + d$

| $a + c$ | $b + d$ | N |

From above, we can rearrange the terms of the probability of a given table (call it "a") as

$$Pr(a) = \frac{(a + b)! \, (c + d)! \, (a + c)! \, (b + d)!}{N! \, a! \, b! \, c! \, d!} \quad \text{and}$$

$$Pr(a + 1) = \frac{(a + b)! \, (c + d)! \, (a + c)! \, (b + d)!}{N! \, (a + 1)! \, (b - 1)! \, (c - 1)! \, (d + 1)!}$$

so that

$$Pr(a + 1) = \frac{bc}{(a + 1)(d + 1)} \times Pr(a) . \tag{1.26}$$

Example 1.20. The following example is taken from Hand et al., dataset 20 and Altman [61, 4], and concerns the relationship between spectacle wearing and delinquency in boys. In a study comparing the proportion of boys who failed a vision test, 1 out of 15 boys who were juvenile delinquents wore eyeglasses whereas 5 out of 7 boys who were not juvenile delinquents wore eyeglasses. The interest is in evaluating whether or not delinquents are more likely to wear glasses than nondelinquents.

Observed Table

	Delinq	Non-delinq	Total
Glasses	5	1	6
No Glasses	2	14	16
Total	7	15	22

Expected Table (Under H_0)

	Delinq	Non-delinq	Total
Glasses	1.909	4.091	6
No Glasses	5.091	10.909	16
Total	7	15	22

1. Notice that the expected counts of two of the cells are < 5.
2. The table is arranged so that the 1st row and 1st column have the smallest margin values.
3. Start with table 0: $\Pr(0) = \binom{7}{0}\binom{15}{6} / \binom{22}{6} = [15!16!]/(9!16!) = 0.06708$. $\Pr(1) = \frac{7 \cdot 6}{1 \cdot 10} \times$ $0.06708 \approx 0.28173$, $\Pr(2) = \frac{6 \cdot 5}{2 \cdot 11} \times 0.28173 \approx 0.38418$, and so on.

Table 0	Table 1	Table 2	Table 3	Table 4	Table 5	Table 6
0 6	1 5	2 4	3 3	4 2	5 1	6 0
7 9	6 10	5 11	4 12	3 13	2 14	1 15

$\Pr(a) = 0.06708$ 0.28173 0.38418 0.21343 0.04925 0.00422 0.00009

Note that the observed table is Table 5 in the list of possible tables given above. The complete distribution of these tables is shown in Figure 1.10.

Figure 1.10 *Hypergeometric Distribution Associated with the Analysis of Spectacle and Delinquency Data*

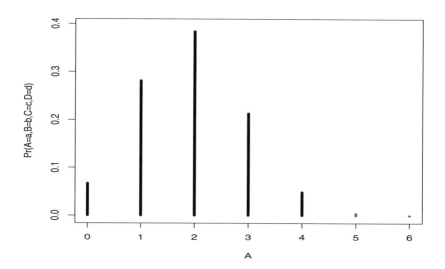

To get a p-value, we're interested in taking the "tail" probability, that is, the sum of the probability of obtaining Table 5 <u>plus</u> all of the probabilities to its right or its left. The sum of the probability of obtaining Table 5 and every table to its right is $0.0001 + 0.00422 \approx 0.0043$. The sum of the probability of obtaining Table 5 and every table to its left is $0.0671 + 0.2817 + 0.3842 + 0.2134 + 0.0493 + 0.0042 = 0.9991$. But which tail should we choose? To calculate a p-value, we take *the more extreme tail*, which is the right tail in this case. The two-sided p-value is $p = 2 \times \left[\min(0.0043, 0.9991) \right] \approx 2 \times 0.0043 = 0.0086$, a statistically significant result. Actually, the calculation is simpler than it appears. If one "tail" probability is *less than* 0.5, then, by definition the other tail has to be *greater than* 0.5. The "more extreme tail" will be the one that is < 0.5. So, in this case, we really only have to sum the right-hand tail because $0.0043 \ll 0.5$. This analysis can be performed in R with the following command: `fisher.test(matrix(c(5, 2, 1, 14),nrow = 2))`. It can also be obtained as part of the output in `proc freq` in SAS.

1.12 The Exponential Distribution

The last distribution we'll consider in this chapter is the *exponential* distribution. As we will see in Chapters 10 and 11, this distribution is widely used in the study of *survival analysis*. If X is distributed as a exponential with parameter, λ, then one formulation of the probability density function for X is

$$f(x) = \begin{cases} \lambda e^{-\lambda x}, & \text{for } x \geq 0 ; \\ 0, & \text{otherwise.} \end{cases} \qquad (1.27)$$

From this, with simple calculus, the cumulative density function can be derived as $F(x) = 1 - e^{-\lambda x}$, $x \geq 0$; 0, elsewhere. Overlayed graphs of the exponential probability density function for $\lambda = 1, 0.75$ and 0.20 are given in Figure 1.11.

The expected value and variance of X are $E(X) = 1/\lambda$ and $\text{Var}(X) = 1/\lambda^2$, respectively. The exponential distribution has the property that it is *memoryless*. What this means is that its conditional probability obeys

$$\Pr(X > x + t \mid X > x) = \Pr(X > t) \text{ for all } x > 0 \text{ and } t \geq 0 . \qquad (1.28)$$

In the context of survival analysis, what this means is that if someone's time to relapse from a disease follows an exponential distribution, then their probability of having a relapse given that they're relapse-free to a given point in time doesn't change based

Figure 1.11 *Graphs of Exponential Probability Density Functions for Different Values of* λ

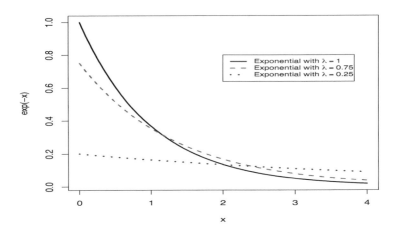

on the point of time being considered. One application where this assumption is apropos is in studies that explore certain types of early stage cancers.

1.13 Exercises

1. Let $G_i = \{$Child i in a family of four is a girl$\}$, $i = 1, \ldots, 4$ and let $Y = \#$ of girls in a family of four children.

    (a) Suppose we are interested in the event that at least one child in a family of four is a girl. Write this event in terms of the G_i and also in terms of Y.

    (b) Repeat part (**a**) for the case that we are interested in the event that exactly two of the four children are girls.

    (c) Find the probability that exactly two of the four children are girls *given that* at least one child is a girl.

2. Assuming that the probability that a given child is a girl is equal to 0.49 and that the sexes of two or more children are independent events, calculate the probabilities associated with the events described in the previous problem.

3. In Example 1.4, calculate the predictive values positive and negative, respectively, given the following information:

    (a) fix the sensitivity and specificity as given in Example 1.4 but vary the population disease prevalence from 1% to 10% in increments of 1%;

    (b) fix the population disease prevalence and specificity as given in Example 1.4 but vary the sensitivity from 80% to 98% in increments of 2%; and

(c) fix the population disease prevalence and sensitivity as given in Example 1.4 but vary the specificity from 80% to 98% in increments of 2%.

(d) From your calculations in parts **(a) – (c)**, which property appears to affect predictive value positive the most? What about predictive value negative?

4. In Example 1.11, use a normal approximation to the Poisson distribution to estimate $\Pr(X \geq 6)$ and compare your answer to the exact result in the example.

5. Create a table for a t-distribution similar to that produced in section 2.3.2 (or as in Fisher and Van Belle [46] and Rosner [124]). However, for the t-distribution, it is customary to find the t-value for a given number of degrees of freedom and for a particular percentile of the distribution as in Rosner [124]. Thus, if $T \sim t_r$, where r is the number of degrees of freedom, then for a given percentile, say, 95, you need to calculate the value of t so that $\Pr(T \leq t) = 0.95$. For your table, you can calculate the 75th, 80th, 95th, 97.5^{th}, 98th, 99th, 99.5^{th} and 99.95^{th} percentiles. For the degrees of freedom, use each integer value between 1 and 30 inclusively and also, 35, 40, 50, 60, 75, 80, 90, 100, 120, and ∞. (Hint: One way to do it is using the qt function in R or S-plus. For example, the 95th percentile of a t-distribution with 10 degrees of freedom (to four decimal places) could be obtained as round(qt(.95,10),4) and would yield the value 1.8125. (For more information, type help(qt) or ?qt in R.)

6. Create a table for a χ^2 distribution similar to that produced in the previous problem. (Hint: You'll replace the qt function with the qchisq function.)

7. For the data in Example 1.17, do a two-sided t-test for comparing the mean dopamine levels between the psychotic and nonpsychotic patients assuming that the two variances cannot be pooled. Is there any difference in the significance levels between the test where the variance is pooled and the test where the variances are not pooled? (Hint: One can easily do the calculations for this by hand or alternatively, use a procedure like PROC TTEST in SAS or t.test in R, which can produce t-tests using both pooled variances and unpooled variances.)

CHAPTER 2

Use of Simulation Techniques

2.1 Introduction

One of the most common techniques for evaluating statistical methods is to perform a series of "numerical experiments," which often involve performing "Monte Carlo" simulations assuming a set of conditions. To do this, the use of "random number generators" are often employed.

2.2 What Can We Accomplish with Simulations?

The development of theoretical statistics in the 20th century has allowed scientists to make powerful inferences on the distributions of populations based both on very large and very small random samples. For large samples, *asymptotic inference* is made by mathematically determining how some value is distributed when the sample size is infinitely large. Hence, such inference will be very accurate if a given sample size is large. Likewise, for sample sizes of data that are very small, often, *exact distributions* can be derived mathematically. A typical example is *Fisher's exact-test* for comparing two proportions as was illustrated in Chapter 1 of this book. In practice, however, many datasets are of *moderate* size, that is, sample sizes of between 10 and 100. This range of sample sizes can be too large for employing exact procedures but may be too small to make one feel comfortable with making "large sample approximations." Simulations allow us to explore the nature of inference in these intermediate-sized datasets and possibly see how well asymptotic inference works in these cases.

Another useful purpose for performing simulations is to help answer the question "What if the assumptions of my analysis are wrong?" For example, in a regression analysis, what effect does having errors that are not normal have on making an inference about the parameter or the prediction of an outcome variable? Or, what if the errors are correlated but the correlation is not properly accounted for? In these cases, if the estimates or resulting inference are not very sensitive to misspecification of the model itself or the underlying distributional properties of the error structure, then the method is said to be *robust*.

A related problem that can be uniquely addressed by simulations is that of understanding the statistical consequences when someone does an "improper" or possibly a proper but new type of statistical analysis of data. For example, in the case of comparing the proportions between two groups, what are the implications of using a t-test of the two proportions instead of the Z or χ^2 tests? Such questions are very difficult or impossible to answer in problems with intermediate-sized samples using purely mathematical arguments. But, by the use of simulations, one can obtain operating characteristics of tests associated with different assumptions and assess the robustness or efficiency of a given method of interest.

A fourth practical problem addressed by simulation studies for analyses of multiple regression data involves situations where some of the predictors of the outcomes or the outcomes themselves are missing. This is a common situation occurring in studies involving animal or human subjects. Simulations help us to understand how sensitive models are to how much missing data there is? Also, they help us understand how the *nature* of the missingness affects modeling the data, e.g., is the missingness related to previous observations? Future observations? Or is the missingness completely at random?

A fifth useful purpose for employing simulations is when a problem of interest is not easily tractable mathematically. Simulations can sometimes help investigators formulate mathematical solutions or to explore what approximate solutions can be employed to solve the particular problem.

2.3 How to Employ a Simple Simulation Strategy

The strategy employed to conduct many statistical simulations is actually quite simple. The idea is to generate data by creating a model and adding some type of random variation or "noise" to the model. The model is then fit and estimates of the parameters or other features of the model are stored. The process is repeated many times so that a distribution of the estimated parameter or feature of interest can be estimated. Often, some summary statistic on the feature such as its mean is computed. A schematic representation of the process is outlined in the Figure 2.1.

A practical example of how to employ the strategy outlined in Figure 2.1 is given in Example 2.1. In this example, we want to examine the robustness of the one-sample t-test to the assumption that the underlying error structure is normal.

Figure 2.1 *Outline of a Simple Simulation Strategy.*

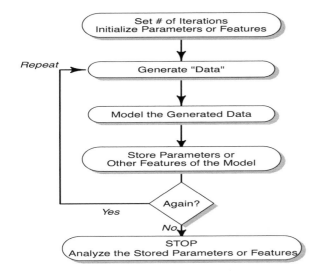

Example 2.1. A simulation for testing the robustness of a t test For a t-test comparing a sample mean to some known value, μ_0, we assume that the sampled data are from a normal distribution. To test the hypothesis, $H_0 : \mu = \mu_0$ versus $H_1 : \mu \neq \mu_0$, we form a test statistic, $T = (\overline{x} - \mu_0)/(s/\sqrt{n})$, where n, $\overline{x}$ and s are the sample size, mean and standard deviation, respectively. The *null hypotheis*, H_0 is rejected if the p-value is less than some prespecified α-level, usually 0.05. Of course, the α-level is the probability of falsely declaring a significant result. Suppose we wish to investigate the effect of a nonnormal error structure on the α level of a one-sample t-test. To make this concrete, we'll set $\mu_0 = 5$ and fix three different sample sizes at 5, 15, and 30. We'll choose the errors to be normally distributed and then compare α-levels to cases where the errors are uniformly distributed, $U(-6, 6)$.

```
> ############################################################
> #    The following is a simple example of how a simulation
> # can be performed in R. In this example, we generate
> # two distributions each with sample size, n.
> # For each realization, one sample t-tests are performed
> #    assuming that
> #    1) the first distribution is normally distributed
> #    2) the second distribution is uniformly distributed
> #
> #    We repeat this process a large number of times and
> # obtain the distributions of the test statistics
> # and p-values.
> ############################################################
> # FIRST, initialize the parameters for the simulation
> N <- 10000  # Input the number of simulations desired
> n <- 5     # Input the number of observations in group 1
> mu <- 0    # Input the mean value for group 1
> sd <- 1  # Input the standard deviation of group 1
> out <- matrix(-9999,nrow=N,ncol=4) # Initialize results
```

```
>
> for (i in 1:N){
+     x <- rnorm(n,mu,sd) # Generate data for group 1
+     y <- runif(n,-6,6) # Generate data for t-test
+     t1.out<-t.test(x,NULL)
+     t2.out<-t.test(y,NULL)
+ #   OUTPUT VARIABLES OF INTEREST
+     out[i,] <- c(t1.out$statistic,t1.out$p.value,
+                  t2.out$statistic,t2.out$p.value)
+ }
> prop1.reject <- sum(ifelse(out[,2]<=0.05,1,0))/N # Test 1 reject prop
> prop2.reject <- sum(ifelse(out[,4]<=0.05,1,0))/N # Test 2 reject prop

OUTPUT
------
SAMPLE SIZE = 5; 10,000 realizations
> prop1.reject
[1] 0.0512
> prop2.reject
[1] 0.0671

SAMPLE SIZE = 15; 10,000 realizations
> prop1.reject
[1] 0.0477
> prop2.reject
[1] 0.056

SAMPLE SIZE = 30; 10,000 realizations
> prop1.reject
[1] 0.05
> prop2.reject
[1] 0.0494
```

As can be seen from the results of this simulation, the one-sample t-test is fairly robust at least with respect to the α-level at the 5% level, to an error structure that is not normal (Gaussian). The average α-levels are very close to the nominal value even when the error structure is not normal. A more thorough examination of the performance of the one-sample t-test with nonnormal errors would involve choosing different α-levels and also examining the *power* of the test under different alternative hypotheses.

2.4 Generation of Pseudorandom Numbers

To begin our discussion of the process of simulating random sequences of numbers, we must recognize that because we are constructing algorithms to create such sequences, in reality, the numbers we create are really not "random" but, rather, *pseudorandom*. For practical purposes, we'll henceforth drop the "pseudo" syllable.

The basis for generating any type of random numbers is contingent on accurately generating a sequence of independent uniform random numbers that display no particular pattern. Recall that the standard uniform distribution has probability density function, $f(x) = 1$, if $0 \leq x \leq 1$, and 0, elsewhere. A set of random numbers from nearly any type of continuous distribution can be generated from a standard uniform distribution. But, how can one generate values that are both as random as possible and are as close to being uniform as possible?

The first step in the process of "random" number generation is to specify a "seed."

In most cases, the seed is fed into an algorithm that produces, as its output, numbers that are uniformly and randomly distributed. The seed is usually an integer that can be specified directly by the user, or can be a default value from some program or which can be identified through a "random process," for example, the number of milliseconds from midnight to a given time. Once the seed is specified, one would want the following properties for the outputted numbers:

1. To be as close to uniformly distributed as possible;

2. To each given number in the sequence of random numbers output to be indepedent of the previous numbers generated; and

3. To not have any type of periodicity existing among the successive numbers generated in a long sequence.

Two efficient methods for reliably generating uniform random numbers involve the use of *multiplicative congruential* generators and *Tausworthe* generators. These methods have been the source of a wealth of mathematical and statistical research and are beyond the scope of this book. However, one can refer to a number of very good references (Deng and Lin [28], Fishman and Moore [48], Gray [59], Kennedy and Gentile [77] Tausworthe [148] and Thompson [151]) to further explore this area.

2.5 Generating Discrete and Continuous Random Variables

Assuming that we can produce a set of numbers that closely follow a uniform distribution, the question becomes "How can we use this ability to produce uniform random numbers (deviates) to produce random numbers that follow other distributions?" Luckily, it turns out many different types of distributions can be produced by using the *Inverse Transform Method*, see Lange [85] and Ross [125]. The idea is that if X is a continuous random variable with cumulative density function, $F(x)$ (see Figure 2.2), then $U = F(X)$ is uniformly distributed on $[0, 1]$. Even if $F(x)$ is not continuous, then $Pr[F(x) \leq t] \leq t$ is still true for all $t \in [0, 1]$. Thus, if $F^{-1}(y)$ is the smallest x having the property that $F(x) \geq y\}$ for any $0 < y < 1$, and if U is uniform on $[0, 1]$, then $F^{-1}(U)$ has cumulative density function $F(x)$.

Using the inverse transform method, we can directly generate random numbers (deviates) of many distributions given that we have already generated (psuedo)random *standard uniform* deviates.

Example 2.2. One discrete distribution that can be easily generated by the use of standard uniform random variables is the *Bernoulli* distribution defined as $Pr\{X = 0\} = 1 - p$; $Pr\{X = 1\} = p$; 0, elsewhere where $p > 0$. One can generate this by generating each U as

$$X = \begin{cases} 1, & \text{if } U \leq p \\ 0, & \text{if } U > p \end{cases}.$$

The code below, which can be run in R or S-plus, produces 100 Bernoulli trials with $p = 0.6$.

Figure 2.2 *Examples of Discrete and Continuous cumulative density functions*

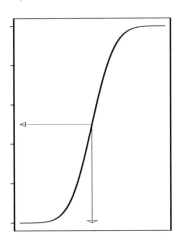
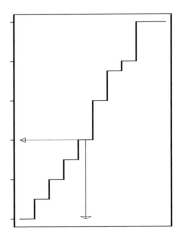

```
> ######################################################
> # Example: Generate 100 Bernoulli trials with p = 0.60
> ######################################################
> NUM.OBS <- 100
> p <- rep(0,NUM.OBS)
> x <- runif(NUM.OBS)
> p[x<=.6] <- 1
> p
  [1] 0 1 1 1 1 0 1 1 1 1 0 0 1 0 0 1 0 0 1 0 1 0 1 1 0 0 0 1 0
 [28] 1 0 1 1 1 0 1 1 0 0 1 0 1 0 0 1 0 0 0 0 1 1 1 0 1 1 1
 [55] 0 1 1 0 1 0 1 0 1 0 1 0 1 1 1 0 1 1 0 1 1 0 1 1 0 1 0 1 0
 [82] 0 1 1 0 1 0 1 1 0 0 1 1 1 1 0 0 1 1 1
> table(p)
p
 0  1
43 57
```

A program in SAS to produce the same type of simulation is given below.

```
options nodate nonumber;
data r (drop=i);
  y=round(mod(1000*time(),1000000));
  num_obs = 100;
  true_p=0.6;
  do i=1 to num_obs;
    u=ranuni(y);
  if u le true_p then p = 1; else p = 0;
  output;
  end;
run;
proc freq; table p; run;
```

yielding the output

The SAS System

The FREQ Procedure

p	Frequency	Cumulative Frequency	Cumulative Percent	Percent
0	43	43	43.00	43.00
1	57	100	57.00	100.00

Of course, for a given sample of 100, we may not observe exactly 60 1's, but over the long run we would expect about 60% of the observations to be 1's. If we sum n independent Bernoulli random variables, we have a *binomial* random variable, that is, if $X \sim b(1, p)$ then $Y = \sum_{i=1}^{n} X_i \sim b(n, p)$.

Ross [125] presents efficient algorithms to produce binomial and multinomial distributions that work well in computer languages or packages that are effective at looping and performing elementwise operations. For statistical packages, such as SAS IML, S-plus or R, that have functions which operate primarily on vectors or matrices, the code above would be efficient to implement the creation of a sample of random variables with a binomial distribution.

One major distribution that cannot be generated by use of the inverse transform method is the normal distribution. However, an efficient methods introduced by Marsaglia and Bray [96] allows one to create a set of normal random numbers from a set of uniform random numbers. This algorithm or other variations of it are built into modern-day statistical packages so that one doesn't have to write any sophisticated code to generate normal random numbers. For example, in R or S-plus, the command, rnorm(1000,2,4) would generate $1,000$ random normal deviates with a mean, $\mu = 2$ and a standard deviation, $\sigma = 4$. In SAS, the following commands would give similar results:

```
DATA random (drop=i);
  DO i=1 to 1000;
    x = 2+ 4*NORMAL();
    OUTPUT;
  END;
RUN;
 proc print; run;
```

Example 2.3. In probability and statistics, the role of a *random walk* has been of great historical and theoretical importance. Feller [40], in his classic 1968 book, *An Introduction to Probability Theory and Its Applications*, describes *binary* random walks in great detail with excellent applications to classic gambler's ruin problems. *Continuous* random walks are of great importance in many applications including smoothing splines and the formulation of early stopping rules in clinical trials.

In this example, we simulate the realization of a continuous random walk at discrete intervals as follows. Suppose we have a set of random variables, $Z_1, Z_2, ..., Z_n$, each of which has a standard normal distribution, that is, $X_i \sim N(0,1)$, $i = 1, \ldots, n$. We also assume that random variables are independent of each other. To form the realization of a continuous random walk, we will create $X_1, X_2, \ldots, X_n$ by adding all of the Z_i's in the sequence up to including the current value. Mathematically, we write this as $X_i = \sum_{k=1}^{i} Z_k$. Hence, each X_i has the properties that $E(X_i) = 0$ and $\text{Var}(X_i) = i$ so that the standard deviation of the process *increases* with iteration number. Also, unlike the Z_i's, the X_i's are correlated.

In this example, we'll plot the sequence of X_i's against each iteration, i. The code to do this for, say, $10,000$ X_i's is implemented very easily in R with the following code.

```
> plot(1:10000,cumsum(rnorm(10000)),type="l",xlab="iteration",ylab="f(x)")
> abline(h=0,lty=2)
```

The graph of this is given in Figure 2.3.

Figure 2.3 *Realization of a One-Dimensional Continuous Random Walk*

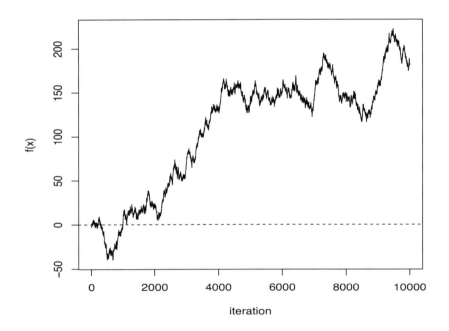

Once an efficient method for generating a set of univariate normal random numbers is established, it is quite easy to generate bivariate or multivariate random numbers. Consequently, one can extend the previous one-dimensional random walk to a two-dimensional random walk as is demonstrated in the example below.

Example 2.4. The *R* code given below created the a discretized version of a two-dimensional continuous random walk. A plot of the two-dimensional random walk is displayed in Figure 2.4.

```
library(mvtnorm) #<-- First must install "mvtnorm" from the web

# Set parameters
N<-1000
mu<-c(0,0)
Sigma<-matrix(c(1,0,0,1),nc=2)

x<-rmvnorm(N,mu,Sigma)
X<-c(0,cumsum(x[,1])); Y<-c(0,cumsum(x[,2]))

# Plot
plot(X,Y,type="l",main="",lwd=2); abline(v=0,lty=2); abline(h=0,lty=2)
points(0,0,pch=8,col="red",lwd=2); text(2,0,"Begin",col="red",cex=.75,lwd=2)
points(X[length(X)],Y[length(Y)],pch=8,col="red",lwd=3);
text(X[length(X)]+2,Y[length(Y)],"End",col="red",cex=.75,lwd=3)
```

Figure 2.4 *Simulation of a Two-Dimensional Continuous Random Walk*

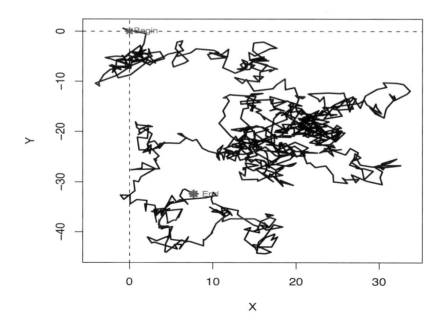

2.6 Testing Random Number Generators

Empirical tests of random number generators usually involve goodness-of-fit tests
and tests of patterns of successive random numbers. In most cases, we can limit our
tests to examining uniform random deviates because deviates from other distributions can bederived from uniform deviates. In the case of using goodness-of-fit tests,
however, we can directly test deviates from any distribution, uniform or not.

2.6.1 Goodness-of-Fit Tests

The two most common goodness-of-fit (GOF) tests are the χ^2 *goodness-of-fit* and
the *one sample Kolmogorov–Smirnov goodness-of-fit test*. The χ^2 GOF test involves
looking at the statistic, $X^2 = \sum_{i=1}^{n} \frac{(f_i - np_i)^2}{np_i}$ where f_i is the frequency of deviates
falling in the ith "cell" (in, say, a histogram), p_i is the *expected proportion* in the ith
cell *if* the deviate is truly from the distribution of interest, k is the number of cells, and
n is the total number of deviates generated, that is, the sample size. Of course, under
the hypothesis that the deviates are truly from a given distribution, then $X^2 \sim \chi^2_{k-1}$.
For the Kolmogorov–Smirnov (K–S) GOF test, we first order the random deviates
from the smallest to the largest, then calculate an *empirical cumulative density func-
tion*, $\widehat{Pr}(X \leq k) = \widehat{F(k)} = \sum_{i=1}^{n} i/n$ and then take $K = \max_{1 \leq k \leq n} \left| \widehat{F(k)} - F(k) \right|$
where $F(k)$ is the cumulative density function of the true distribution. The value
of K is then compared to a critical value of a known distribution. The functions,
`chisq.gof` and `ks.gof` in *S*-plus can be used to perform the tests above (see examples below). (We will meet the *two-sample* Kolmogorov–Smirnov test later in
Chapter 9.)

2.6.2 Tests of Serial Correlation Using Time Series and Other Methods

Another class of tests useful for testing randomness in a sequence of numbers is that
which identifies particular types of *serial correlation*. This topic is thoroughly discussed in Box and Jenkins [12], Harvey [64], and Shumway and Stouffer [139]. One
common way to identify serial correlation involves the examination of the *autocorrelation function (ACF)* of successive numbers. The underlying assumption made
when estimating the ACF is that a series of random variables, $X_1, X_2, \ldots, X_t, \ldots$
is *weakly stationary*, that is, $E(X_t) = \mu$ for every t. Often, since deviates or other
data are collected over *time*, the sequences, $X_1, X_2, \ldots, X_t, \ldots$, are referred to as a
time series. If, indeed, a time series is weakly stationary, then we define the underlying *autocovariance* as $\gamma(h, 0) = \gamma(h) = E\left[(X_{t+h} - \mu)(X_t - \mu)\right]$. $\gamma(h, 0)$ can
be shortened to "$\gamma(h)$" because the autocovariance of a weakly stationary process is
only a function of the time separation or *lag*, h. Another common way to write the
autocovariance function is $\gamma(s - t) = E\left[(X_s - \mu)(X_t - \mu)\right]$. In the case of the
generation of standard uniform deviates, we would expect to observe *white noise*,
that is, $E\left[(X_s - \frac{1}{2})(X_t - \frac{1}{2})\right] = E\left[(X_{t+h} - \frac{1}{2})(X_t - \frac{1}{2})\right] = 0$ for all s, t, and h.

Also, if $h = 0$ then $E\left[(X_{t+h} - \mu)(X_t - \mu)\right] = E\left[(X_t - \mu)^2\right] = \sigma^2$ if σ^2 is constant for every t.

The *autocorrelation* of the underlying weakly stationary time series is defined to be

$$\rho(h) = \frac{\gamma(h)}{\gamma(0)} = \frac{E\left[(X_{t+h} - \mu)(X_t - \mu)\right]}{E\left[(X_t - \mu)^2\right]} \tag{2.1}$$

and is *estimated as*

$$\hat{\rho}(h) = \frac{\sum_{i=1}^{n-h}\left\{(x_{i+h} - \overline{x})(x_i - \overline{x})\right\}}{\sum_{i=1}^{n}(x_i - \overline{x})^2}. \tag{2.2}$$

The estimated autocorrelation function (ACF) is merely the $\hat{\rho}(h)$ as a function of the lags, h. Thus, in most statistical packages, we can get plots of the $\hat{\rho}(h)$ versus the h. In the example below, we illustrate an ACF for $h = 0, \ldots, 40$ associated with a simulation in which $10,000$ (pseudo)random numbers were generated in S-plus.

Example 2.5. Example of χ^2 and K–S GOF tests and graphical technique for testing a random sequence of numbers. The program below generates $10,000$ random standard uniform numbers. The sequence is then tested for goodness-of-fit and sequential serial correlation.

```
> # S-plus program that 1) generates "random" uniform numbers;
> # 2) uses GOF tests to test for uniformity; and
> # 3) displays 40 lags of an ACF of the distribution of numbers generated.
> N<-10000
> u<-runif(N)
> chisq.gof(u, dist = "uniform", min = 0, max = 1)

Chi-square Goodness of Fit Test

data:  u
Chi-square = 63.632, df = 79, p-value = 0.8959
alternative hypothesis:
    True cdf does not equal the uniform Distn. for at least one sample point.

> ks.gof(u, dist = "uniform", min = 0, max = 1)

One-sample Kolmogorov-Smirnov Test
Hypothesized distribution = uniform

data:  u
ks = 0.0094, p-value = 0.3374
alternative hypothesis:
    True cdf is not the uniform distn. with the specified parameters
> hist(u,ylab="Frequency")
> abline(h=500)
> acf(u)
```

Figure 2.5 *Histogram of 10,000 Uniform Random Numbers*

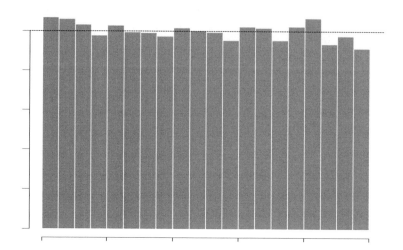

As can be seen from Figure 2.5 and the two corresponding GOF tests, the $10,000$ numbers generated show variability around the flat (uniform) distribution but do not differ significantly from the uniform distribution. The ACF for 40 lags are displayed in Figure 2.6. Also displayed are the 95% confidence intervals about 0. From this, one can see that no lag other than lag 0 (the variance scaled to 1) has an autocorrelation significantly different from 0. This is what we would expect with white noise, that is, no obvious pattern of serial correlation. However, to truly ensure that a sequence of numbers is random is a very difficult task as many very subtle patterns can be present in any given sequence. For example, one could test for "runs" of numbers or sequences of numbers in a long string of generated deviates (see, for example, Draper and Smith [30]).

2.7 A Brief Note on the Efficiency of Simulation Algorithms

In the implementation of any statistical or mathematical computation, whether it is a simulation or not, there are four major considerations for determining its efficiency:

1. efficiency of the underlying simulation algorithm;
2. economy of computer time used in the computations;
3. economy of computer memory used in the computations; and
4. economy of code used to implement the computations.

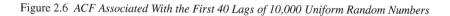

Figure 2.6 *ACF Associated With the First 40 Lags of 10,000 Uniform Random Numbers*

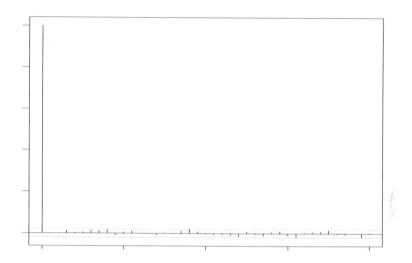

These four considerations are usually interrelated although, sometimes, economy of code, economy of computer memory and economy of computation time are difficult to achieve at the same time. In this book, we primarily emphasize establishing efficiency based on the second item above although, most good algorithms are efficient in all of the above four items. The speed of an algorithm is primarily dependent on the number of operations required. Other factors such as the type of hardware or computer language used also have a bearing on the speed of an algorithm. Furthermore, factors such as how many times a program "reads" and "writes" can affect computational efficiency.

The number of operations in some statistical computations can be reduced by the use of recursive algorithms [see, for example, equations (1.9) and (1.13) for calculating the cumulative density functions of the binomial and Poisson distributions, respectively]. In the computation of statistical simulations, the efficiency issue is magnified as computations are repeated a large number of times. In depth discussions of computational efficiency can be found in Atkinson [9], Gray [59], Lange [85], Ross [125], and Thompson [151]. One issue not often emphasized in these discussions, however, is how various different computer languages or statistical packages compare with respect to the speed of computations. A useful way to get a feel for how fast a particular simulation or other computation is on a particular computer is to calculate the elapsed computer time required. An example of how to do that in R is given below.

Example 2.6. Determining elapsed time for the implementation of a simulation.

```
> #################################################################
> #     The following is a simple example of how a simulation
> # can be performed in R. In this example, we generate
> # two samples of standard uniform random variables.
> # We use a unpaired t-test to compare these and output
> # the test statistic & the p-value.
> # We repeat this process a large number of times and
> # count the number of times that we reject the null hypothesis
> #   to try to estimate the alpha level.
> # This simulation examines one aspect of determining how
> # robust a 2-sample t-test is to the nonnormality of the data.
> #################################################################
> N <- 100000   # Input the number of simulations to be performed
> n1 <- 20      # Input sample size for treatment 1
> n2 <- 20      # Input sample size for treatment 2
> true.alpha <- .01 # Input alpha level
> sd1 <- 1    # Input the standard deviation of the noise (Treatment 1)
> sd2 <- 1    # Input the standard deviation of the noise (Treatment 2)
>
> p.value <- matrix(-1,nrow=N,ncol=1) # Initialize results
>
> simul.time<-
+ system.time(for (i in 1:N){
+     x <- runif(n1,0,sd1)
+     y <- runif(n2,0,sd2)
+     t.xy <- t.test(x,y)
+     p.value[i,]<- t.xy$"p.value"
+ }

> total.number <- length(p.value[is.na(p.value)==F])
> reject <- ifelse(p.value<=true.alpha,1,0)
> number.rejected<-length(p.value[p.value<=true.alpha&is.na(p.value)==F])
> prop.rejected <- number.rejected / total.number
> print(paste(N,"simulations using uniform distributions of sizes",
+ n1,"and",n2))
[1] "1e+05 simulations using uniform distributions of sizes 20 and 20"
>
> print(cbind(total.number,number.rejected,prop.rejected,true.alpha))
       total.number number.rejected prop.rejected true.alpha
[1,]          1e+05            1074       0.01074       0.01

> elapsed.time<-simul.time[3]
> print(elapsed.time)
 elapsed.time
[1] 118.67
```

The elapsed time (in seconds) is the sum of the system Central Processing Unit (CPU) time, the time it takes to execute the commands and the time for reading and writing output. In SAS, one can determine the cpu time by examining the ".LOG" file, which gives this information along with other aspects of how the computation performed. The "real time" in SAS is similar to the "elapsed.time" in R. As can be seen from the output above, the t-test seems to be reasonably robust to non-normality with respect to extreme α-levels. To further investigate the robustness, one would need to assess the ability of the t-tests to reject H_0 under various alternative hypotheses, H_1. This could be accomplished by varying the means of the two distributions and plotting the proportion of rejected tests against the difference of the mean values. By comparing these empirical values against those expected if the distribution were normal would allow one to determine how the nonnormality affects the *power* to detect a difference between two means. "Power" here refers to the probability of detecting a difference between means with a particular statistical test given that an underlying difference between means actually exists.

2.8 Exercises

1. Use your favorite program or package to generate $10,000$ standard uniform deviates using the preferred generator in that package (e.g., runif() in R or S-plus, gen r=unif() in Stata or the UNIFORM function in SAS). Use the following methods to check the validity of the distributional properties:

(a) Plot a histogram of your deviates;

(b) Use a Kolmogorov one-sample test to test whether or not your sample is significantly different from uniformity;

(c) Use a chi-square goodness-of-fit test to test whether or not your sample is significantly different from uniformity;

(d) Make a plot of the ACF versus the first 40 lags; and

(e) State your conclusions from parts **(a)** - **(d)**.

2. Construct three multiplicative congruential generators of the form $k_{i+1} \equiv ak_i \bmod m$ to generate $10,000$ potentially standard uniform [i.e., $\sim U(0,1)$] random deviates. This can be done in your favorite package. The parameters for the three generators should be as follows:

- Generator # 1: $a = 7^5$, $b = 0$, $m = 2^{31} - 1$
- Generator # 2: $a = 69069$, $b = 23606797$, $m = 2^{32}$
- Generator # 3: $a = 501$, $b = 0$, $m = 2^{10}$

Use any random seed you want but use the *same* random seed for all three generators. Use the tests given in parts **(a)**–**(e)** of Exercise 2.1 to evaluate your generators. [Hint: See the "%%" functions in R or S-plus or "MOD(x,d);" (x and d integers) in SAS.]

3. Consider again the simulation in Example 2.1. In that simulation, we compared one-sample t-tests comparing $H_0 : \mu = 0$ versus $H_1 : \mu \neq 0$ applied to data that were distributed $N(0,1)$ and as $U(-6,6)$.

(a) Why were parameters in the uniform distribution chosen to be -6 and 6?

(b) Reproduce the results in Example 2.1 but with only $N = 2000$ replications in the simulation.

(c) Repeat part **(b)** but with distributions changed to $N(0.2,1)$ and $U(-5.9,6.1)$.

(d) Repeat part **(c)** but with distributions changed to $N(0.4,1)$ and $U(-5.8,6.2)$.

(e) Overlay your results for the normal and uniform simulated distributions on a single plot for parts (b) – (d).

## CHAPTER 3

The Central Limit Theorem

3.1 Introduction

One of the fundamental notions of statistical reasoning is the idea that one can make reasonable inferences from some sample of observations randomly drawn from a population of potential observations. Figure 3.1 depicts a crude schema of how statistical inference is made.

Figure 3.1 *An Overview of How a Population and a Random Sample are Linked Through Inferential Statistics*

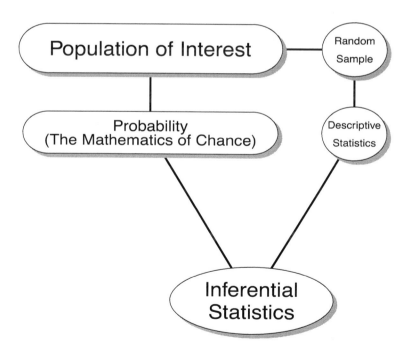

To start a process of making statistical inference, one must first define a *population* from which a *sample* of observations can be drawn. The underlying population has some distribution that can usually be characterized by a mean, a standard deviation, skewness, kurtosis, and other parameters. One attempts to draw a *random* sample of observations from the population that theoretically involves a scheme in which all observations in the population have equal chances of being drawn for the sample. It is assumed that each sample represents only one realization of the infinite possible samples that could be drawn from an infinitely large population and that other possible samples would yield observations that would vary depending on the particular sample that is drawn. Then, using, for example, the *sample* mean and standard deviation together with the laws of probability, one can make *statistical inferences* about population parameters.

One of the most important distributions used in statistical inference is the normal (Gaussian or "bell-shaped") distribution. One reason that the normal distribution has a central role in statistics is because of a remarkable property of the arithmetic mean of a sample that is described by what is known as the *Central Limit Theorem*. (Of course, when a concept is titled with a word like "central" or "fundamental" in it, then one should usually have the idea that the notion is of high importance.) The Central Limit Theorem essentially states if one takes many different samples of size n where n is reasonably large from any distribution with a known mean and variance, then the *mean values of those samples will approximate a normal distribution* even though the original samples may be drawn from a population with a distribution that is highly skewed or otherwise not bell-shaped.

In the remainder of this chapter, the Strong Law of Large Numbers and the Central Limit Theorem will be stated and we will visualize how they work with simulated data. We will also demonstrate a situation in which the Central Limit Theorem does not apply. Finally, some of the resulting implications of the Strong Law of Large Numbers and the Central Limit Theorem will be summarized.

3.2 The Strong Law of Large Numbers

An important prelude to the formulation of the Central Limit Theorem was the discovery of what is now known as the "Strong Law of Large Numbers," which essentially states that the chances that the arithmetic mean of a random sample will differ substantially from the population mean will be close to 0 as the sample gets large. This can be formulated mathematically as follows: if $X_1, X_2, \ldots, X_n$ are random variables, each taken from a distribution with a common mean, μ, and if $\overline{X}_n = \frac{X_1 + \ldots + X_n}{n}$, then

$$\Pr\left\{ |\overline{X}_n - \mu| > \epsilon \right\} \to 0 \text{ as } n \to \infty \tag{3.1}$$

for any $\epsilon > 0$ no matter how small it is (Feller [40], Hsu [72]). In other words, the probability that the sample mean differs from the population mean by even a small amount will go to 0 as the size of the sample gets very large. This property can be visualized by simulating the proportion of heads one would get if she or he were flipping a coin many times.

Example 3.1. Consider a set of Bernoulli trials, each of which, has a probability of success equal to 0.5. That is, $X_i \sim b(1, 0.5)$ for $i = 1, \ldots, n$. Recall that Bernoulli trials have values of 1 for trials that are "successes" and 0 for trials that are "failures." Each trial is also independent of all other trials. We can think of each trial as being a coin flip with a success being a "Heads" and a failure being a "Tails." In this case, we have a *unbiased* coin. If we take the sum of the values of these Bernoulli trials and divide by the number of trials, we get an estimate of the sample mean or equivalently, an estimate of the proportion of successes. In our example, the sample proportion, according to the Strong Law of Large Numbers, should go to the true proportion of 0.5. Summarized in Figure 3.2 is a single realization of 2000 Bernoulli trials whose mean values are plotted against the number of trials. Notice that the mean (or proportion of successes) gets closer to 0.5 as the number of trials increases. The variation of the estimated proportion also decreases as the number of trials increases.

Figure 3.2 *Proportion of Successes by the Number of Trials for 2000 Bernoulli Trials*

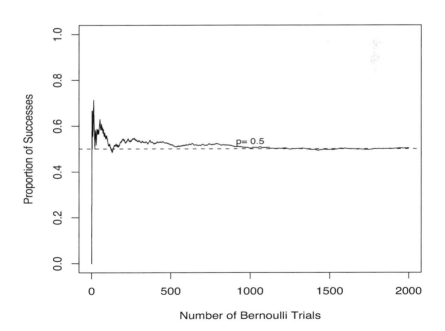

3.3 The Central Limit Theorem

3.3.1 The Statement of the Central Limit Theorem

If $X_1, X_2, \ldots, X_n$ represent a sample of n independently and identically distributed (i.i.d.) random variables, each with mean μ and variance $\sigma^2 < \infty$, then the distribution of the *arithmetic* mean of the X_i's is *approximately normally* distributed as n gets large with mean μ and variance σ^2/n. This property is true *even if* the individual X_i's *are not normally distributed*. More formally, this is written as follows:

If $X_1, X_2, \ldots, X_n$ are independently and identically distributed $D(\mu, \sigma^2)$ for any distribution D, then

$$\frac{\sum_{j=1}^{n} X_i}{n} \; = \; \overline{X} \; \dot{\sim} \; N(\mu, \sigma^2/n) \text{ as } n \to \infty, \tag{3.2}$$

where, " $\dot{\sim}$ " means "approximately distributed as."

3.3.2 A Demonstration of How the Central Limit Theorem Works

One way to see how this important theoretical result works in practice is through simulation. A simulation allows us to observe *many* realizations of samples instead of only one. To do this, we will generate different samples from a distribution that is *not* normal and then take the mean values of each of the samples. A histogram of those mean values should look somewhat like a normal distribution. This is demonstrated in the example given below.

Example 3.2. To demonstrate one example of how the Central Limit Theorem works, we generated 100 samples, each of size 30, of random deviates that have a standard uniform distribution. Recall that a standard uniform distribution is a continuous and flat distribution that takes on values between 0 and 1, inclusively. The expected value (mean) of a uniform distribution is $1/2$ and the variance is $1/12$. Thus, both μ and σ^2 exist and thus, the Central Limit Theorem should hold. Listed below are the means of the 100 samples of 30 uniform distributions.

```
> round(mxu,3)
  [1] 0.519 0.550 0.501 0.572 0.632 0.558 0.519 0.473
  [9] 0.544 0.582 0.516 0.616 0.393 0.456 0.515 0.494
 [17] 0.452 0.532 0.476 0.472 0.597 0.558 0.468 0.572
 [25] 0.525 0.440 0.491 0.499 0.546 0.469 0.477 0.467
 [33] 0.376 0.509 0.497 0.430 0.461 0.507 0.606 0.510
 [41] 0.487 0.464 0.419 0.467 0.482 0.519 0.512 0.418
 [49] 0.490 0.513 0.526 0.535 0.583 0.512 0.434 0.494
 [57] 0.533 0.585 0.629 0.521 0.527 0.545 0.516 0.452
 [65] 0.455 0.547 0.473 0.534 0.487 0.598 0.464 0.387
 [73] 0.401 0.421 0.538 0.399 0.518 0.505 0.376 0.600
 [81] 0.515 0.467 0.453 0.511 0.485 0.539 0.515 0.641
 [89] 0.529 0.487 0.500 0.482 0.477 0.567 0.511 0.424
 [97] 0.487 0.412 0.533 0.529
```

A summary of the 100 means is given with the following R (or S-plus) or command:

```
> summary(mxu)
  Min. 1st Qu. Median  Mean 3rd Qu.  Max.
 0.376   0.468  0.508 0.503   0.533 0.641
> stdev(mxu)
[1] 0.057079
```

Notice that the mean value is near the mean value of a standard uniform distribution and that standard deviation of the mean value is near the *standard error of the mean* for 30 uniform random variables, that is, $\sqrt{\frac{1/12}{30}} \approx \sqrt{0.0027778} \approx 0.052705$.

The 100 histograms in Figure 3.3 allow us to visualize the distributions of each of the 100 samples. In Figure 3.4, a histogram of the mean values of the 100 samples is displayed. Notice that the histogram of the means displayed in Figure 3.4 looks roughly like a normal distribution. With each of the n sample sizes being as large as 30, one expects to see the distribution of means to look more normal as n increases.

Figure 3.3 *100 Samples of Size 30 of Uniform Random Deviates*

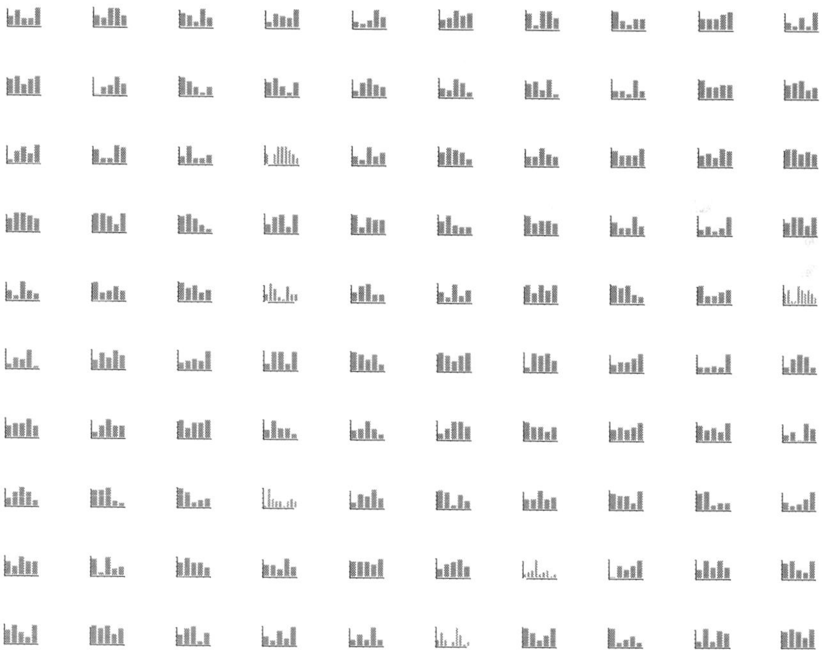

Figure 3.4 *Histogram of 100 Sample Means of Uniform Distributions Each of Sample Size 30*

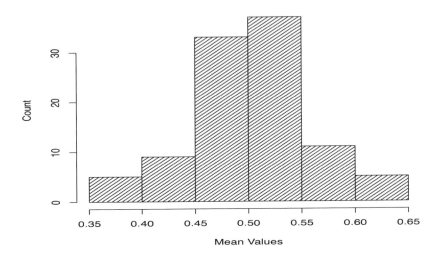

3.3.3 A Case When the Central Limit Theorem "Fails"

There are situations, however, when the Central Limit Theorem fails, for example, when either the mean or the standard deviation are infinite. A particular case where the central limit theorem fails involves the *Cauchy* distribution, which is equivalent to a t-distribution with one degree of freedom. The probability density function of the standard Cauchy distribution is

$$f(x) = \frac{1}{\pi(1+x^2)}, \quad -\infty < x < \infty .$$

Neither the mean nor the variance exists for the Cauchy distribution. For example, $\int_{-\infty}^{\infty} f(x)\, dx = 1$ *but* $\int_{-\infty}^{\infty} x\, f(x)\, dx = \infty$ and $\int_{-\infty}^{\infty} x^2\, f(x)\, dx = \infty$.

Example 3.3. We generated 100 samples, each of size 30, of random deviates that had a t-distribution with one degree of freedom (Cauchy distribution). Histograms of the 100 individual samples and the means of those 100 samples are given in Figures 3.5 and 3.6, respectively. A listing and summary statistics of the 100 means of the Cauchy samples are given below.

```
> round(mxt1,3)
  [1]    2.104 -21.258  -0.737   4.564   3.463  37.269   0.521  -8.589
  [9]   -0.462  -0.284   0.267  -1.012   0.674   5.871  -2.822   1.393
 [17]  -12.379  26.779  -0.615   3.845  -1.720   1.560  -0.377   0.135
 [25]   -2.651  -1.151  -0.240  -0.568   1.264  -0.197  -1.374   0.050
 [33]   -1.128  10.950  -0.319   2.092  -0.413  -0.610   0.722  -0.914
 [41]    1.679   0.414   7.571  -0.195  -0.328  -0.763   0.400  -2.115
```

```
[49]    3.661   -0.047    1.889   -1.499    7.221   -5.528    1.540   -0.379
[57]    0.596   -1.449   -1.329    1.098    1.202   -1.790    0.654    1.119
[65]    3.753   -1.721    2.797    2.057   -6.082    0.424    3.233   -1.615
[73]    5.662   -0.388    3.051    1.031    0.992   -1.322    2.173    0.837
[81]    1.323   -3.033   -1.273    1.450    2.501    1.208   -1.235   -0.076
[89]    0.723    4.633    1.718   -1.720   -1.688    0.479    0.778    5.343
[97]    0.867   -0.505   -1.560   -0.239


> summary(mxt1)
      Min.   1st Qu.    Median      Mean   3rd Qu.      Max.
-21.25771  -1.04086   0.33366   0.81896   1.58981  37.26908

> stdev(mxt1)
[1] 5.827644
```

Notice that even though the mean of the 100 mean values is 0.819, the range of the 100 mean values is $37.26908 - (-21.25771) = 58.52679 \approx 58.5$ and the standard deviation is over 5, which is no smaller than many of the individual sample standard deviations. The histogram of the 100 means looks symmetric but has values that are far from the center of the distribution. Thus, although the means of the samples are somewhat symmetric, they don't appear to converge to any particular distribution.

Figure 3.5 *100 Samples of Size 30 of Cauchy Random Deviates*

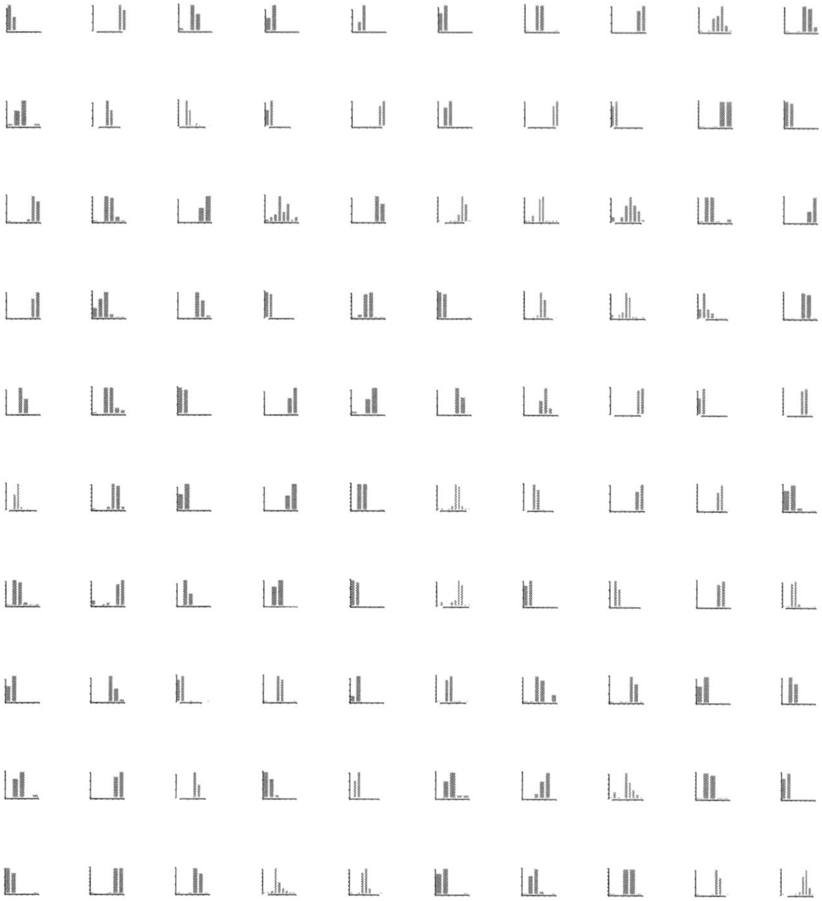

Figure 3.6 *Distribution of 100 Means of Cauchy Samples Each of Sample Size 30*

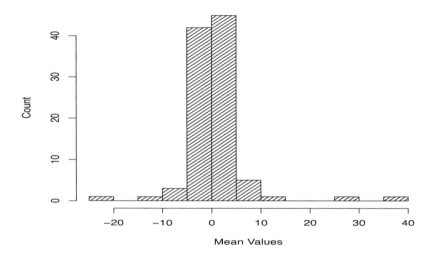

3.4 Summary of the Inferential Properties of the Sample Mean

We have seen from the Strong Law of Large Numbers and the Central Limit Theorem that the arithmetic mean of a random sample is useful for making statistical inferences about a population mean. Summarized below are several properties of a sample mean and their relationship to making a statistical inference about a population mean.

1. The Strong Law of Large Numbers and the Central Limit Theorem both tell us about the relationship between sample size and the accuracy of our sample estimate of the population mean. Namely, if we take the arithmetic means, $\overline{X}$, of larger and larger samples of identically distributed random variables, then we will expect the mean to be a more accurate estimate of the the the *true* population mean, μ. Consequently, the *variance* of $\overline{X}$ gets smaller and smaller as n gets larger and larger. For example, suppose $X_1, X_2, \ldots X_n$ are independently and identically distributed $\sim D$ with mean, μ, and variance, σ^2. Then

$$\text{Var}(\overline{X}) = \text{Var}\left(\frac{1}{n}\sum_{j=1}^{n} X_i\right) = \frac{1}{n^2}\left[\text{Var}(X_1) + \ldots \text{Var}(X_n)\right]$$

$$= \frac{1}{n^2}\underbrace{\left[\sigma^2 + \ldots + \sigma^2\right]}_{n \text{ times}} = \frac{1}{n^2}(n\sigma^2) = \frac{\sigma^2}{n}. \tag{3.3}$$

Thus, $\text{Var}(\overline{X}) = \frac{\sigma^2}{n}$. The *square root* of this quantity is $\sigma/\sqrt{n}$ and is called the *standard error of the mean*. The standard error of the mean (s.e.m.) is *estimated* from a sample of observations by $\hat{\sigma}_{\overline{x}} = \hat{\sigma}/\sqrt{n} = s/\sqrt{n}$ where

$$s = \sqrt{\frac{1}{n-1}\left[\sum_{j=1}^{n}(x_i - \overline{x})^2\right]} = \sqrt{\frac{1}{n-1}\left[\sum_{j=1}^{n}x_i^2 - \frac{1}{n}(\sum_{j=1}^{n}x_i)^2\right]}.$$

The standard deviation and the standard error of the mean differ in that the standard deviation measures the variability of *individual* observations about the mean whereas the standard error of the mean measures the variability of the *mean* of n observations and so, estimates the *population* variation.

2. The Central Limit Theorem further indicates that if we take the mean of a *large* number of observations that come from even a very highly skewed distribution, then that mean is approximately normally distributed. As was indicated earlier, the mean of a *random* sample of n observations, $\overline{x}$, estimates the population mean, μ. In the case where the individual random variables, that is, the X_i are i.i.d. $N(\mu, \sigma^2)$ $i = 1, \ldots, n$, then the mean of these random variables, $\overline{X}$ is said to be an *unbiased estimator* of μ, that is, $E(\overline{X}) = \mu$. $\overline{X}$ is also said to be a *minimum variance estimator* of μ, i.e., $\text{Var}(\overline{X}) \leq \text{Var}(f(X_1, \ldots, X_n))$ for any function, f, of the X_i's. The *sample* estimate, $\overline{x}$, is said to be a *point estimate* of μ.

3. Knowing (1) and (2) allows us to make approximate probability statements about an arithmetic mean from almost any type of distribution (i.e., those with known means and variances) when the sample size, n, is large,

$$\frac{\overline{X} - \mu}{\sigma/\sqrt{n}} \ \dot{\sim}\ N(0,1) \implies \Pr\left\{-1.96 < \frac{\overline{X} - \mu}{\sigma/\sqrt{n}} < 1.96\right\} \approx .95 . \qquad (3.4)$$

If the original underlying distribution is normal, then for all n, equation (3.4) will hold exactly, that is, if the X_i are i.i.d. $N(\mu, \sigma^2)$, $i = 1, \ldots, n$, then

$$\frac{\overline{X} - \mu}{\sigma/\sqrt{n}} \sim N(0,1) \implies \Pr\left\{-1.96 < \frac{\overline{X} - \mu}{\sigma/\sqrt{n}} < 1.96\right\} = .95 . \qquad (3.5)$$

Equations (3.4) and (3.5) can be used for constructing confidence intervals and performing hypothesis tests for a population mean value, μ.

3.5 Appendix: Program Listings

The following listings are programs which allow one to create the plots in Figures 4.2–4.6. It should be noted that other R programs such as that by John Verzani [154] (see `simple.sim` in the library `UsingR`) can be used to do similar tasks.

```
##########################################################
# Create a series of Bernoulli processes, sum them
# and then divide each sum by the number of trials
#
```

```
#    This program can be run both in R and in S-plus
#      Program by Stewart J. Anderson, 2/3/2006
#########################################################
#

Initialize the parameters of interest
N <- 2000 #<-- Number of Bernoulli processes to be generated
p <- .5 #<--- Pick a parameter for the Bernoulli process

x <- rbinom(N,1,p)
sum.x <- cumsum(x)
number.x <- 1:N
p.hat <- sum.x/number.x
plot(number.x,p.hat,type="l",xlab="Number of Bernoulli trials",
     ylab="Proportion of Successes",ylim=c(0,1))
abline(h=p,lty=2)
text(N/2,p+.03,paste("p =",p))



#############################################################
#   This program creates n samples of pseudo-random numbers,
#   each of size, ss.  In this case, the samples come from
#   a standard uniform distribution.  Each sample is plotted
#   separately and the means of the samples are also plotted.
#   This program will run in S-plus but needs a few
#   modifications to run in R.
#
#      Program by Stewart J. Anderson, 7/9/2005
#############################################################
n <- 100  # No. of replications
ss <- 30  # No. of deviates per sample
tot <- ss*n
gdim <- sqrt(n)
xu <- matrix(rep(0,tot),byrow=T,ncol=n)
mxu <- c(rep(0,n))
par(mfrow=c(gdim,gdim),oma=c(3,0,3,0),xaxs="i")
for (i in 1:n) {
  xu[,i] <- runif(ss) # ran uniform dist [rt(ss,1) gives Cauchy]
  mxu[i] <- mean(xu[,i])
  hist(xu[,i],sub="")
#  title(paste("Simulation",i))
  }
mtext(paste(n,"samples of size",ss,
                " of uniform random deviates\n"),
side=3,cex=1.25,outer=T)

graphsheet()
par(mfrow=c(1,1))
hist(mxu,xlab="Mean Value")
#title(paste("Distribution of",n,"means from
# uniform distributions each of size",ss),cex=1)
```

3.6 Exercises

1. Write a program to

(a) produce 5000 independent Bernoulli trials with $p = 0.1$.

(b) For the output in part (a), write a program to produce a figure similar to Figure 3.2.

(c) What would the standard error of the mean of p be in this problem? (HINT: If $X \sim b(1,p)$ then $\text{Var}(X) = pq$ and $\text{Var}\left(\left[\sum_i^n X_i\right]/n\right) = n \cdot (pq/n^2) = \frac{pq}{n}$. Hence, the standard error of the mean in this case would be $\sqrt{\frac{pq}{n}}$.)

2. Write a program to produce the *mean values* for

(a) 150 samples, which each have a sample size of $n = 5$ and are from a t-distribution with 2 degrees of freedom;

(b) 150 samples, which each have a sample size of $n = 20$ and are from a t-distribution with 2 degrees of freedom;

(c) 150 samples, which each have a sample size of $n = 50$ and are from a t-distribution with 2 degrees of freedom.

(d) For the three resultant samples of 150 created in parts **(a)** – **(c)**, calculate the mean, standard deviation, and quartiles of your distributions. Also, for each case, produce a histogram of the distribution of 150 mean values.

(e) Does the central limit theorem appear to hold for a t-distribution with 2 degrees of freedom?

3. Write a program to produce a distribution of *median* values of

(a) 150 samples, which each have a sample size equal to 10, and are from an exponential distribution with $\lambda = 1$; and

(a) 150 samples, which each have a sample size equal to 50, and are from an exponential distribution with $\lambda = 1$; and

(c) For the two resultant distributions created in **(a)** and **(b)**, calculate the mean, standard deviation, and quartiles of your distributions. Also, for each case, produce a histogram of the distribution of 150 median values.

(d) Does the distribution of medians of an exponential distribution appear to converge towards a normal distribution as the sample size gets larger?

4. Repeat the steps in the previous problem but use the first quartile of each distribution instead of the median.

## CHAPTER 4

Correlation and Regression

4.1 Introduction

In many medical, biological, engineering, economic and other scientific applications, one wishes to establish a linear relationship between two or more variables. If there are only two variables, X and Y, then there are two ways a linear relationship can be characterized: (1) using the *correlation coefficient*; and (2) using *linear regression*. One would typically use a correlation coefficient to quantify the strength and direction of the linear association. If neither variable is used to predict the other, then *both* X and Y are assumed to be random variables, which makes inference more complicated. Linear regression is useful for answering the question: Given a value of X, what is the predicted value of Y? For answering this type of question, values of X are assumed to be fixed (chosen) while values of Y are assumed to be random.

In this chapter, we first review the Pearson correlation coefficient and then tackle simple, multiple, and polynomial regression models. Our primary approach for the presentation of the regression models is to use the general linear model involving matrices. We provide a short appendix at the end of the chapter to review matrix algebra. Our strategy for the presentation of regression in this chapter allows us to use the same approach for the different types of regression models. Also included in this chapter are strategies for visualizing regression data and building and assessing regression data. In the final section, we introduce two smoothing techniques, namely, the loess smoother and smoothing polynomial splines. To facilitate the discussion of the techniques covered in this chapter, we provide numerical examples with hand calculations to demonstrate how to fit simple models and also provide programs in SAS and R to demonstrate the implementation of calculations in more complex models.

4.2 Pearson's Correlation Coefficient

Correlation is a scaled version of *covariance*, which describes the "co-variation" of two random variables, X and Y. Mathematically, covariance is defined as

$$\text{Cov}(X,Y) \equiv E\big[(X - \mu_X)(Y - \mu_Y)\big] = E(XY) - \mu_X \mu_Y, \tag{4.1}$$

where μ_X and μ_Y are the population means for the random variables X and Y and "E" is expected value as defined in Chapter 1. The correlation coefficient, ρ, is scaled so that $-1 \leq \rho \leq 1$. Its mathematical definition is as follows:

$$\rho \equiv \frac{\text{Cov}(X,Y)}{\sigma_X \sigma_Y} = \frac{E\left[(X - \mu_X)(Y - \mu_Y)\right]}{\sigma_X \sigma_Y} = \frac{E(XY) - \mu_X \mu_Y}{\sigma_X \sigma_Y}, \quad (4.2)$$

where σ_X and σ_Y are the population standard deviations for the random variables X and Y. Given data pairs, $(x_1, y_1), \ldots, (x_n, y_n)$, the correlation coefficient, ρ, is *estimated* as

$$r \equiv \widehat{\rho} = \frac{\sum_{j=1}^{n}(x_i - \overline{x})(y_i - \overline{y})}{\sqrt{\sum_{j=1}^{n}(x_i - \overline{x})^2 \sum_{j=1}^{n}(y_i - \overline{y})^2}}$$

$$= \frac{\sum_{j=1}^{n} x_i y_i - n\overline{x}\overline{y}}{\sqrt{\left(\sum_{j=1}^{n} x_i^2 - n\overline{x}^2\right)\left(\sum_{j=1}^{n} y_i^2 - n\overline{y}^2\right)}}. \quad (4.3)$$

In Figure 4.1, four scatterplots depicting various estimated correlations are presented. Datasets 1 and 2 show perfect positive and negative correlations, respectively. Datasets 3 and 4 depict situations where the correlation coefficients are reasonably high and reasonably low, respectively. The pattern associated with a high estimated correlation tends to be football shaped.

For the purposes of making an inference, we'll assume that both X and Y are normally distributed. Unfortunately, since both X and Y are random, the *estimated* correlation coefficient, r, is not distributed in a convenient way for doing statistical analyses. Therefore, in order to perform statistical analyses, we have to *transform* r so that it is in a convenient form.

To test the hypothesis, $H_0 : \rho = 0$ versus $H_1 : \rho \neq 0$, form

$$T_\rho = \frac{r\sqrt{n-2}}{\sqrt{1-r^2}}, \quad (4.4)$$

which, under H_0, is approximately distributed as t_{n-2}.

<u>Rule: Two-sided hypothesis test for $\rho = 0$</u>

$$\begin{cases} \text{If } |T_\rho| \geq t_{n-2,1-\frac{\alpha}{2}} & \text{then } \textit{reject } H_0 \text{ ;} \\ \text{Otherwise} & \textit{do not} \text{ reject } H_0, \end{cases}$$

where $t_{n-2,1-\frac{\alpha}{2}}$ denotes the $100\% \times (1-\frac{\alpha}{2})^{th}$ percentile of the central t-distribution with $n-2$ degrees of freedom. For a one-sided alternative, $H_1 : \rho < 0$ or $H_1 : \rho > 0$, replace $t_{n-2,1-\frac{\alpha}{2}}$ by $t_{n-2,1-\alpha}$.

Figure 4.1 *Different Patterns of Data Pairs and Their Corresponding Correlations*

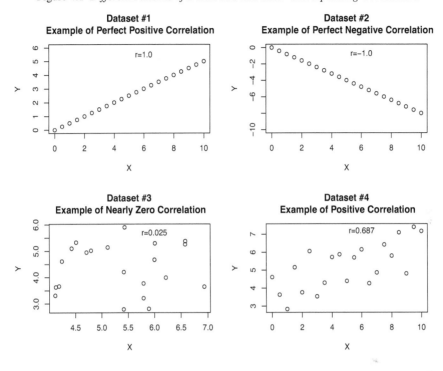

Example 4.1. Suppose blood pressure in sibling pairs is measured to determine whether or not there is a correlation between blood pressure levels in siblings. Specifically, suppose that $r = .61$ is based on 12 sibling pairs. Suppose that we're testing $H_0: \rho = 0$ versus $H_1: \rho > 0$. Form

$$T_\rho = \frac{.61\sqrt{10}}{\sqrt{1 - (.61)^2}} \approx \frac{1.929}{0.792} \approx 2.434$$

which, under H_0, is approximately distributed as t_{25}. Since $t_{10,.99} \approx 2.764 \Longrightarrow |T_\rho| < 2.764$ $\Longrightarrow$ we cannot reject H_0 at the $\alpha = .01$ level. In this case the p-value is approximately 0.0176. Thus, the one-sided statistic is close to being significant at the $\alpha = .01$ level.

In some cases, one wants to test the hypothesis that ρ is equal to some value not equal to 0, for example, $H_0 : \rho = .5$. The pecularities of the correlation coefficient require that a separate test be performed in this case. Consequently, in general, to test $H_0: \rho = \rho_0, \rho_0 \neq 0$ versus $H_1 : \rho \neq \rho_0$, form

$$Z = \frac{1}{2} \ln \left[\frac{1+r}{1-r} \right], \text{ which under } H_0, Z \dot{\sim} N \left(\frac{1}{2} \ln \left[\frac{1 + \rho_0}{1 - \rho_0} \right], \frac{1}{n-3} \right),$$

where "ln" denotes the natural logarithm. The above transformation is known as

Fisher's "Z" transformation. Also, form

$$Z_0 = \frac{1}{2} \ln \left[\frac{1 + \rho_0}{1 - \rho_0} \right],$$

where ρ_0 is the value of ρ under the null hypothesis. Finally, form

$$W_1 = (Z - Z_0)\sqrt{n - 3} \sim N(0, 1), \text{ under } H_0. \tag{4.5}$$

<u>Rule: Two-sided hypothesis test for $\rho = \rho_0$ where $\rho_0 \neq 0$</u>

$$\begin{cases} \text{If } |W_1| \geq Z_{1-\frac{\alpha}{2}} & \text{then } \textit{reject } H_0 \text{ ;} \\ \text{Otherwise} & \textit{do not } \text{reject } H_0. \end{cases}$$

For a one-sided alternative, $H_1 : \rho < \rho_0$ or $H_1 : \rho > \rho_0$, replace $Z_{1-\frac{\alpha}{2}}$ by $Z_{1-\alpha}$. From the relationships in equation (4.5), one can show that a two-sided $(1 - \alpha)\%$ confidence interval for ρ is given by

$$\left(\frac{e^{2z_1} - 1}{e^{2z_1} + 1}, \frac{e^{2z_2} - 1}{e^{2z_2} + 1} \right), \tag{4.6}$$

where $z_1 = z - Z_{1-\frac{\alpha}{2}} \sqrt{1/(n-3)}$, $z_2 = z + Z_{1-\frac{\alpha}{2}} \sqrt{1/(n-3)}$ and $z = \frac{1}{2} \ln \left(\frac{1+r}{1-r} \right)$.

Example 4.2. Suppose that, in Example 4.1, we're interested in testing $H_0 : \rho = .2$, versus $H_1 : \rho \neq .2$. For the data in Example 4.1, $z = 0.7089$, $z_0 = 0.2027$. Thus, $|W_1| = |.7089 - .2027|\sqrt{9} \approx 0.5062 \times 3 \approx 1.5186 < 2.5758 = Z_{.995} \Longrightarrow$ do not reject H_0 at the $\alpha = .01$ level. To get corresponding 99% confidence intervals for ρ, we have

$$z_1 = 0.7089 - 2.5758 \left(\frac{1}{\sqrt{9}} \right) \approx -0.1497, \quad z_2 = 0.7089 + 2.5758 \left(\frac{1}{\sqrt{9}} \right) \approx 1.5675$$

$$\Longrightarrow \left(\frac{e^{2(-0.1497)} - 1}{e^{2(-0.1497)} + 1}, \frac{e^{2(1.5675)} - 1}{e^{2(1.5675)} + 1} \right) \approx \left(\frac{-0.2587}{1.7413}, \frac{21.99}{23.99} \right) \approx \left(-0.1486, 0.9166 \right).$$

Finally, assume we have correlation coefficients from *two samples* and we wish to test $H_0 : \rho_1 = \rho_2$ versus $H_1 : \rho_1 \neq \rho_2$. To do this, we'll first calculate Fisher's Z transformation for each sample:

$$Z_1 = \frac{1}{2} \ln \left(\frac{1 + r_1}{1 - r_1} \right), \quad Z_2 = \frac{1}{2} \ln \left(\frac{1 + r_2}{1 - r_2} \right).$$

$$W_2 = \frac{Z_1 - Z_2}{\sqrt{\frac{1}{n_1 - 3} + \frac{1}{n_2 - 3}}} \sim N(0, 1) \text{ under } H_0, \tag{4.7}$$

where n_1 and n_2 are the number of *pairs* in each of the two samples.

<u>Rule: Two-sided hypothesis test for $\rho_1 = \rho_2$</u>

$$\begin{cases} \text{If } |W_2| \geq Z_{1-\frac{\alpha}{2}} & \text{then } \textit{reject } H_0 \text{ ;} \\ \text{Otherwise} & \textit{do not } \text{reject } H_0. \end{cases}$$

For a one-sided alternative, $H_1 : \rho_1 < \rho_2$ or $H_1 : \rho_1 > \rho_2$, replace $Z_{1-\frac{\alpha}{2}}$ by $Z_{1-\alpha}$.

Example 4.3. Suppose that the cholestorol for members of 75 families who have at least one natural child and 23 families who have at least one adopted child is measured. Comparing correlations of cholestorol between the mother and the eldest child for natural parents versus adopted parents, we get the following data:

$$r_1 = .41, \; n_1 = 50, \text{ and}$$

$$r_2 = .08, \; n_2 = 20,$$

where n_1 and n_2 represent the number of mother–child pairs in each group. Test $H_0 : \rho_1 = \rho_2$ versus $H_1 : \rho_1 \neq \rho_2$. Compute

$$Z_1 = \frac{1}{2} \ln\left(\frac{1 + .41}{1 - .41}\right) \approx .4356, \quad Z_2 = \frac{1}{2} \ln\left(\frac{1 + .08}{1 - .08}\right) \approx .0802$$

$$W_2 = \frac{.4356 - .0802}{\sqrt{\frac{1}{72} + \frac{1}{20}}} \approx 1.41 < 1.96 = Z_{.975}$$

$\implies$ do not reject H_0 at the $\alpha = .05$ level $\implies$ the correlation coefficients for cholesterol between mother and child are *not significantly* greater for natural parents than for adopted parents. The p-value here is about .08.

4.3 Simple Linear Regression

Suppose that we believe that a linear relationship exists between two variables, X and Y. Suppose further that we're interested in *predicting* a value for Y given a *fixed* value of a single variable, X. Such a problem is solved via *linear regression*. If the model involving Y and X is a line, then it is called a *simple* linear regression. Recall that the mathematical formula for a line requires an *intercept* and a *slope* to be specified. Define the intercept to be β_0 and the slope to be β_1. If we have a sample of n pairs of observations, $(x_1, y_1), \ldots, (x_n, y_n)$, we can write a linear regression model as follows:

$$y_i = \beta_0 + \beta_1 x_i + \epsilon_i, \; i = 1, \ldots, n \tag{4.8}$$

<u>Assumptions</u>

1. The x_i are *fixed and known* for all $i = 1, \ldots, n$; and
2. β_0 and β_1 are *fixed* <u>but</u> *unknown* (to be estimated from the data); and
3. y_i and ϵ_i are random; for $i = 1, \ldots, n$. [In order to make an inference, we also assume that the $\epsilon_i \sim N(0, \sigma^2)$].

Figure 4.2 *Example of Data Fitted by Using a Simple Linear Regression Model*

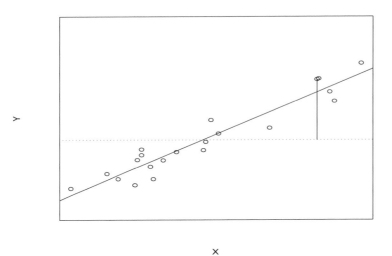

In regression terminology, Y is called the *dependent variable* whereas X is called the *independent variable* or *predictor variable*. A visualization of a simple linear regression can be seen in Figure 4.2. In the figure, the data are represented as single points overlayed by a fitted line. The dotted horizontal line represents the mean of the observed outcomes, that is, $\bar{y}$. The solid vertical line in the graph represents the deviation of a typical data point from the $\bar{y}$. One can see that this consists of two components: (1) the deviation between $\bar{y}$ and the fitted curve; and (2) the deviation between the fitted curve and the observed value of Y, that is, y. More on this subject will be covered in sections 4.3.2 and 4.3.5.

To obtain the formulas necessary to describe a regression line, one must calculate the *sums of squares* of the predictor variable, the independent variable around their own mean values. Also needed are the sums of squares of the cross products adjusted for mean values. To denote these and other quantities needed for regression, we use the notation given below.

$$SS_{xx} = \sum_{j=1}^{n}(x_i - \bar{x})^2 = \sum_{j=1}^{n} x_i^2 - \frac{\left(\sum_{j=1}^{n} x_i\right)^2}{n},$$

$$SS_{yy} = \sum_{j=1}^{n}(y_i - \bar{y})^2 = \sum_{j=1}^{n} y_i^2 - \frac{\left(\sum_{j=1}^{n} y_i\right)^2}{n},$$

$$\text{and } SS_{xy} = \sum_{j=1}^{n}(x_i - \bar{x})(y_i - \bar{y}) = \sum_{j=1}^{n} x_i y_i - \frac{\left(\sum_{j=1}^{n} x_i\right)\left(\sum_{j=1}^{n} y_i\right)}{n} ;$$

$$s_x^2 = \frac{SS_{xx}}{n-1}, \quad s_y^2 = \frac{SS_{yy}}{n-1}, \quad \text{and} \quad s_{xy} = \frac{SS_{xy}}{n-1}.$$

Estimated residuals will be denoted as $e_i = \hat{\epsilon}_i = y_i - \hat{y}_i$, for all $i = 1, \ldots, n$.

A general approach to regression involves the use of *matrix algebra*. For a brief review of basic matrix algebra, see section 4.9. Using matrices, equation (4.8) can be rewritten as

$$\mathbf{y} = \mathbf{X}\boldsymbol{\beta} + \boldsymbol{\epsilon}, \tag{4.9}$$

where

$$\mathbf{y} = \begin{pmatrix} y_1 \\ y_2 \\ \vdots \\ y_n \end{pmatrix}, \quad \mathbf{X} = \begin{pmatrix} 1 & x_1 \\ 1 & x_2 \\ \vdots & \vdots \\ 1 & x_n \end{pmatrix}, \quad \boldsymbol{\beta} = \begin{pmatrix} \beta_0 \\ \beta_1 \end{pmatrix} \quad \text{and} \quad \boldsymbol{\epsilon} = \begin{pmatrix} \epsilon_1 \\ \epsilon_2 \\ \vdots \\ \epsilon_n \end{pmatrix}.$$

Notice that when we multiply $\underbrace{\mathbf{X}}_{n \times 2}$ by $\underbrace{\boldsymbol{\beta}}_{2 \times 1}$, the resulting matrix is $n \times 1$, which matches the dimensions of the $\mathbf{y}$ and $\boldsymbol{\epsilon}$ "matrices" (vectors). The usual assumption is that the error vector is distributed as a *multivariate* normal distribution with mean $\mathbf{0}$ and variance $\mathbf{I}\sigma^2$.

4.3.1 Parameter Estimation

Once the linear (regression) model is written, we must solve for $\boldsymbol{\beta}$ so that we obtain our "best-fitting" model. The typical way to do this is via the use of *least squares* solutions. The goal in the least squares problem for simple linear regression is to minimize the residual sums of squares with respect to β_0 and β_1 given the data, that is minimize

$$\text{Resid SS} = \sum_{j=1}^{n}(y_i - \hat{y}_i)^2 = \sum_{j=1}^{n}(y_i - \hat{\beta}_0 - \hat{\beta}_1 x_i)^2$$

$$= (\mathbf{y} - \mathbf{X}\hat{\boldsymbol{\beta}})'(\mathbf{y} - \mathbf{X}\hat{\boldsymbol{\beta}}) = \mathbf{y}'\mathbf{y} - \hat{\boldsymbol{\beta}}'\mathbf{X}'\mathbf{y}, \tag{4.10}$$

where $'$ denotes transpose, with respect to the components of $\hat{\boldsymbol{\beta}}$, that is, $\hat{\beta}_0$ and $\hat{\beta}_1$, respectively. In equation (4.10), to denote that the *estimates* come from the data, we change the notation from β_0 and β_1, to $\hat{\beta}_0$ and $\hat{\beta}_1$, respectively.

The least squares estimates can be easily derived using standard methods of calculus by set taking partial derivatives of (4.10) with respect to the parameters and setting the resulting equations equal to 0 and solving for $\hat{\beta}_0$ and $\hat{\beta}_1$. The resulting least squares solution for $\boldsymbol{\beta}$, is

$$\hat{\boldsymbol{\beta}} = \begin{pmatrix} \hat{\beta}_0 \\ \hat{\beta}_1 \end{pmatrix} = (\mathbf{X}'\mathbf{X})^{-1}\mathbf{X}'\mathbf{y}, \tag{4.11}$$

where $'$ denotes transpose. The beauty of this solution is that it holds *for every linear or multiple regression model no matter how many independent $x's$ there are in the equation.* Thus, instead of having to deal with different formulas for cases where the number of independent variables vary, we can use one formula for *all* situations. In the special case of *simple* linear regression, the least squares estimates are

$$\mathbf{c}_0'\widehat{\boldsymbol{\beta}} = \widehat{\beta}_0 = \bar{y} - \widehat{\beta}_1\bar{x}, \text{ and } \mathbf{c}_1'\widehat{\boldsymbol{\beta}} = \widehat{\beta}_1 = \frac{SS_{xy}}{SS_{xx}}, \tag{4.12}$$

where $\mathbf{c}_0' = \begin{pmatrix} 1 & 0 \end{pmatrix}$ and $\mathbf{c}_1' = \begin{pmatrix} 0 & 1 \end{pmatrix}$.

4.3.2 Testing the Adequacy of the Regression Model

To test the adequacy of the model, it is convenient to set up an Analysis of Variance (ANOVA or AOV) table. An ANOVA table usually consists of the following: (1) The source of each component of variation (Source); (2) The degrees of freedom (df) associated with each component; (3) The sums of squares for each component; (4) The mean square (MS) for each component; and (5) An F-statistic used for making inference about the components of variation. Listed below is a typical ANOVA table for a simple linear regression. In addition, an equivalent matrix formulation of the same table is given below.

ANOVA table for Simple Linear Regression

Source	df	SS	MS	F
Regression	1	$\sum_{j=1}^{n}(\widehat{y}_i - \bar{y})^2 = \frac{SS_{xy}^2}{SS_{xx}}$	$\frac{SS_{reg}}{1}$	$\frac{MS_{reg}}{MS_{res}}$
Residual	$n-2$	$\sum_{j=1}^{n}(y_i - \widehat{y}_i)^2 = SS_{yy} - \frac{SS_{xy}^2}{SS_{xx}}$	$\frac{SS_{res}}{n-2}$	
Total	$n-1$	$\sum_{j=1}^{n}(y_i - \bar{y})^2 = SS_{yy}$		

Equivalent ANOVA table for Simple Linear Regression with Matrix Notation

Source	df	SS	MS	F
Regression	1	$\widehat{\boldsymbol{\beta}}'\mathbf{X}'\mathbf{y} - n\bar{y}^2$	$\frac{\widehat{\boldsymbol{\beta}}'\mathbf{X}'\mathbf{y} - n\bar{y}^2}{1}$	$\frac{MS_{reg}}{MS_{res}}$
Residual	$n-2$	$\mathbf{y}'\mathbf{y} - \widehat{\boldsymbol{\beta}}'\mathbf{X}'\mathbf{y}$	$\frac{\mathbf{y}'\mathbf{y} - \widehat{\boldsymbol{\beta}}'\mathbf{X}'\mathbf{y}}{n-2}$	
Total	$n-1$	$\mathbf{y}'\mathbf{y} - n\bar{y}^2$		

In either formulation of the ANOVA table, since there are n pairs of data, we start with n degrees of freedom. In simple linear regression, the regression coefficients account for two degrees of freedom. However, because both the regression and total sums of squares are "adjusted for the mean," that is, in both, $\bar{y}$ is subtracted off, we remove one degree of freedom in each case. Thus, the regression and total sums of squares have 1 and $n-1$ degrees of freedom, respectively. This leaves $n-2$ degrees of freedom for the "errors" or "residuals."

Another property unique to simple linear regression is that the test for the "adequacy of the model" is equivalent to testing the question of whether or not there is a signifi-cant linear relationship between X and Y or testing $H_0 : \beta_1 = 0$ versus $H_1 : \beta_1 \neq 0$. This particular two-sided test is done via an F-test. Thus, for a *simple* linear regres-sion, form

$$F = \frac{\text{Reg MS}}{\text{Res MS}} = \frac{(\text{Reg SS})/1}{(\text{Res SS})/(n-2)} = \frac{SS_{xy}^2/SS_{xx}}{\left[\frac{SS_{yy} - \left(SS_{xy}^2/SS_{xx}\right)}{n-2}\right]}. \qquad (4.13)$$

Under the null hypothesis, $H_0 : \beta_1 = 0$, $F \sim F_{1,n-2}$. Thus, if the F statistic we calculate from our data is such that $F \geq F_{1,n-2,1-\alpha}$ then we reject H_0 and declare X to be a *significant predictor* of Y; otherwise we do not reject H_0.

4.3.3 Hypothesis Testing for the Parameters of the Regression Model

We can also *directly* test for the significance of each coefficient (β_0 or β_1). Suppose, for example, that we would like to test $H_0 : \beta_1 = 0$ versus $H_1 : \beta_1 \neq 0$. Similar to a simple t-test, the strategy here is to divide $\hat{\beta}_1$ by its standard error and compare the result with some prespecified value of an appropriately constructed t-distribution. We already have an expression for $\hat{\beta}_1$ so now we must calculate $\text{Var}(\hat{\beta}_1)$ and its square root. Now, if Z is any *univariate* random variable and a is any *scalar*, then $\text{Var}(aZ) = a^2 \text{Var}(Z)$. The "variance" of a vector is obtained by calculating the covariance of every other component of the vector (including the component itself). These covariances are arranged in a square matrix called the *dispersion matrix* or the *variance–covariance matrix* or sometimes, just the *covariance matrix*. Suppose that $\mathbf{Z} = \begin{pmatrix} z_1 \\ z_2 \end{pmatrix}$ and that $\text{Var}(z_1) = \text{Cov}(z_1, z_1) = 16$, $\text{Var}(z_2) = \text{Cov}(z_2, z_2) = 9$, and $\text{Cov}(z_1, z_2) = 4$. Then,

$$\mathbf{Var(Z)} = \begin{pmatrix} 16 & 4 \\ 4 & 9 \end{pmatrix}. \qquad (4.14)$$

Sometimes $\mathbf{Var(Z)}$ is denoted as $\mathbf{\Sigma_Z}$. Notice that if the dimensions of $\mathbf{Z}$ are 2×1, then $\mathbf{\Sigma_Z}$ has dimensions, 2×2. The matrix analog of the variance rule for re-scaling univariate variables is that if $\mathbf{Z}$ is any *random* vector and $\mathbf{A}$ is any compatible matrix with *fixed* values, then

$$\mathbf{Var(AZ)} = \mathbf{A Var(Z) A'} = \mathbf{A \Sigma_Z A'} \qquad (4.15)$$

Simple matrix algebra will show that if $\mathbf{Var(e)} = \mathbf{I}\sigma^2$, then

$$\mathbf{Var}(\widehat{\boldsymbol{\beta}}) = \mathbf{\Sigma}_{\widehat{\boldsymbol{\beta}}} = (\mathbf{X'X})^{-1}\sigma^2. \qquad (4.16)$$

Since we don't know σ^2, we must estimate it by $\hat{\sigma}^2 = \text{Residual SS}/(n-p)$ where the number of parameters, $p = 2$, in this case. To calculate $\text{Var}(\hat{\beta}_1)$, we must first observe that $\hat{\beta}_1 = \mathbf{c}_1'\widehat{\boldsymbol{\beta}}$ where $\mathbf{c}_1' = (\,0 \quad 1\,)$. The vector, $\mathbf{c}_1'$, is nothing more than a matrix expression, which in this case allows us to "pick off" the second row of the estimated parameter vector, $\widehat{\boldsymbol{\beta}}$, that is, the *slope* associated with the fitted line. The variance of $\hat{\beta}_1$ is given by

$$\widehat{\text{Var}}(\hat{\beta}_1) = \mathbf{c}_1'(\mathbf{X'X})^{-1}\mathbf{c}_1\,\hat{\sigma}^2 \qquad (4.17)$$

The seemingly difficult matrix expression in equation (4.17) is nothing more than an algebraic ploy for picking off the correct element of the variance–covariance matrix, that is, the second column of the second row, which is the variance of $\widehat{\beta}_1$. Consequently, we can conduct a t-test for testing the significance of β_1 by observing that, under the null hypothesis, $H_0 : \beta_1 = 0$,

$$T_{\beta_1} = \frac{\widehat{\beta}_1}{\widehat{se}(\widehat{\beta}_1)} = \frac{\mathbf{c}_1'\widehat{\boldsymbol{\beta}}}{\sqrt{\mathbf{c}_1'(\mathbf{X}'\mathbf{X})^{-1}\mathbf{c}_1\,\widehat{\sigma}^2}} = \frac{SS_{xy}/SS_{xx}}{\sqrt{\frac{s_{y \cdot x}^2}{SS_{xx}}}} \sim t_{n-2}, \qquad (4.18)$$

where $s_{y \cdot x}^2$ is the Residual Mean Squared or Res MS as is used in equation (4.13), t_{n-2} denotes the t-distribution with $n - 2$ degrees of freedom and everything else is defined above. The quantity, $s_{y \cdot x}^2$ in equation (4.18) is also called the "error mean square" or "mean squared error (MSE)." As was remarked upon earlier, the number of degrees of freedom for this type of test statistic is $n - p$ but, in the particular case of the simple linear regression model, $p = 2$. Both the numerator and the denominator of T are scalar values so that simple division is appropriate. We also note that equation (4.18) hinges on the assumption that $\boldsymbol{\epsilon} \sim N(\mathbf{0}, \mathbf{I}\sigma^2)$. The hypothesis test for β_1 is performed by using the following rule:

<u>Rule: Two-sided hypothesis test for β_1 in a simple linear regression model</u>

$$\begin{cases} \text{If } |T_{\beta_1}| \geq t_{n-2,1-\frac{\alpha}{2}} & \text{then } reject\ H_0\ ; \\ \text{Otherwise} & do\ not\ \text{reject } H_0. \end{cases}$$

For a one-sided alternative, $H_1 : \beta_1 < 0$ or $H_1 : \beta_1 > 0$, replace $t_{n-2,1-\frac{\alpha}{2}}$ by $t_{n-2,1-\alpha}$.

It is of interest here that, in the case of simple linear regression, where a single predictor variable, X, is used to predict an outcome, Y, the <u>two-sided</u> t-test for testing whether or not β_1 is equal to 0 is <u>equivalent</u> to the F-test for adequacy of the model. The direct method (e.g., using t-tests) allows us to use one-sided as well as two-sided inferences on both β_1 and β_0. However, the test for model adequacy is useful for multiple regression as well as simple linear regression.

A similar strategy is used to test whether or not the intercept, β_0 is equal to 0. Hence, to test $H_0 : \beta_0 = 0$ versus, say, $H_1 : \beta_0 \neq 0$, form

$$T_{\beta_0} = \frac{\widehat{\beta}_0}{\widehat{se}(\widehat{\beta}_0)} = \frac{\mathbf{c}_0'\widehat{\boldsymbol{\beta}}}{\sqrt{\mathbf{c}_0'(\mathbf{X}'\mathbf{X})^{-1}\mathbf{c}_0\,\widehat{\sigma}^2}} = \frac{\bar{y} - \widehat{\beta}_0\bar{x}}{\sqrt{s_{y \cdot x}^2\left(\frac{1}{n} + \frac{\bar{x}^2}{SS_{xx}}\right)}}, \qquad (4.19)$$

where $\mathbf{c}_0' = (\,1 \quad 0\,)$ and the other quantities are as defined above. Under H_0, $T_{\beta_0} \sim t_{n-2}$. The subsequent rule for rejecting or not rejecting H_0 is as follows:

<u>Rule: Two-sided hypothesis test for β_0 in a simple linear regression model</u>

$$\begin{cases} \text{If } |T_{\beta_0}| \geq t_{n-2,1-\frac{\alpha}{2}} & \text{then } reject\ H_0\ ; \\ \text{Otherwise} & do\ not\ \text{reject } H_0. \end{cases}$$

For a one-sided alternative, $H_1 : \beta_0 < 0$ or $H_1 : \beta_0 > 0$, replace $t_{n-2,1-\frac{\alpha}{2}}$ by $t_{n-2,1-\alpha}$.

4.3.4 Confidence Intervals for the Regression Parameters

Two-sided $(1 - \alpha)\%$ confidence intervals (CI) for the parameters (or linear combinations of parameters) can be obtained as

$$\mathbf{c}_0'\widehat{\boldsymbol{\beta}} \mp t_{n-p,1-\frac{\alpha}{2}} \sqrt{\mathbf{c}_0'(\mathbf{X}'\mathbf{X})^{-1}\mathbf{c}_0\,\widehat{\sigma}^2} = \widehat{\beta}_0 \mp t_{n-2,1-\frac{\alpha}{2}}\,\widehat{se}\left(\widehat{\beta}_0\right) \tag{4.20}$$

for β_0, where $\mathbf{c}_0' = (\,1 \quad 0\,)$; and

$$\mathbf{c}_1'\widehat{\boldsymbol{\beta}} \mp t_{n-p,1-\frac{\alpha}{2}} \sqrt{\mathbf{c}_1'(\mathbf{X}'\mathbf{X})^{-1}\mathbf{c}_1\,\widehat{\sigma}^2} = \widehat{\beta}_1 \mp t_{n-2,1-\frac{\alpha}{2}}\,\widehat{se}\left(\widehat{\beta}_1\right) \tag{4.21}$$

for β_1, where $\mathbf{c}_1' = (\,0 \quad 1\,)$. The above expressions are the shortest $(1 - \alpha)\%$ confidence intervals that can be obtained for β_0 and β_1, respectively.

4.3.5 Goodness-of-Fit Measure: R^2

The R^2 statistic is used for quantifying how much of the total variability in Y is due to the regression on X. The vertical line from one of the points to the fitted line in Figure 4.2 allows us to visualize the total variation as compared to the variation due to the the model for a typical residual. The total variation is measured from the mean Y value to the designated point whereas the variation due to the model is measured from the fitted line to the point. R^2 is formally defined as

$$R^2 \equiv \frac{\text{Reg SS}}{\text{Total SS}} = \frac{\sum_{j=1}^{n}(\widehat{y}_i - \overline{y})^2}{\sum_{j=1}^{n}(y_i - \overline{y})^2} \tag{4.22}$$

R^2 has the property that $0 \leq R^2 \leq 1$. If R^2 is near 0, then very little of the variability in Y is due to X whereas if R^2 is near 1, then nearly all of the variability in Y is due to X. R^2 values can sometimes be quite low *even if* the test for the adequacy of the model is significant. This result means that either the process of interest is inherently noisy or that there is another set of independent variables, which could be included to explain the dependent variable, Y.

4.3.6 Confidence Intervals for y Given x

There are two types of situations one can encounter when calculating confidence intervals for y:

1. x is an individual point not necessarily taken from the sample of x's, i.e., it is a "new" sample point and we're interested in predicting an individual y; or

2. x is a point taken from the existing sample of x's and we're interested in projecting the average y given the x.

In each case, there is an expression for the standard error of $\widehat{y}$:

1. $\widehat{se}_1\left(\widehat{y}\right) = \sqrt{s_{y \cdot x}^2 \left[1 + \frac{1}{n} + \frac{(x-\overline{x})^2}{SS_{xx}}\right]}$ $\implies$ two-sided $(1 - \alpha) \times 100\%$ confidence interval for y is given by $\widehat{y} \mp t_{n-2,1-\frac{\alpha}{2}} \widehat{se}_1\left(\widehat{y}\right)$

2. $\widehat{se}_2\left(\widehat{y}\right) = \sqrt{s_{y \cdot x}^2 \left[\frac{1}{n} + \frac{(x-\overline{x})^2}{SS_{xx}}\right]}$ $\implies$ two-sided $(1 - \alpha) \times 100\%$ confidence interval for y is given by $\widehat{y} \mp t_{n-2,1-\frac{\alpha}{2}} \widehat{se}_2\left(\widehat{y}\right)$

Because there is more variability in predicting a y for a "new" value of x, $\widehat{se}_1\left(\widehat{y}\right)$ is larger than $\widehat{se}_2\left(\widehat{y}\right)$. The associated confidence intervals are also wider.

Example 4.4. The data below are taken from a subset of 20 of a group 44 healthy male subjects. Measurements of duration (in seconds) on a treadmill until VO_2 $_{MAX}$ (X) were recorded along with the value of the VO_2 $_{MAX}$ in ml/kg/min (Y) (see Fisher and van Belle, pp. 347–348 [46] for details of the experiment).

Subject	1	2	3	4	5	6	7	8	9	10
Duration (s)	706	732	930	900	903	976	819	922	600	540
VO_2 $_{max}$	41.5	45.9	54.5	60.3	60.5	64.6	47.4	57.0	40.2	35.2

Subject	11	12	13	14	15	16	17	18	19	20
Duration(s)	560	637	593	719	615	589	478	620	710	600
VO_2 $_{max}$	33.8	38.8	38.9	49.5	37.1	32.2	31.3	33.8	43.7	41.7

For these data, we will **(a)** estimate β_0 and β_1; **(b)** project the VO_2 $_{MAX}$ for a 45-second treadmill duration; **(c)** create an ANOVA table; **(d)** test whether or not the model is adequate; **(e)** calculate the R^2 associated with the fit; **(f)** do hypothesis tests, at the $\alpha = .05$ level, on β_1 and β_0; and **(g)** calculate 95% confidence intervals for β_0 and β_1.

(a) First, we'll fit $E(Y) = \beta_0 + \beta_1 X$ to the data by estimating β_1 and β_0.

$$\widehat{\beta}_1 = \frac{SS_{xy}}{SS_{xx}} \approx \frac{656085 - \frac{1}{20}(14149)(887.9)}{10439699 - \frac{1}{20}(14809)^2} \approx \frac{28093.57}{432133.2} \approx 0.06498$$

$$\implies \widehat{\beta}_0 = \overline{y} - \widehat{\beta}_1 \overline{x} \approx \frac{887.9}{20} - 0.06497\left(\frac{14149}{20}\right) \approx 44.395 - 45.969 = -1.574 \ .$$

[*NOTICE* that β_1 was solved for first!!] Thus, $\widehat{Y} = -1.574 + .065X$.

(b) Predicted VO_2 $_{MAX}$ for a subject who was on the treadmill for a 45 second period $=$ $\widehat{y}(45) \approx -1.612 + 0.065(45) = 1.35$ ml/kg/min.

(c) The Analysis of Variance table is as follows:

ANOVA table

Source	df	SS	MS	F
Regression	1	$\frac{27940.14^2}{429989.0} \approx 1815.5$	1815.5	$\frac{1815.5}{10.062} \approx 180.44$
Residual	18	$1996.6 - 1815.5 = 181.1$	$\frac{181.2}{18} \approx 10.062$	
Total	19	$41414.95 - \frac{788366.4}{20} \approx 1996.6$		

(d) The test for the overall adequacy of the model results in

$$F = \frac{1815.5}{10.062} \approx 180.4 \sim F_{1,18} \text{ under } H_0$$

Therefore, $F = 180.4 \gg 15.38 = F_{1,18,.999} > F_{1,18,.999} \implies F > F_{1,18,.999} \implies$ reject H_0 at the $\alpha = .001$ level. (From S-plus, the p-value $\approx 8.05 \times 10^{-11}$.) Hence, duration until $V0_{2\ MAX}$ is a *highly* significant predictor of the value of $V0_{2\ MAX}$.

(e) The goodness-of-fit test, $R^2 = \frac{\text{Reg SS}}{\text{Total SS}} = \frac{1815.5}{1996.6} \approx 0.9093$. The high value of R^2 indicates that a large amount of the variation in $V0_{2\ MAX}$ is due to the regression on Duration. Hence, as the R^2 and the plot indicate, the fit of $V0_{2\ MAX}$ using Duration is quite good $(r = \sqrt{.9093} = 0.9536)$!!

(f) To test $H_0 : \beta_1 = 0$ versus $H_1 : \beta_1 \neq 0$ at the $\alpha = .05$ level, set $\mathbf{c}_1' = (\,0 \quad 1\,)$ and form

$$T_{\beta_1} = \frac{\widehat{\beta}_1}{\widehat{se}(\widehat{\beta}_1)} = \frac{\mathbf{c}_1'\widehat{\boldsymbol{\beta}}}{\sqrt{\mathbf{c}_1'(\mathbf{X}'\mathbf{X})^{-1}\mathbf{c}\widehat{\sigma}}} = \frac{SS_{xy}/SS_{xx}}{\sqrt{\frac{s_{y\cdot x}^2}{SS_{xx}}}} \approx \frac{.0650}{\sqrt{\frac{10.0619}{429989.0}}} \approx \frac{.0650}{.004837} \approx 13.433$$

which, under H_0, is $\sim t_{n-2} = t_{18}$. $|T_{\beta_1}| = 13.433 \gg 2.101 = t_{18,.975} \implies$ reject H_0 at the $\alpha = .05$ level. *Notice that* $T_{\beta_1}^2 = (13.433)^2 \approx 180.4 \approx F$ *in* **(d)**.

Now test $H_0 : \beta_0 = 0$ versus $H_1 : \beta_0 \neq 0$ at the $\alpha = .05$ level. So, form

$$T_{\beta_0} = \frac{\widehat{\beta}_0}{\widehat{se}(\widehat{\beta}_0)} = \frac{\overline{y} - \widehat{\beta}_1\overline{x}}{\sqrt{s_{y\cdot x}^2\left(\frac{1}{n} + \frac{\overline{x}^2}{SS_{xx}}\right)}} \approx \frac{44.395 - .065(707.45)}{\sqrt{10.062\left(\frac{1}{20} + \frac{(705.19)^2}{432133.2}\right)}} \approx \frac{-1.574}{3.495} \approx -0.4504$$

which, under H_0, is $\sim t_{n-2} = t_{19}$. $|T_{\beta_0}| = 0.4504 < 2.101 = t_{18,.975} \implies$ do <u>not</u> reject H_0 at the $\alpha = .05$ level.

(g) Two-sided 95% confidence intervals:

for $\beta_0 : \ -1.574 \mp 2.101 \cdot 3.4949 \approx -1.574 \mp 7.343 \implies (-8.917, 5.768)$

for $\beta_1 : \ 0.0650 \mp 2.101 \cdot 0.004837 \approx .0650 \mp .0102 \implies (0.0548, 0.0751)$

OBSERVE that the two-sided confidence interval for β_1 does *not* contain 0, which is consistent with the hypothesis tests done in **(d)** and **(f)** and the two-sided confidence interval for β_0 *contains* 0, which is consistent with the hypothesis test done in **(f)**.

4.4 Multiple Regression

A linear regression model with more than two predictors is called a *multiple* linear regression model. In such models, the formulas introduced in the section above understandably become more complex. However, as was demonstrated earlier, using straightforward matrix concepts, one can develop a very general approach to multiple regression problems. For example, as was developed in equations (4.9–4.11) we can develop a small set of formulas that are applicable to *any* linear regression model, which has one or more predictors.

Example 4.5. Suppose we are given the data below (Source: Rosner, *Fundamentals of Biostatistics*, 4th edition [122], Example 11.24, pp. 470). The purpose of the investigation was to predict systolic blood pressure in infants from the birthweight of the infant and the age (in days) of the infant. We can write the multiple regression model as

$$
\begin{pmatrix} 89 \\ 90 \\ 83 \\ 77 \\ 92 \\ 98 \\ 82 \\ 85 \\ 96 \\ 95 \\ 80 \\ 79 \\ 86 \\ 97 \\ 92 \\ 88 \end{pmatrix}
=
\begin{pmatrix}
1 & 135 & 3 \\
1 & 120 & 4 \\
1 & 100 & 3 \\
1 & 105 & 2 \\
1 & 130 & 4 \\
1 & 125 & 5 \\
1 & 125 & 2 \\
1 & 105 & 3 \\
1 & 120 & 5 \\
1 & 90 & 4 \\
1 & 120 & 2 \\
1 & 95 & 3 \\
1 & 120 & 3 \\
1 & 150 & 4 \\
1 & 160 & 3 \\
1 & 125 & 3
\end{pmatrix}
\begin{pmatrix} \beta_0 \\ \beta_1 \\ \beta_2 \end{pmatrix}
+
\begin{pmatrix} \epsilon_1 \\ \epsilon_2 \\ \vdots \\ \epsilon_{15} \\ \epsilon_{16} \end{pmatrix}.
\tag{4.23}
$$

The least squares solution to this problem is

$$
\widehat{\boldsymbol{\beta}} = \begin{pmatrix} \hat{\beta}_0 \\ \hat{\beta}_1 \\ \hat{\beta}_2 \end{pmatrix} = (\mathbf{X'X})^{-1}\mathbf{X'y}.
\tag{4.24}
$$

To solve for $\widehat{\boldsymbol{\beta}}$ in equation (4.24), we must calculate $\mathbf{X'X}$ and $\mathbf{X'Y}$, which themselves contain much important information about the regression variables. In our example,

$$
\mathbf{X'X} =
\begin{pmatrix}
n & \sum_{j=1}^{16} x_{1j} & \sum_{j=1}^{16} x_{2j} \\
\sum_{j=1}^{16} x_{1j} & \sum_{j=1}^{16} x_{1j}^2 & \sum_{j=1}^{16} x_{1j}x_{2j} \\
\sum_{j=1}^{16} x_{2j} & \sum_{j=1}^{16} x_{1j}x_{2j} & \sum_{j=1}^{16} x_{2j}^2
\end{pmatrix}
=
\begin{pmatrix}
16 & 1925 & 53 \\
1925 & 236875 & 6405 \\
53 & 6405 & 189
\end{pmatrix}
$$

$$
\text{and } \mathbf{X'y} =
\begin{pmatrix}
\sum_{j=1}^{16} y_j \\
\sum_{j=1}^{16} x_{1j}y_j \\
\sum_{j=1}^{16} x_{2j}y_j
\end{pmatrix}
=
\begin{pmatrix} 1409 \\ 170350 \\ 4750 \end{pmatrix}.
\tag{4.25}
$$

The element in the upper left-hand corner of $\mathbf{X'X}$ is the product of an $1 \times n$ row vector of 1's multiplied by an $n \times 1$ column vector of 1's. The result is a scalar value $n = \sum_{i=1}^{16} 1 = 16$ (that is, the sum of 16 ones is 16). Thus, this particular element tells us how many data points were included in the analysis; in this case, 16 "data triplets," (x_{1i}, x_{2i}, y_i), were observed.

The $(1,2)^{th}$ element of $\mathbf{X'X}$ is the product of an $1 \times n$ row vector containing the values of the infant birthweights multiplied by the $n \times 1$ column vector of 1's. The resulting product is the sum of all $n = 16$ of the birthweights, i.e., $\sum_{j=1}^{16} x_{1j} = 1925$. Likewise, all of the elements of both $\mathbf{X'X}$ and $\mathbf{X'y}$ represent sums, sums of squares, or sums of cross products of the variables used in the regression.

The matrices in equation (4.25) are very convenient for summarizing the descriptive statistics for the variables of interest in a regression. Most statistical packages provide the user with an option to print these matrices (i.e., $\mathbf{X'X}$ and $\mathbf{X'y}$) out. (See, for example, the option XPX in the SAS procedure PROC REG.)

Once we have $\mathbf{X'X}$ and $\mathbf{X'y}$ from equation (4.25), we can obtain the least squares solution for $\boldsymbol{\beta}$ in (4.24) as

$$\widehat{\boldsymbol{\beta}} = \begin{pmatrix} 16 & 1925 & 53 \\ 1925 & 236875 & 6405 \\ 53 & 6405 & 189 \end{pmatrix}^{-1} \begin{pmatrix} 1409 \\ 170350 \\ 4750 \end{pmatrix}$$

$$\approx \begin{pmatrix} 3.3415265 & -0.021734 & -0.200517 \\ -0.021734 & 0.0001918 & -0.000406 \\ -0.200517 & -0.000406 & 0.0752777 \end{pmatrix} \begin{pmatrix} 1409 \\ 170350 \\ 4750 \end{pmatrix} \approx \begin{pmatrix} 53.450 \\ 0.1256 \\ 5.8877 \end{pmatrix}. \qquad (4.26)$$

Two important quantities for hypothesis testing and for constructing an analysis of variance table for multiple regression are the *regression sums of squares* and the *residual* or *error sums of squares*. The _unadjusted_ regression sums of squares can be calculated as

$$(\text{Regression SS})_u = \widehat{\boldsymbol{\beta}}'\mathbf{X'y} \approx 1409(53.45) + 170350(0.1256) + 4750(5.888) \approx 124671.1 \,.$$

The *regression sums of squares adjusted for the mean* can be calculated as

$$(\text{Regression SS})_a = \widehat{\boldsymbol{\beta}}'\mathbf{X'y} - n\bar{y}^2 \approx 124671.1 - 124080.06 \approx 591.0356 \,.$$

In our example, the unadjusted regression sums of squares has $p = 3$ degrees of freedom whereas the regression sums of squares adjusted for the mean has $p - 1 = 2$ degrees of freedom. *It is much more common to use adjusted sums of squares than to use unadjusted sums of squares.*

The *residual sums of squares* for our example can be calculated as

$$\text{Residual SS} = (n-p)\widehat{\sigma}^2 = (n-3)\widehat{\sigma}^2 = \text{Residual SS} = \mathbf{y'y} - \widehat{\boldsymbol{\beta}}'\mathbf{X'y}$$

$$= (\,89 \quad 90 \quad \ldots \quad 92 \quad 88\,) \begin{pmatrix} 89 \\ 90 \\ \vdots \\ 92 \\ 88 \end{pmatrix} - (\text{Regression SS})_u$$

$$= (89^2 + 90^2 + \ldots + 92^2 + 88^2) - (\text{Regression SS})_u$$

$$\approx 124751 - 124671.1 \approx 79.9019 \,.$$

SAS program for regression problem using a matrix approach

```
/* -----------------------------------------------------------------
 | The purpose of this program is to show how to
 | 1) Read design and response matrices into PROC IML,
 | 2) Use PROC IML to calculate the least squares estimate of
 | the parameter vector in a linear model.
 ----------------------------------------------------------------- */

PROC IML;

* Form the DESIGN matrix by entering a column of 1's followed
  by the two predictor variables;

X = {
1 135 3,
1 120 4,
1 100 3,
1 105 2,
1 130 4,
1 125 5,
1 125 2,
1 105 3,
1 120 5,
1  90 4,
1 120 2,
1  95 3,
1 120 3,
1 150 4,
1 160 3,
1 125 3};

* Form Y and NOTE that Y is a 16 by 1 COLUMN vector;

Y = {89, 90, 83, 77, 92, 98, 82, 85, 96, 95, 80, 79, 86, 97, 92, 88};
PRINT Y X;

* Now, get estimate of parameter vector;
BETAHAT = INV(X` * X) * X` * Y;
print 'Parameter Vector';
print BETAHAT;
Title1 'Example of Multiple Linear Regression Problem Using Matrices';

* Find regression and residual sums of squares, print results;
REGSS = BETAHAT` * X` * Y - NROW(Y) * Y(|:,|)##2;
* Note that Y(|:,|) takes the mean value of all of the Y's;
RESIDSS = Y` * Y - BETAHAT` * X` * Y;
print 'Regression Sums of Squares';
print REGSS;
print 'Error Sums of Squares';
print RESIDSS;
SIGHAT2 = RESIDSS / (NROW(Y) - NROW(BETAHAT));
c = {0,0,1};
VARB = INV(X`*X) * SIGHAT2;
VARB2 = c` * INV(X`*X) * c * SIGHAT2;
T = c` * BETAHAT / SQRT(VARB2);
print 'T statistic';
print T;
```

Note that in the above example, by running the SAS PROC IML program, we obtain

$$
\widehat{\Sigma}_{\widehat{\beta}} = (\mathbf{X'X})^{-1}\widehat{\sigma}^2 \approx
\begin{pmatrix}
3.3415265 & -0.021734 & -0.200517 \\
-0.021734 & 0.0001918 & -0.000406 \\
-0.200517 & -0.000406 & 0.0752777
\end{pmatrix} \times 6.1462969
$$

$$\approx \begin{pmatrix} 20.538014 & -0.133581 & -1.23244 \\ -0.133581 & 0.001179 & -0.002495 \\ -1.23244 & -0.002495 & 0.462679 \end{pmatrix}$$

Thus,

$$T_{\widehat{\beta}_2} = \frac{\widehat{\beta}_2}{\sqrt{\widehat{\mathrm{Var}}(\widehat{\beta}_2)}} = \frac{\mathbf{c}'\widehat{\boldsymbol{\beta}}}{\sqrt{\mathbf{c}'(\mathbf{X}'\mathbf{X})^{-1}\mathbf{c}\widehat{\sigma}^2}} \approx \frac{5.8877}{\sqrt{0.462679}} \approx \frac{5.8877}{0.6802} \approx 8.6558 .$$

Since $8.6558 \gg 4.221 = t_{13,.9995}$, we have that β_2 is highly significantly different from 0, $p < .001$. That is, an infant's age in days is a highly significant predictor of his/her systolic blood pressure (even after adjusting for birthweight).

One can verify the results from the above programs by running the following SAS program:

```
DATA ROSNER;
 INPUT SYSBP BIRTHWT AGE;
 datalines;
89 135 3
90 120 4
83 100 3
77 105 2
92 130 4
98 125 5
82 125 2
85 105 3
96 120 5
95  90 4
80 120 2
79  95 3
86 120 3
97 150 4
92 160 3
88 125 3
 ; run;

Title 'Example of Multiple Linear Regression Problem using PROC REG';
Title2 'Example 11.24 from Rosner's 4th edition, page 470';

PROC REG DATA=ROSNER/XPX;
MODEL SYSBP = BIRTHWT AGE; run;
```

Confidence intervals (CI) for the parameters (or linear combinations of parameters) are given by

$$\mathbf{c}'\widehat{\boldsymbol{\beta}} \mp t_{n-p,1-\frac{\alpha}{2}} \sqrt{\mathbf{c}'(\mathbf{X}'\mathbf{X})^{-1}\mathbf{c}\widehat{\sigma}^2} , \tag{4.27}$$

where $\mathbf{c}' = (0 \quad 0 \quad 1)$, $n = 16$ and $p = 3$, so that, for example, the shortest 95% CI for β_2 is $5.8877 \mp t_{13,.975}(0.6802) \approx 5.89 \mp 2.16(0.680)$, which yields the interval $(4.42, 7.36)$.

The beauty of the general linear models approach to regression is that t-tests and general analysis of variance problems comparing, for example, several means can all be viewed as regression, that is, using the model given in equation (4.9). This concept is reinforced in the following example.

Example 4.6. To demonstrate the applicability of the general liner model, we consider an example of a two-sample t-test. Suppose that a small study is conducted where the mean body mass index (BMI) [that is, mass/(height)^2 measured in kg/m^2] for four slightly hypertensive women with BMIs equal to 25.3, 22.1, 33.0, and 29.2 is compared to the mean BMI of three other normotensive women with BMIs equal to 16.8, 23.7, and 15.3. Assuming these data are independent and normal, we can write a regression model as $y_{ij} = \mu_j + \epsilon_{ij}$ where $i = 1, 2$ denotes the two groups and $j = 1, \ldots, n_i$ denotes the individual observations within group i. By stacking the observations, we can write

$$
\mathbf{y} = \begin{pmatrix} 25.3 \\ 22.1 \\ 33.0 \\ 29.2 \\ 16.8 \\ 23.7 \\ 15.3 \end{pmatrix} = \begin{pmatrix} 1 & 0 \\ 1 & 0 \\ 1 & 0 \\ 1 & 0 \\ 0 & 1 \\ 0 & 1 \\ 0 & 1 \end{pmatrix} \begin{pmatrix} \mu_1 \\ \mu_2 \end{pmatrix} + \begin{pmatrix} \epsilon_{11} \\ \epsilon_{12} \\ \epsilon_{13} \\ \epsilon_{14} \\ \epsilon_{21} \\ \epsilon_{22} \\ \epsilon_{23} \end{pmatrix}
$$

so that $\mathbf{X'X} = \begin{pmatrix} 4 & 0 \\ 0 & 3 \end{pmatrix}$ and $\widehat{\boldsymbol{\mu}} = \begin{pmatrix} \widehat{\mu}_1 \\ \widehat{\mu}_2 \end{pmatrix} = (\mathbf{X'X})^{-1}\mathbf{X'y} = \begin{pmatrix} \frac{1}{4}\sum_{j=1}^{4} y_{1j} \\ \frac{1}{3}\sum_{j=1}^{3} y_{2j} \end{pmatrix}$

$= \begin{pmatrix} \bar{y}_{1\bullet} \\ \bar{y}_{2\bullet} \end{pmatrix} = \begin{pmatrix} 27.4 \\ 18.6 \end{pmatrix}$. By choosing $\mathbf{c'} = \begin{pmatrix} 1 & -1 \end{pmatrix}$ and noting that

$$
\widehat{\sigma}^2 = \frac{\mathbf{y'y} - \widehat{\boldsymbol{\mu}}'\mathbf{X'y}}{n-p} = \frac{\sum\limits_{j=1}^{4} y_{1j}^2 - n_1\bar{y}_{1\bullet}^2 + \sum\limits_{j=1}^{3} y_{2j}^2 - n_2\bar{y}_{2\bullet}^2}{(n-p)} = \frac{(n_1-1)s_1^2 + (n_2-1)s_2^2}{n_1+n_2-2},
$$

we can then write the usual t statistic as

$$
T = \frac{\mathbf{c'}\widehat{\boldsymbol{\mu}}}{\sqrt{\mathbf{c'}(\mathbf{X'X})^{-1}\mathbf{c}\widehat{\sigma}^2}} = \frac{\bar{y}_{1\bullet} - \bar{y}_{2\bullet}}{\sqrt{\left(\frac{1}{n_1} + \frac{1}{n_2}\right)\widehat{\sigma}^2}} \approx \frac{27.4 - 18.6}{\sqrt{\left(\frac{1}{4} + \frac{1}{3}\right)21.448}} \approx \frac{8.8}{3.5371} \approx 2.488
$$

which is slightly less than $t_{n-p,.975} = t_{5,.975} \approx 2.57$ since $n - p = n_1 + n_2 - 2 = 5$. The result is that with these two very small samples of women, the mean BMIs are not significantly different at a 5% level. What we can see, therefore, is that the test of equality of coefficients of this "regression" model is equivalent to a simple two-sample t-test testing the means of two populations.

4.5 Visualization of Data

One trap that a busy data analyst can fall into is to merely analyze data without visualizing them either before or after a model has been fit. As we shall see, failure to perform this practical step in analyzing regression data can lead to some puzzling results. In the example below, we demonstrate very simple plots for visualizing regression data. A very comprehensive survey of graphical and diagnostic techniques in regression analysis are given in Cook and Weisberg [21] and Harrell [62].

Example 4.7. Consider the following four datasets originally constructed by F. Anscombe [7] and later reproduced by Draper and Smith [30]. Each dataset consists of 11 observations. For each of the four datasets, a simple linear regression (in Minitab) is fit.

Dataset											
# 1	X	10	8	13	9	11	14	6	4	12	7	5
	Y	8.04	6.95	7.58	8.81	8.33	9.96	7.24	4.26	10.84	4.82	5.68
# 2	X	10	8	13	9	11	14	6	4	12	7	5
	Y	9.14	8.14	8.74	8.77	9.26	8.10	6.13	3.10	9.13	7.26	4.74
# 3	X	10	8	13	9	11	14	6	4	12	7	5
	Y	7.46	6.77	12.74	7.11	7.81	8.84	6.08	5.39	8.15	6.42	5.73
# 4	X	8	8	8	8	8	8	8	8	8	8	19
	Y	6.58	5.76	7.71	8.84	8.47	7.04	5.25	5.56	7.91	6.89	12.50

Dataset #1

```
The regression equation is
Y1 = 3.00 + 0.500 X1

Predictor        Coef         StDev           T       P
Constant        3.000         1.125        2.67    0.026
X1              0.5001        0.1179        4.24    0.002

S = 1.237       R-Sq = 66.7%      R-Sq(adj) = 62.9%

Analysis of Variance

Source      DF        SS       MS           F       P
Regression   1    27.510    27.510       17.99   0.002
Error        9    13.763     1.529
Total       10    41.273
```

Dataset #2

```
The regression equation is
Y2 = 3.00 + 0.500 X1

Predictor        Coef         StDev           T       P
Constant        3.001         1.125        2.67    0.026
X1              0.5000        0.1180        4.24    0.002

S = 1.237       R-Sq = 66.6%      R-Sq(adj) = 62.9%

Analysis of Variance

Source      DF        SS       MS          F          P
Regression   1    27.500    27.500      17.97     0.002
Error        9    13.776     1.531
Total       10    41.276
```

Dataset #3

```
The regression equation is
Y3 = 3.00 + 0.500 X1

Predictor        Coef         StDev           T       P
Constant        3.003         1.134        2.67    0.026
```

```
X1                 0.4997         0.1179        4.24     0.002

S = 1.236          R-Sq = 66.6%      R-Sq(adj) = 62.9%

Analysis of Variance
  Source         DF         SS        MS       F        P
  Regression      1      27.470    27.470    17.97     0.002
  Error           9      13.756     1.528
  Total          10      41.226

Unusual Observations
  Obs      X1        Y3     Fit StDev Fit   Residual   St Resid
   3     13.0     12.740   9.499     0.601     3.241      3.00R

R denotes an observation with a large standardized residual
```

Dataset #4

```
The regression equation is
Y4 = 3.00 + 0.500 X2

  Predictor        Coef       StDev        T       P
  Constant        3.002       1.124      2.67     0.026
  X2              0.4999      0.1178     4.24     0.002

S = 1.236          R-Sq = 66.7\%      R-Sq(adj) = 63.0\%

Analysis of Variance
  Source         DF         SS        MS          F         P
  Regression      1      27.490    27.490     18.00     0.002
  Error           9      13.742     1.527
  Total          10      41.232

Unusual Observations
  Obs   X2     Y4      Fit    StDev Fit   Residual    St Resid
   11  19.0 12.500  12.500      1.236     -0.000       * X

X denotes an observation whose X value gives it large influence.
```

4.5.1 Visualization before Model Fitting

Simple scatterplots of datasets 1–4 shown in Figure 4.3 reveal four quite distinct patterns of data. However, when examining the regression models given above, we note that the parameter estimates, standard errors of parameter estimates and analysis of variance tables are virtually identical!! Thus, *had we not visualized the data first*, we would have concluded that the linear pattern held for each of the four datasets, an obvious falsehood. For example, a model that was quadratic in the X's would have been more appropriate for dataset #2.

What is clear from this example is that from the output, one sees almost identical results, which without any further analyses or visualization of the data, would lead an investigator to falsely conclude that the datasets were nearly equivalent.

Figure 4.3 *Scatterplots of Datasets 1–4*

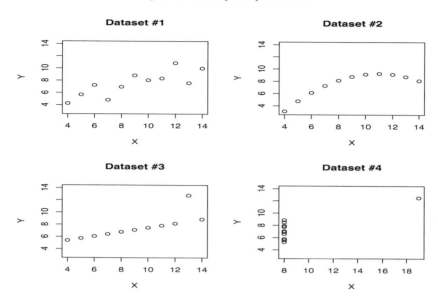

4.5.2 *Visualization after Model Fitting*

Another useful role for visualization as part of the model building process is to see how well the model actually fits the data. Three plots that can be useful in this process are:

1. A scatterplot of the *fitted* Y_i's, that is, $\widehat{Y}_i$, $i = 1, \ldots, n$ overlayed onto the *actual* Y_i's both plotted versus the X_i's $i = 1, \ldots, n$;
2. A scatterplot of the *fitted residuals*, $\widehat{e}_i$'s, $i = 1, \ldots, n$, plotted versus the X_i's $i = 1, \ldots, n$;
3. A scatterplot of the *fitted residuals*, $\widehat{e}_i$'s, $i = 1, \ldots, n$, plotted versus the *fitted* Y_i's, that is, $\widehat{Y}_i$, $i = 1, \ldots, n$.

Plots 1 and 2 are especially useful when there is *one* predictor variable. In particular, one can use these plots for detecting a misspecified model, for example, misspecify-ing the model as being linear in the X's instead of quadratic. Both types of plots can also indicate whether or not there is lack of fit, that is, whether or not the fitted model tends to explain the overall variability of the observed outcomes.

Plot 3 is always useful but is uniquely useful when there are *more than one* predictor variables. Again, it can be used for getting a general feel for model misspecification or lack of fit. Plots 1 and 3 for the four datasets are displayed in Figures 4.4 and 4.5.

Figure 4.4 *Fitted Linear Regression Lines to Four Datasets*

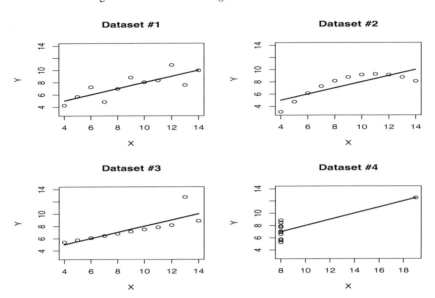

Figure 4.5 *Residual Plots versus Predicted Y's for Datasets 1–4*

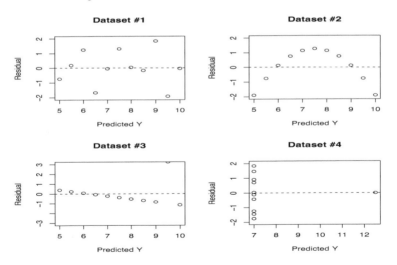

4.6 Model Assessment and Related Topics

As was emphasized in the previous section, some of the most effective ways of assessing a model is by *visualizing* various residual plots after fitting the regression.
However, the act of visualization doesn't allow us to draw a formal inference about

the adequacy of models. Arriving at a "best model" is often quite complex because of the fact that the predictor variables may be related with *each other* as well as being related to the outcome variable. In fact, there are many situations, where several competing models may be essentially equivalent. In such cases, one must use his/her *scientific judgment* (rather than to *only* rely on statistical inference) in order to create a hierarchy of plausible models. Nevertheless, proper statistical assessment will almost always enhance one's ability to assess the competing models under consideration.

4.6.1 Methods for Assessing Model Adequacy in Regression

Listed below are seven methods commonly used to assess model adequacy. These are not exhaustive as there are a plethora of such methods.

1. Assess the significance of the regression coefficients, i.e., test whether or not each of the p coefficients, $\beta_0, \beta_1, \ldots, \beta_{p-1}$ are significantly different from 0.

2. Examine *studentized* residuals. Studentized residuals (sometimes referred to as *standardized* residuals) are obtained by dividing each estimated residual, $\hat{e}_i$ by its standard error, $se(\hat{e}_i)$, $i = 1, \ldots, n$ (see Figure 4.6). Strictly speaking, each "studentized" residual should be compared to a t-distribution with $n - p$ degrees of freedom in order to assess whether it is a significant outlier. The α-level for the assessments of residuals is usually set to some small value such as .01. A practical rule of thumb widely used is to declare any datum whose standardized residual is > 3 as being a significant outlier (see Draper and Smith [30] for further discussion).

3. Assessment of the R^2 ("goodness-of-fit") statistic. (The square root of R^2 is called the *multiple correlation coefficient*.) R^2 *always increases* as the number of variables in the model increases. However, only very modest increases in R^2 will be observed for some variables, which indicate that they contribute very little to the overall variation of the data. One can use the R^2 statistic to test whether or not a subset of variables simultaneously are different from 0. For example, suppose k variables, $X_1, X_2, \ldots X_j, X_{j+1}, \ldots X_k$ are fitted in a regression model. We can first calculate R_2^2 corresponding to the hypothesis $H_{0,2}: \beta_1 = \ldots = \beta_j = \beta_{j+1} = \ldots = \beta_k = 0$, that is, the hypothesis that all k variables are equal to 0. If we are interested in testing whether or not the coefficients associated with a <u>subset</u> of the k variables, say, $X_{j+1}, \ldots, X_k$ taken together are significantly different from 0, then we would be interested in testing $H_0: \beta_{j+1} = \ldots = \beta_k = 0$ versus H_1: at least one $\beta_i \neq 0$, $i = j + 1, \ldots, k$. If the test *fails* to reject, then we can discard the subset of variables from the model. To do this test, we must also regress Y on the first j variables and then calculate R_1^2 associated with the hypothesis $H_{0,1}$: $\beta_1 = \ldots = \beta_j = 0$. (The hypothesis $H_{0,1}: \beta_1 = \ldots = \beta_j = 0$ is said to be *nested within* $H_{0,2}: \beta_1 = \ldots = \beta_j = \beta_{j+1} = \ldots = \beta_k = 0$.) The statistic for testing $H_0: \beta_{j+1} = \ldots = \beta_k = 0$ versus H_1: at least one $\beta_i \neq 0$, $i = j + 1, \ldots, k$, is

written as

$$F = \frac{(R_2^2 - R_1^2)/(k - j)}{(1 - R_2^2)/(n - k - 1)},$$

which, under H_0: $\beta_{j+1} = \ldots = \beta_k = 0$ is distributed as $F_{k-j,n-k-1}$.

Example 4.8. Suppose we have a sample size of 25 in a regression and originally fit six variables to predict the outcome of interest. The value of multiple R^2 for this fit is $R_2^2 = 0.6$. R_2^2 would be associated with the hypothesis $H_{0,2}$: $\beta_1 = \beta_2 = \beta_3 = \beta_4 = \beta_5 = \beta_6 = 0$. Suppose further we wish to test whether or not variables 1 and 5 can be discarded. We would then calculate R_1^2 associated with the hypothesis $H_{0,1}$: $\beta_2 = \beta_3 = \beta_4 = \beta_6 = 0$, that is, the four variables we wish to keep in the model. Suppose that $R_1^2 = 0.5$. Thus, in our situation, $n = 25$, $k = 6$ and $j = 4$. Then, to test H_0: $\beta_1 = \beta_5 = 0$ we calculate

$$F = \frac{(0.6 - 0.5)/(6 - 4)}{(1 - 0.6)/(25 - 6 - 1)} = \frac{.05}{.0222} = 2.25 < F_{2,18,.95} = 3.55.$$

The corresponding p-value is $p \approx 0.134$. Thus, because we do *not* reject H_0, we can discard both X_1 and X_5 from the model.

---

4. Assessment of Mallow's C_p statistic. The idea behind this assessment is to balance the reliability of obtaining a "good fit" (that is, one without much bias) using a lot of variables with the practicality of obtaining the simplest ("most parsimonious") model available. The C_p statistic is obtained by first fitting a model with all the variables considered (i.e., *saturated* model) and calculating the mean squared error, s^2. Then for each *subset* of variables being considered, the following statistic is calculated: $C_p = RSS_p/s^2 - (n - 2p)$; where RSS_p is the residual sum of squares from the model fit for the given subset. If there is no lack of fit, then the *expected value* of C_p, i.e., $E(C_p)$ is equal to p. Thus, for each subset of p variables (including the intercept), we would look for the model whose C_p statistic is closest to p. In the saturated model, C_p is *always equal to* p. In that case, however, we are usually able to discard a number of variables due to the fact that they are not *significant* predictors of the outcome. Thus, we want to find the model with the smallest value of p with the constraint that the value of C_p is close to p.

5. Assessment of influence points. Probably the most common way of doing this is by examining the Cook's D (Distance) statistic. The idea is to drop each point and obtain a new parameter vector, $\boldsymbol{\beta}(i) = (\beta_0(i) \quad \beta_1(i) \quad \ldots \quad \beta_{p-1}(i))^T$ for the regression equation. The "(i)" indicates that the ith data point is dropped. Then, for each (dropped) point, a function of the *squared* distance between the parameter vector with and without that data point is calculated. Under the null hypothesis of no lack of fit, each Cook's D is approximately distributed as $F_{p,n-p}$. Thus, in order to do formal testing, one would compare the Cook's D for each point to $F_{p,n-p,1-\alpha}$ for some α. In Figure 4.7, plots of Cook's Distance versus predicted Y's for datasets # 1–4 are given. In dataset 4, the last point has an extremely large influence (indicated by "↑" in the plot).

6. Assessment of the *correlation matrix* of the estimated parameter vector. Two or more variables being highly correlated can result in very *unreliable* parameter

estimates. One can look at the correlation matrix of the estimated parameter vector to assess this. If the correlation between two coefficients is very high, say over .95, then either a regression adjustment should be made, e.g., use of methods such as ridge regression or one should drop one of the variables from the model.

Example 4.9. Consider again the data in Example 4.5, where we were looking at a regression which used birthweight and age to predict the systolic blood pressure of an infant. The variance–covariance matrix obtained for example is as follows:

$$\widehat{\Sigma}_{\widehat{\beta}} = \begin{pmatrix} 20.538014 & -0.133581 & -1.23244 \\ -0.133581 & 0.001179 & -0.002495 \\ -1.23244 & -0.002495 & 0.462679 \end{pmatrix}$$

The corresponding correlation matrix is

$$\begin{pmatrix} 1.0000 & -0.8584 & -0.3998 \\ -0.8584 & 1.0000 & -0.1068 \\ -0.3998 & -0.1068 & 1.0000 \end{pmatrix}$$

The correlation between the coefficients for birthweight and age in this model is -0.1068, which is obtained by dividing the covariance between the coefficients of birthweight and age by the square root of the product of their variances, i.e.,

$$-0.1068 \approx \frac{-0.002495}{\sqrt{(0.001179) \cdot (0.462679)}}.$$

---

7. Assessment of the Durbin–Watson statistic for first-order serial correlation (see Draper and Smith, 1981 [30] for more details). This is a very effective method for determining whether or not a model has a "lack of fit." In many models that do have a lack of fit, consecutive residuals are highly correlated. Such correlation between consecutive residuals (e_{i-1} and e_i, $i = 2, \ldots, n$) is called *first-order serial correlation*. The Durbin–Watson test formally tests the hypothesis, H_0 : $\rho_s = 0$ versus H_1 : $\rho_s = \rho^s$ ($0 < \rho < 1$) by forming the test statistic $d = \sum_{i=2}^{n} (e_i - e_{i-1})^2 / \sum_{i=1}^{n} e_i^2$ and comparing these to certain tabled values. The significance of a two-sided test is determined as follows: if $d < d_L$ or $4 - d < d_L$ then conclude that d is significant at level 2α; if $d > d_U$ and $4 - d > d_U$ then conclude that d is *not* significant at level 2α; otherwise, the test is said to be inconclusive.

Figure 4.6 *Plots of Studentized Residual Versus the Predicted Y's for Datasets 1–4*

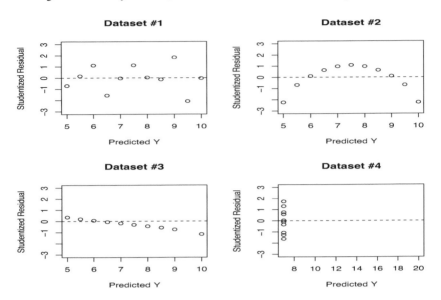

4.6.2 Picking the Best Models in Multiple Regression

A practical consideration when fitting many variables to an outcome is how to arrive at the "best" set of predictors. By "best," we mean the smallest set of predictors
that are most highly associated with the outcome of interest and generally satisfy
the model adequacy criteria outlined above. Two common strategies for stepwise
regression are "forward elimination" or "step-up" procedures and "backwards elimination" or "step-down" procedures. A third procedure uses a combination of both of
these methods. The forward elimination procedure starts with variables either one at
a time or a few at a time and adds potential predictors at each step. The backwards
elimination procedure starts with many or all of the variables (if feasible) and then
eliminates variables that are either not sufficiently related to the outcome of interest
or are redundant (possibly colinear with other variables) in predicting the outcome
of interest. Excellent discussions of these procedures are given in Fisher and van
Belle [46], Seber and Lee [136] and Draper and Smith [30].

4.6.3 Partial Correlation Coefficient

Sometimes when many variables, say, $Y, X_1, \ldots, X_k$ are being measured in a study,
one is interested in determining whether or not a linear association exists among two

Figure 4.7 *Plots of Cook's Distance Versus the Predicted Y's for Datasets 1–4*

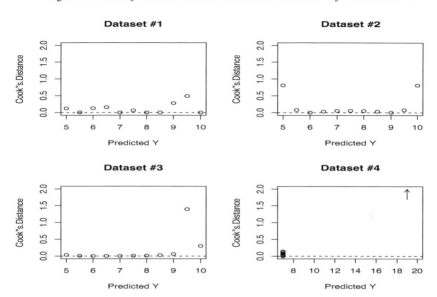

variables, say Y and X_1, *after adjusting for a number of other variables*. Such a measure is referred to a *partial correlation coefficient* and is denoted as $\rho_{Y X_1 \bullet X_2 \ldots, X_k}$. The general formula for calculating an estimate, $r_{Y X_1 \bullet X_2 \ldots, X_k}$, of the partial correlation coefficient requires an iterative algorithm. However, in the <u>*special case*</u> where two variables, Y and X_1 are related to a third variable, X_2, one can calculate the estimate of the partial correlation coefficient between Y and X_1 *while adjusting for* X_2 as

$$r_{Y X_1 \bullet X_2} = \frac{r_{Y X_1} - r_{Y X_2} r_{X_1 X_2}}{\sqrt{\left(1 - r_{Y X_2}^2\right)\left(1 - r_{X_1 X_2}^2\right)}},$$

where each "r" in the right-hand side of the formula represents an estimate of the simple Pearson correlation coefficient.

Example 4.10. Suppose, in the systolic blood pressure data for infants, that we were more interested in characterizing the correlation between systolic blood pressure and birthweight while *adjusting for* the age in days. For this application, define Y = systolic blood pressure, X_1 = birthweight and X_2 = age in days. The simple Pearson correlation coefficients are $r_{Y X_1} = 0.4411$, $r_{Y X_2} = 0.87084$ and $r_{X_1 X_2} = 0.10683$,

$$r_{Y X_1 \bullet X_2} \approx \frac{0.4411 - 0.8708 \cdot 0.1068}{\sqrt{\left(1 - 0.8708^2\right)\left(1 - 0.1068^2\right)}} \approx \frac{0.3481}{\sqrt{0.2416 \cdot 0.9886}} \approx \frac{0.3481}{0.4888} \approx 0.712 \,.$$

Partial correlation coefficients are meaningful *only when* the outcome variable, Y, _and_ all of the Z's of interest are approximately _normally distributed_. However, in a general regression setting, there is no requirement that the _predictor_ variables be normally distributed. In contrast, the _outcome_ variable, Y, conditional on conditional on the X's (predictors), should be at least approximately normally distributed in a multiple linear regression. Therefore, the calculation of partial correlation coefficients is not always appropriate in a given regression setting.

4.7 Polynomial Regression

Often the relationship between an outcome variable, Y, and a predictor variable, X, is not linear. A common relationship can be described using _polynomial_ functions of X, that is,

$$E(y_i) = \beta_0 + \beta_1 X_i + \beta_2 X_i^2 + \ldots + \beta_{p-1} X_i^{p-1}, i = 1, \ldots, n \qquad (4.28)$$

Notice the the relationship with regard to the _parameters_ is linear. Consequently, one can view polynomial regression as being a special case of multiple linear regression. Thus, system of equations described in (4.28) can be written in matrix form as

$$E(\mathbf{y}) = E \begin{pmatrix} y_1 \\ y_2 \\ \vdots \\ y_n \end{pmatrix} = \begin{pmatrix} 1 & X_1 & X_1^2 & \cdots & X_1^{p-1} \\ 1 & X_2 & X_2^2 & \cdots & X_2^{p-1} \\ \vdots & \vdots & \vdots & \cdots & \vdots \\ 1 & X_n & X_n^2 & \cdots & X_n^{p-1} \end{pmatrix} \begin{pmatrix} \beta_0 \\ \beta_1 \\ \vdots \\ \beta_{p-1} \end{pmatrix} = \mathbf{X}\boldsymbol{\beta} \qquad (4.29)$$

Notice that this model is a _linear function_ of the parameter vector, $\boldsymbol{\beta}$.

4.8 Smoothing Techniques

In regression analysis, the relationship between n data points, (t_i, y_i), is described by a fixed functional form, $y_i = f(t_i) + e_i$, $i = 1, \ldots, n$, where the e_i are usually independently and identically distributed $N(0, \sigma^2)$. This functional form is usually specified prior to fitting the data. In some cases, however, no particular functional form can be readily be found to adequately describe a relationship between two observed variables. In this case, it is appropriate to use _smoothing techniques_ to characterize such a relationship. One advantage to this approach is that no prespecified functional form must be used. Moreover, one can explore functional forms with smoothing techniques. The two most common smoothing methods are _smoothing polynomial splines_ (Wahba [155], Craven and Wahba [23], Kohn and Ansley [81]), and _lowess_ (Cleveland [16, 17]).

A _polynomial spline_ of degree $2m - 1$ as a piecewise polynomial of degree $2m - 1$ in each of the intervals between adjacent points (t's) with the polynomials grafted together so that the first $2m - 2$ derivatives are continuous. The points where the spline is grafted are called _knots_ in the spline literature. A _smoothing_ polynomial spline is then the smoothest spline that achieves a certain degree of fidelity for a given dataset. "Fidelty" here refers to how well the predicted curve follows the data. The knots for smoothing polynomial splines are located at every unique t value. Schoenberg [131] showed that the smoothing spline, $g_{n,\lambda}(t_i)$, of degree $2m - 1$ minimizes the quantity

$$r(t) = n^{-1} \sum_{i=1}^{n} [g(t_i) - y_i]^2 + \lambda \int_0^1 (g^{(m)}(t))^2 dt, \tag{4.30}$$

where $g^{(m)}(t)$ denotes the mth derivative of $g(t)$. Notice that if $\lambda = 0$, this criterion is the same as the least squares criterion. Since the curve is *flexible* (not fixed), the least squares criterion when $\lambda = 0$ results in fitting an *interpolating* spline. On the other hand, if $\lambda \to \infty$, then the criterion is equivalent to fitting a polynomial regression model of degree m. Wahba [155] showed that the stochastic process

$$g_{n,\lambda}(t) = \sum_{k=0}^{m-1} \alpha_k \frac{(t-a)^k}{k!} + R^{1/2}\sigma \int_a^t \frac{(t-h)^{m-1}}{(m-1)!} dW(h), \tag{4.31}$$

where $dW(h) \sim N(0,1)$ is the solution to minimizing $r(t)$ in equation (4.30) with $\lambda = 1/\sigma$. This is conditional on letting $\alpha = (\alpha_0, \alpha_1, \ldots, \alpha_{m-1})'$ have a diffuse prior, that is, assuming that the variance of each component is infinitely large. Smoothing polynomial splines have an advantage in that they are equivalent to many fixed functional forms. Thus, they can be used whether or not an underlying function is known for a particular process. Also, the smoothing or "bandwidth" parameters are often chosen automatically using fixed statistical criteria (Wahba[155] and Kohn and Ansley [81]).

Locally weighted polynomial methods like *lowess* are based on the concept of fitting low degree polynomials around neighborhoods of each of the x values in a set of n (x, y) pairs. The algorithm in R allows the degrees of polynomials to be 0,1 or 2. The user determines how smooth the fitted lowess curve will be by specifying the smoothing parameter, α. The value of α is the proportion of the dataspan of x $[= \max(x) - \min(x)]$ used in each fit. Thus, α controls how flexible or "wiggly" the lowess curve will be. Large values of α produce the smoothest curves whereas small values are associated with curves that fluctuate a lot. Values near 0 would interpolate the data. Unless one has knowledge of the nature of the underlying function to be fit, a small value of the smoothing parameter is often not desirable. In the R implementation of lowess, the default value of the smoothing parameter (denoted by f in that program) is $2/3$. Another related algorithm implemented in R, called *loess*, uses a slightly different formulation to arrive at the fitted curve.

Example 4.11. To demonstrate the ability of smoothing and lowess splines to model different patterns of data, we simulated three different sine curves: one with a high "signal to noise" ratio, one with a medium "signal to noise" ratio, and one with a low "signal to noise" ratio. Smoothing splines are fit with no prior specification of smoothing parameters and loess curves are fit with the bandwidth of the smoothing parameter roughly 1/8 of the length of the data span. The program and outputted graphs are displayed in Figure 4.8. The solid lines are the

true sine curves. As can be seen by the dashed lines of the two predicted curves, both the
smoothing spline and loess fits tend to track the data well with low signal to noise ratios, but
fluctuate somewhat due to local influence points with medium signal to noise ratios. With high
signal to noise ratios, the signal tends to be lost and the two predictors oversmooth somewhat.

```
x<-1:100/10
f.x<-sin(pi*x)
e1<-rnorm(length(x),0,.25); e2<-rnorm(length(x),0,1); e3<-rnorm(length(x),0,2)

Y1<-f.x+e1; Y2<-f.x+e2; Y3<-f.x+e3

par(mfrow=c(3,1)) # Set to three frames per plot
plot(x,Y1,lwd=1); lines(x,f.x,lty=1,lwd=1)
lines(lowess(x,Y1,f=1/8),lty=3,col="blue",lwd=2)
lines(smooth.spline(x,Y1),lty=4,col="red",lwd=3)
title("Sine wave with low noise")
plot(x,Y2,lwd=1); lines(x,f.x,lty=1,lwd=1)
lines(lowess(x,Y2,f=1/8),lty=3,col="blue",lwd=2)
lines(smooth.spline(x,Y2),lty=4,col="red",lwd=3)
title("Sine wave with medium noise")
plot(x,Y3,lwd=1); lines(x,f.x,lty=1,lwd=1)
lines(lowess(x,Y3,f=1/8),lty=3,col="blue",lwd=2)
lines(smooth.spline(x,Y3),lty=4,col="red",lwd=3)
title("Sine wave with high noise")

par(mfrow=c(1,1)) # Set back to one frame per plot
```

Figure 4.8 *Fitted Smoothing Splines to Simulated Sine Wave Data*

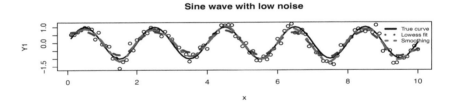

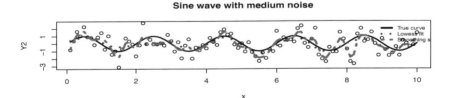

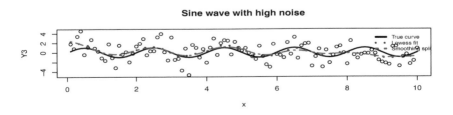

4.9 Appendix: A Short Tutorial in Matrix Algebra

In this section, we first review some very simple matrix concepts and show how matrix calculations can be performed in SAS IML or R (S-plus). Similar types of programs can also be easily implemented in other statistical packages that have built-in matrix algorithms.

A matrix is an *array of numbers*. Examples of matrices are as follows:

$$\begin{pmatrix} 1 & 0 \\ 0 & 1 \end{pmatrix}, \quad (1 \quad 1 \quad 1 \quad 1), \quad \begin{pmatrix} -1 & 3 & 2 \\ 7 & 5 & 4 \\ 0 & -2 & 1 \\ 4 & -10 & \frac{1}{2} \end{pmatrix}.$$

Matrices have *dimensions*. The dimensions of a matrix tell us how many rows and columns, respectively, the matrix has. For example, the matrix $\begin{pmatrix} 1 & 0 \\ 0 & 1 \end{pmatrix}$ is a 2×2 matrix (2 rows and 2 columns) whereas $(1 \quad 1 \quad 1 \quad 1)$ is a 1×4 matrix (1 row and 4 columns). A matrix with only one row is also called a *row vector*. Likewise, a matrix with only one column is called a *column vector*. A matrix with one row <u>and</u> one column is called a *scalar*. Thus, the number 7 can be thought of as either a 1×1 matrix or as a scalar.

Matrices are useful for describing *systems of equations*. For example, a simple linear regression model can be written as follows:

$$y_i = \beta_0 + \beta_1 x_i + \epsilon_i, \quad i = 1, \ldots, n, \qquad (4.32)$$

where the x_i are *fixed* and *known*, β_0 and β_1 are fixed and *unknown* and the ϵ_i are *random variables*. Equation (4.32) actually represents <u>only one</u> of a total of n equations for a given dataset. The *system* of n equations can be written as follows:

$$y_1 = \beta_0 + \beta_1 x_1 + \epsilon_1$$

$$y_2 = \beta_0 + \beta_1 x_2 + \epsilon_2$$

$$\vdots \qquad (4.33)$$

$$y_n = \beta_0 + \beta_1 x_n + \epsilon_n$$

Another way to write this system of equations is

$$\begin{pmatrix} y_1 \\ y_2 \\ \vdots \\ y_n \end{pmatrix} = \begin{pmatrix} 1 & x_1 \\ 1 & x_2 \\ \vdots & \vdots \\ 1 & x_n \end{pmatrix} \begin{pmatrix} \beta_0 \\ \beta_1 \end{pmatrix} + \begin{pmatrix} \epsilon_1 \\ \epsilon_2 \\ \vdots \\ \epsilon_n \end{pmatrix}. \qquad (4.34)$$

To understand equation (4.34), we must first define arithmetic operations for matrices. Addition (or subtraction) of two matrices is obtained by simply adding (or subtracting) the corresponding elements of the matrices. For example,

$$\begin{pmatrix} 2 & 4 \\ \frac{-1}{2} & 5 \end{pmatrix} + \begin{pmatrix} -2 & 5 \\ \frac{1}{2} & 3 \end{pmatrix} = \begin{pmatrix} 2+(-2) & 4+5 \\ \frac{-1}{2}+\frac{1}{2} & 5+3 \end{pmatrix} = \begin{pmatrix} 0 & 9 \\ 0 & 8 \end{pmatrix}. \tag{4.35}$$

Note that addition and subtraction of two or more matrices are *only defined when the dimensions of the matrices are equal*. Thus, it makes no sense to, say, subtract a 2×3 matrix from a 3×3 matrix.

A *null* matrix is defined as a matrix whose elements are all equal to 0. A null matrix, $\mathbf{N}$, has the property that when added or subtracted from another matrix, $\mathbf{A}$, the sum or difference will equal $\mathbf{A}$. For example,

$$\begin{pmatrix} 1 & 2 \\ 3 & 4 \end{pmatrix} + \begin{pmatrix} 0 & 0 \\ 0 & 0 \end{pmatrix} = \begin{pmatrix} 1 & 2 \\ 3 & 4 \end{pmatrix}.$$

Matrix multiplication is defined in a little different way than matrix addition and subtraction. First, assume that two matrices, $\mathbf{A}$ and $\mathbf{B}$ have dimensions $n \times p$ and $p \times r$, respectively. Each element, C_{ij}, of $\mathbf{C} = \mathbf{AB}$ is obtained by summing the products of the ith row of $\mathbf{A}$ by the jth column of $\mathbf{B}$. For example,

$$\begin{pmatrix} 1 & 2 \\ 3 & 4 \end{pmatrix} \begin{pmatrix} -3 & 1 \\ -1 & 0 \end{pmatrix} = \begin{pmatrix} (1 \times -3) + (2 \times -1) & (1 \times 1) + (2 \times 0) \\ (3 \times -3) + (4 \times -1) & (3 \times 1) + (4 \times 0) \end{pmatrix}$$

$$= \begin{pmatrix} -2+(-3) & 1+0 \\ -9+(-4) & 3+0 \end{pmatrix} = \begin{pmatrix} -5 & 1 \\ -13 & 3 \end{pmatrix}.$$

Another example is

$$(1 \quad 1 \quad 1) \begin{pmatrix} 2 & -1 \\ -1 & 3 \\ 4 & -1 \end{pmatrix} = (2 - 1 + 4 \quad -1 + 3 - 1) = (5 \quad 1). \tag{4.36}$$

Notice that the dimensions of the first matrix on the left-hand side of equation (4.36) are 1×3 whereas the dimensions of the second matrix on the left-hand side are 3×2. The resulting matrix product (on the right-hand side of the equation) is 1×2. In matrix multiplication, the number of *columns* of the first member of the product *must equal* the number of *rows* in the second member of the product. In cases where this property does not hold, i.e., the number of columns of the first member of the product does *not equal* the number of rows in the second member, matrix multiplication *is not defined*. The only exception to this rule, however, is when a scalar is multiplied by a matrix. For example, if

$$k = 2 \text{ and } \mathbf{A} = \begin{pmatrix} 1 & 2 \\ 3 & 4 \\ -1 & 1 \\ 0 & 2 \end{pmatrix}$$

then

$$kA = \begin{pmatrix} 2(1) & 2(2) \\ 2(3) & 2(4) \\ 2(-1) & 2(1) \\ 2(0) & 2(2) \end{pmatrix} = \begin{pmatrix} 2 & 4 \\ 6 & 8 \\ -2 & 2 \\ 0 & 4 \end{pmatrix}. \tag{4.37}$$

An *identity* matrix, $\mathbf{I}$, has the property that when multiplied by another matrix, $\mathbf{A}$, the product will equal $\mathbf{A}$. The identity matrix is *always a square matrix*, i.e., has equal row and column dimensions. The 2×2 identity matrix is

$$\mathbf{I}_{2\times 2} = \begin{pmatrix} 1 & 0 \\ 0 & 1 \end{pmatrix}. \tag{4.38}$$

The "identity" property can be illustrated with the following two examples:

$$\begin{pmatrix} 1 & -2 \\ \frac{-1}{3} & \frac{1}{3} \end{pmatrix} \begin{pmatrix} 1 & 0 \\ 0 & 1 \end{pmatrix} = \begin{pmatrix} 1 & 0 \\ 0 & 1 \end{pmatrix} \begin{pmatrix} 1 & -2 \\ \frac{-1}{3} & \frac{1}{3} \end{pmatrix} = \begin{pmatrix} 1 & -2 \\ \frac{-1}{3} & \frac{1}{3} \end{pmatrix}, \tag{4.39}$$

$$\begin{pmatrix} 1 & 2 & 3 \\ -1 & 0 & 1 \end{pmatrix} \begin{pmatrix} 1 & 0 & 0 \\ 0 & 1 & 0 \\ 0 & 0 & 1 \end{pmatrix} = \begin{pmatrix} 1 & 2 & 3 \\ -1 & 0 & 1 \end{pmatrix}. \tag{4.40}$$

Notice that in equation (4.40), that we cannot *premultiply* by the identity matrix because the multiplication is not defined. However, we can *postmultiply* by the identity matrix.

The analog to division in matrix operations is obtained by calculating the *inverse* of a matrix. The *inverse* of a matrix is that matrix when multiplied by the original matrix yields the identity matrix. In symbols, we write the inverse of a matrix, $\mathbf{A}$ as $\mathbf{A}^{-1}$. $\mathbf{A}^{-1}$ has the property that

$$\mathbf{A}^{-1}\mathbf{A} = \mathbf{A}\mathbf{A}^{-1} = \mathbf{I}, \tag{4.41}$$

where $\mathbf{I}$ is the identity matrix. For example,

$$\text{if } \mathbf{A} = \begin{pmatrix} 13 & 5 \\ 5 & 2 \end{pmatrix} \text{ then } \mathbf{A}^{-1} = \begin{pmatrix} 2 & -5 \\ -5 & 13 \end{pmatrix} \text{ because}$$

$$\begin{pmatrix} 13 & 5 \\ 5 & 2 \end{pmatrix} \begin{pmatrix} 2 & -5 \\ -5 & 13 \end{pmatrix} = \begin{pmatrix} 2 & -5 \\ -5 & 13 \end{pmatrix} \begin{pmatrix} 13 & 5 \\ 5 & 2 \end{pmatrix} = \begin{pmatrix} 1 & 0 \\ 0 & 1 \end{pmatrix}. \tag{4.42}$$

Only square matrices have unique inverses. Also, there are square matrices that do not have unique inverses. $\begin{pmatrix} 1 & 1 \\ 2 & 2 \end{pmatrix}$ does not have a unique inverse, for example. One can obtain an inverse (by hand calculations) for a given matrix as follows:

1. *Augment* original matrix by the identity matrix. If, for instance, the matrix to be inverted is $\begin{pmatrix} 5 & 6 \\ 4 & 5 \end{pmatrix}$ then the *augmented* matrix is:

$$\begin{pmatrix} 5 & 6 & | & 1 & 0 \\ 4 & 5 & | & 0 & 1 \end{pmatrix}. \tag{4.43}$$

If the original matrix had dimension $p \times p$, the augmented matrix will have dimension $p \times 2p$.

2. Perform row operations on the augmented matrix with the strategy of eventually obtaining the identity matrix on the left-hand side.

Subtract row 2 from row 1 in equation (4.43) and put the result in row 1:

$$\left(\begin{array}{cc|cc} 1 & 1 & 1 & -1 \\ 4 & 5 & 0 & 1 \end{array} \right).$$

Now, subtract $4\times$ (row 1) from row 2 and put the result in row 2:

$$\left(\begin{array}{cc|cc} 1 & 1 & 1 & -1 \\ 0 & 1 & -4 & 5 \end{array} \right).$$

Next, subtract row 2 from row 1 and place the result in row 1:

$$\left(\begin{array}{cc|cc} 1 & 0 & 5 & -6 \\ 0 & 1 & -4 & 5 \end{array} \right).$$

The complete process can be written mathematically as follows:

$$\left(\begin{array}{cc|cc} 5 & 6 & 1 & 0 \\ 4 & 5 & 0 & 1 \end{array} \right) \sim \left(\begin{array}{cc|cc} 1 & 1 & 1 & -1 \\ 4 & 5 & 0 & 1 \end{array} \right)$$

$$\sim \left(\begin{array}{cc|cc} 1 & 1 & 1 & -1 \\ 0 & 1 & -4 & 5 \end{array} \right) \sim \left(\begin{array}{cc|cc} 1 & 0 & 5 & -6 \\ 0 & 1 & -4 & 5 \end{array} \right).$$

3. The matrix on the right-hand side of the augmented matrix is the inverse of the original matrix. In our example, $\left(\begin{array}{cc} 5 & -6 \\ -4 & 5 \end{array} \right)$ is the inverse of $\left(\begin{array}{cc} 5 & 6 \\ 4 & 5 \end{array} \right)$. We can check this by multiplying the matrices together and obtaining the 2×2 identity matrix. The above mentioned technique is not the only technique for inverting matrices. Another method uses *determinants* and *cofactor* matrices to invert a given matrix. Still another technique uses *pivot points* as a starting point for the inversion. For interested readers, further elaboration of this technique can be found in Harville [66]

The inversion of a matrix can be quite tedious if the matrix has large dimensions. Luckily, matrix inversions can be easily performed by computer programs like SAS and R.

Another matrix operation needed for regression is the *transpose* operation. If $\mathbf{A}$ is any matrix, then the transpose of $\mathbf{A}$, denoted as $\mathbf{A}'$ or $\mathbf{A}^{\mathrm{T}}$, is obtained by writing each column of $\mathbf{A}$ as a row of $\mathbf{A}'$. For instance, if

$$\mathbf{A} = \left(\begin{array}{ccc} 1 & 2 & 3 \\ 4 & 5 & 6 \end{array} \right) \text{ then } \mathbf{A}' = \left(\begin{array}{cc} 1 & 4 \\ 2 & 5 \\ 3 & 6 \end{array} \right).$$

If $\mathbf{A} = \mathbf{A}'$, then $\mathbf{A}$ is said to be a *symmetric* matrix. An example of a symmetric matrix is $\mathbf{A} = \left(\begin{array}{cc} 5 & 4 \\ 4 & 7 \end{array} \right)$.

A useful number associated with a square matrix is called a *determinant*. The determinant of a 2×2 matrix, $\begin{pmatrix} a_{11} & a_{12} \\ a_{21} & a_{22} \end{pmatrix}$, is denoted $\begin{vmatrix} a_{11} & a_{12} \\ a_{21} & a_{22} \end{vmatrix}$ and equal to $a_{11}a_{22} - a_{12}a_{21}$. A *submatrix* of a square matrix, $\mathbf{A}$, is itself a rectangular array obtained by deleting rows and columns of $\mathbf{A}$. A *minor* is the determinant of a square submatrix of $\mathbf{A}$. The *minor of a_{ij}* is the determinant obtained by deleting row i and column j of $\mathbf{A}$. The *cofactor* of a_{ij} is $(-1)^{i+j} \times ($ minor of $a_{ij})$. The value of a determinant can be calculated by moving down any *single* row or column of a matrix and multiplying each a_{ij} in that row or column by the cofactor of a_{ij} and summing the results of those products.

Example 4.12. Let $\mathbf{A} = \begin{pmatrix} 1 & 2 & 3 \\ 4 & 5 & 6 \\ 7 & 8 & 9 \end{pmatrix}$. The cofactor of a_{11} is $(-1)^2 \begin{vmatrix} 5 & 6 \\ 8 & 9 \end{vmatrix} = (1)[45 - 48] = -3$.

$$\begin{vmatrix} 1 & 2 & 3 \\ 4 & 5 & 6 \\ 7 & 8 & 9 \end{vmatrix} = a_{11}(\text{cofactor of } a_{11}) + a_{21}(\text{cofactor of } a_{21}) + a_{31}(\text{cofactor of } a_{31})$$

$$\left(1 \times (-1)^2 \times \begin{vmatrix} 5 & 6 \\ 8 & 9 \end{vmatrix} \right) + \left(4 \times (-1)^3 \times \begin{vmatrix} 2 & 3 \\ 8 & 9 \end{vmatrix} \right) + \left(7 \times (-1)^4 \times \begin{vmatrix} 2 & 3 \\ 5 & 6 \end{vmatrix} \right)$$

$$= 1(45 - 48) - 4(18 - 24) + 7(12 - 15) = -3 + 24 - 21 = 0 .$$

Two properties of determinants are (1) $\det(\mathbf{AB}) = \det(\mathbf{A}) \cdot \det(\mathbf{B}) = \det(\mathbf{B}) \cdot \det(\mathbf{A}) = \det(\mathbf{BA})$ and (2) $\det(\mathbf{A}) = \det(\mathbf{A}')$. Determinants are particularly useful for assessing the *rank* of a matrix that is related to whether or not a matrix, $\mathbf{A}$, is *invertible*. If, for example, $|\mathbf{A}| = 0$ then there is no unique inverse of $\mathbf{A}$. Also, if A is $n \times n$ then if $|\mathbf{A}| = 0$, rank$(A) < n$. Another very useful number associated a square matrix is its *trace*, which is nothing more than the sum of its diagonal elements. Hence, the trace of an $n \times n$ matrix, $\mathbf{A}$, is $\text{tr}(\mathbf{A}) = \sum_{i=1}^{n} a_{ii}$. Finally, a square matrix, $\mathbf{A}$, is called *idempotent* if and only if $\mathbf{A}^2 = \mathbf{A} \cdot \mathbf{A} = \mathbf{A}$. An identity matrix, $\mathbf{I}$, is idempotent because $\mathbf{I}^2 = \mathbf{I}$. In the case that a matrix is idempotent, $\text{tr}(\mathbf{A}) = \text{rank}(\mathbf{A})$. This is very important because the ranks of certain idempotent matrices are associated with the degrees of freedom used for statistical tests.

Example 4.13. Consider the model, $\mathbf{y} = \mathbf{X}\boldsymbol{\beta} + \boldsymbol{\epsilon}$, where $\boldsymbol{\epsilon} \sim N(\mathbf{0}, \mathbf{I}\sigma^2)$. The *least squares* estimate for $\boldsymbol{\beta}$ is $\widehat{\boldsymbol{\beta}} = (\mathbf{X}'\mathbf{X})^{-1}\mathbf{X}'\mathbf{y}$. Then the predicted outcomes at the sampled points are given by $\widehat{\mathbf{y}} = \mathbf{X}\widehat{\boldsymbol{\beta}} = \mathbf{X}(\mathbf{X}'\mathbf{X})^{-1}\mathbf{X}'\mathbf{y} = \mathbf{Hy}$. The matrix, $\mathbf{H} = \mathbf{X}(\mathbf{X}'\mathbf{X})^{-1}\mathbf{X}'$ is called the "hat matrix" because $\mathbf{H}$ is the transformation that maps $\mathbf{y}$ into $\widehat{\mathbf{y}}$, that is, the *observed y* values to the *predicted y* values. $\mathbf{H}$ is *idempotent* because $\mathbf{H}^2 = \left[\mathbf{X}(\mathbf{X}'\mathbf{X})^{-1}\mathbf{X}' \right] \left[\mathbf{X}(\mathbf{X}'\mathbf{X})^{-1}\mathbf{X}' \right] = \mathbf{X}(\mathbf{X}'\mathbf{X})^{-1}\mathbf{X}'\mathbf{X}(\mathbf{X}'\mathbf{X})^{-1}\mathbf{X}' = \mathbf{X}(\mathbf{X}'\mathbf{X})^{-1}\mathbf{X}' = \mathbf{H}$. This particular property of $\mathbf{H}$ is *crucial* to regression analysis because of its *scale preserving* quality, that is, the least squares regression predictor is *guaranteed* to be on the same scale as is the original data.

Matrix calculations, while mostly quite straightforward conceptually, can be quite tedious or even insurmountable when the dimensions are large. Luckily, most computer
languages have a vast array of matrix functions that are very efficient for making desired matrix calculations. Some matrix calculations in SAS and R (or S-plus) are
demonstrated in the following example.

Example 4.14. The following programs in SAS and R illustrate how to:

1. Create a matrix from scratch;

2. Transpose a matrix;

3. Invert a matrix;

4. Add matrices;

5. Multiply matrices;

6. Calculate the determinant of a matrix; and

7. Print all results.

SAS program

```
PROC IML;
  A = {1.5 1.7, -2.0 5.3}; AT = A';
  A_INV = INV(A);
  B = {19.5 -1.7, -1.7 15.8}; C = A + B;  D = A * B; detA = det(A);
  PRINT A; PRINT AT;   PRINT A_INV;
  PRINT B; PRINT C; PRINT D; PRINT detA;
run;
```

R (S-plus) program

```
A<-matrix(c(1.5,1.7,-2,5.3),byrow=T,nr=2);
A.t<-t(A);A.inv<-solve(A);
B<-matrix(c(19.5,-1.7,-1.7,15.8),byrow=T,nr=2)
# Note: C & D are reserved words in R, S-plus & hence, CC & DD are used instead
CC<-A+B; DD<- A%*%B
A; A.t; A.inv; B; CC; DD; det(A)
```

More details about how SAS and R can be used are included in the last two chapters of the
book.

Two other matrix properties that are very important to development of statistical
techniques are *eigenvalues* and *eigenvectors* (also known as *characteristic roots* and
characteristic vectors). These properties only apply to square matrices. The *eigenvalues* of a square matrix, $\mathbf{B}$ are defined the roots of the *characteristic equation*

$$P(\lambda) = |\mathbf{B} - \lambda\mathbf{I}| = 0. \tag{4.44}$$

Example 4.15. Let $\mathbf{B} = \begin{pmatrix} 6 & 2 \\ 2 & 3 \end{pmatrix}$. Then $\left| \mathbf{B} - \lambda \mathbf{I} \right| = 0 \Longrightarrow (6-\lambda)(3-\lambda) - 4 = \lambda^2 - 9\lambda + 14$

$= (\lambda - 7)(\lambda - 2) = 0 \Longrightarrow \lambda_1 = 7$ and $\lambda_2 = 2$ are the characteristic roots of $\mathbf{B}$.

Eigenvalues are important because if one takes the eigenvalues of a proper covariance matrix, then they should all be > 0. This property is associated with *positive definite* matrices. A matrix, $\mathbf{B}$, that is positive definite has the property that $\mathbf{x}'\mathbf{Bx} > 0$ for all $\mathbf{x} \neq \mathbf{0}$. Consequently, if $\mathbf{B}$ represents a covariance matrix then the property of positive definiteness ensures that any linear combination of a vector has a positive variance. Another important property of eigenvalues applied to statistical applications is that, in certain settings, eigenvalues are proportional to the different independent components of variance associated with multivariate observations.

If λ_i is a eigenvalue of $\mathbf{B}$, then a vector, $\mathbf{x}_i \neq \mathbf{0}$, satisfying $(\mathbf{B} - \lambda_i \mathbf{I})\mathbf{x}_i = \mathbf{0}$ is called a *eigenvector* of the matrix, $\mathbf{B}$, corresponding to the eigenvalue, λ_i. Any scalar multiple of $\mathbf{x}_i$ is also a eigenvector.

Example 4.16. In the last example, for $\lambda_1 = 7$, $(\mathbf{B} - 7\mathbf{I})\mathbf{x}_i = \begin{pmatrix} -1 & 2 \\ 2 & -4 \end{pmatrix} \mathbf{x}_1 = \mathbf{0} \Longrightarrow$

$\mathbf{x}_1 = \begin{pmatrix} 2 \\ 1 \end{pmatrix}$ is a characteristic vector. For $\lambda_2 = 2$, $(\mathbf{B} - 2\mathbf{I})\mathbf{x}_2 = \begin{pmatrix} 4 & 2 \\ 2 & 1 \end{pmatrix} \mathbf{x}_2 = \mathbf{0} \Longrightarrow$

$\mathbf{x}_2 = \begin{pmatrix} 1 \\ -2 \end{pmatrix}$ is a characteristic vector. Note that $(2 \quad 1)(-1 \quad 2) = 0$ which implies that the two eigenvectors are *orthogonal* to each other. This is a general property of eigenvectors.

4.10 Exercises

1. For each of the following sums, write an equivalent expression as products of a row vector times a column vector, that is, find $\mathbf{x}$ and $\mathbf{y}$ so that $\mathbf{x}'\mathbf{y}$ is equivalent to each sum.

(a) $1 + 2 + 3 + 4 + 5$ (Hint: You can use a "sum vector" for either $\mathbf{x}'$ or $\mathbf{y}$.)

(b) $1 + 1 + 1 + 1 + 1 + 1$

(c) $\displaystyle\sum_{i=1}^{5} x_i$

(d) $\displaystyle\sum_{i=1}^{5} x_i^2$

(e) $\displaystyle\sum_{i=1}^{5} x_i y_i$

2. Let $\mathbf{y} = \begin{pmatrix} y_1 \\ y_2 \\ \vdots \\ y_n \end{pmatrix}$ and $\mathbf{A} = \begin{pmatrix} 1 - \frac{1}{n} & -\frac{1}{n} & \cdots & -\frac{1}{n} \\ -\frac{1}{n} & 1 - \frac{1}{n} & \cdots & -\frac{1}{n} \\ \vdots & & \ddots & \vdots \\ -\frac{1}{n} & \cdots & -\frac{1}{n} & 1 - \frac{1}{n} \end{pmatrix}$. Calculate $\mathbf{y'Ay}$.

Comment on your answer.

3. Write an R program to replicate the results in Example 4.5.

4. Download the dataset called BODYFAT.DAT from the following website:
http://www.sci.usq.edu.au/staff/dunn/Datasets/applications/health/bodyfat.html. This dataset has four variables measured on 20 healthy females aged 20 to 34: body fat, triceps skinfold thickness, thigh circumference, and midarm circumference in that order. Body fat, which is cumbersome to obtain, can be predicted by the triceps skinfold thickness, the thigh circumference, and the midarm circumference This data was originally presented in the book by Neter et al. [107]

(a) A useful procedure for comparing goodness-of-fit and other model qualities for all possible subsets of predictors is provided in PROC RSQUARE. (You can get details of this procedure by googling "proc rsquare." Use PROC RSQUARE in SAS to produce R^2 statistics for all possible subsets of the three predictors (triceps skinfold thickness, thigh circumference, and midarm circumference).

(b) Use a backwards elimination procedure to arrive at a "best model" for the body-fat outcome. (Hint: To do this, you can use a "/selection=backward" option in the model statement of PROC REG. The model statement should include all three predictors.)

(c) Use a forwards elimination to arrive at a "best" model for the bodyfat outcome. (Hint: To do this, you can use a /selection=forward option in the model statement of PROC REG. The model statement should include all three predictors.)

(d) Comment on the "best" models you arrived at in parts **(b)** and **(c)**.

5. Consider the following data (Source: *Small Data Sets*, Hand, et al., Chapman & Hall, dataset 102; original source Lea, AJ [86]), which lists mortality rates by mean annual temperature in certain regions of Great Britain, Norway, and Sweden. Using Average temperature as the abscisssa (X value) and Mortality Index as the ordinate (Y value), fit a line, a quadratic, and a cubic polynomial to the data. Then, fit a smoothing spline to the data. Which of these curves do you think provides the best fit of the data? State your reasons for coming to this conclusion. (Hint: Google "Hand small dataset 102"; pick "Index of /data/hand-daly-lunn-mcconway-ostrowsk" and then pick the dataset called BREAST.DAT.)

6. Repeat the similation given in Example 4.11 but use the function,

$$f(x) = \begin{cases} 1 & \text{if } 0 \le x < 5 \\ 0 & \text{if } 5 \le x \le 10 \end{cases}.$$

Use the same error structure as was used in that example. In this example, you may wish to adjust the default value of the smoothing parameter in the program. Comment on your results.

7. In this exercise, we wish to scratch the surface through simulation about how robust least square estimators are with respect to the underlying distribution of the errors. For each simulation, you will generate data from a line, $y = \beta_0 + \beta_1 t + \epsilon$ where the t's are equally spaced integers from 0 to 10 and the ϵ will be distributed according to a specified distrbution. For the simulation study, you'll set $\beta_0 = 1$ and $\beta_1 = 1/2$. Thus, for each realization there will be 11 data points of the form, $y = 1 + (1/2) t + \epsilon$. After generating the data, you'll fit a least squares line and obtain the *estimates* of β_0 and β_1, i.e., $\widehat{\beta}_0$ and $\widehat{\beta}_1$. These last two values will be saved at each iteration. Each simulation will be repeated 1000 times and the interest is comparing the mean value of the $\widehat{\beta}_0's$ and $\widehat{\beta}_1's$ to their "true values" of 1 and 1/2, respectively. We'll also be interested in other distributional properties of the $\widehat{\beta}_0$s and $\widehat{\beta}_1$s.

(a) For the first simulation, assume $\epsilon \sim N(0,1)$. (Thus, for each of the 1000 iterations, you'll generate a new set of 11 random deviates that are $N(0,1)$ and add them to the vector of $Y = 1 + 1/2t$. If you use R, type ?rnorm at the R console.)

    (I) Calculate the mean, standard deviation, minimum, maximum of your 1000 values of $\widehat{\beta}_0$ and $\widehat{\beta}_1$.

    (II) Provide histograms of the $\widehat{\beta}_0$ and $\widehat{\beta}_1$ (you can put both of the histograms on one page if you wish).

(b) For the second simulation, repeat exactly what you did in part **(a)** but assume now that $\epsilon \sim t_{10}$, where t_{10} is a central t-distribution with 10 degrees of freedom. (If you use R, type ?rt at the R console.) Answer the same two questions, **(I)** and **(II)**, as in part **(a)**.

(c) For the third simulation, repeat exactly what you did in part **(a)** but assume now that $\epsilon \sim t_1$, where t_1 is a central t-distribution with 1 degrees of freedom. (This is the same as a standard *Cauchy* distribution.) Answer the same two questions, **(I)** and **(II)**, as in part **(a)**.

(d) Comment on the accuracy and variability of your estimates, $\widehat{\beta}_0$ and $\widehat{\beta}_1$. From these simulations, can you make a statement on the robustness of least squares regression estimators based on different distributional properties of the errors?

8. Show that if one fits the model, $\widehat{\mathbf{y}} = \mathbf{X}\widehat{\beta}$, the matrix that is used to calculate the fitted residuals, $\mathbf{y} - \widehat{\mathbf{y}} = \mathbf{y} - \mathbf{X}\widehat{\beta}$ are idempotent. Thus, show that the residuals are guaranteed to be on the same scale as the original data. [Hint: See Example 4.13 and note that $\mathbf{y} - \mathbf{X}\widehat{\beta} = \mathbf{y} - \mathbf{X}(\mathbf{X}'\mathbf{X})^{-1}\mathbf{X}'\mathbf{y} = (\mathbf{I} - \mathbf{H})\mathbf{y}$. Show that $\mathbf{I} - \mathbf{H}$ is idempotent.]

## CHAPTER 5

Analysis of Variance

5.1 Introduction

In Chapter 1, we reviewed several methods regarding making inferences about the means of one or two groups or populations. The methods all involved the use of *t*-tests. The inference about means of *more* than two groups requires an analytic technique known as *analysis of variance* (ANOVA). Analysis of variance techniques provide a very general framework for making inferences on mean values. In this chapter, we'll explore three types of ways that ANOVA can be used to make inferences about means. First, we'll consider the analysis of variance associated with a *one-way* design that involves comparing means according to *one factor* which may have many levels. Because we are interested in comparing means for the *many* levels, we may run into a problem with falsely declaring the differences to be statistically significant. Accordingly, we will discuss several methods to address this *multiple comparisons* issue, Next, we consider *two-way* designs that generalizes making inferences about means according to the levels associated with *two* factors. Such methods, of course, could be extended to many factors. Finally, we consider a generalization of the regresssion methods introduced in Chapter 4 by considering inferences about means for a factor with many levels while *adjusting for a continuous variable*. The methods for handlings such data involve the use of *analysis of covariance* (ANCOVA).

5.2 One-Way Analysis of Variance

In this section, we consider making inferences about $g \geq 2$ group means. The groups are arranged according to one factor and are "parallel" in that the observations across groups are independent. In medical or biological applications, the factor involved is usually different types of treatment, different levels of demographic or biological characteristics, or different modes of intervention.

Example 5.1. The following example is from Kurban. Mehmetoglu, and Yilmaz [82]. They studied the effect of diet oils on lipid levels of triglyceride in the brain of rats. Seventy-two rats were divided into six groups and given diets consisting of components of different kinds of oils. The outcome of interest was the triglyceride level in the brains of the rats. The results of the study are summarized in the table below.

Brain triglyceride levels of control and oil groups (*mg/g* protein)

Group	n	Triglyceride (mean ± SD)
Control [C]	12	32.57 ± 3.71
Sunflower oil [Su]	12	36.14 ± 4.59
Olive oil [O]	12	43.62 ± 4.92
Margarine [M]	12	30.68 ± 4.32
Soybean oil [So]	12	39.73 ± 4.32
Butter [B]	12	35.56 ± 2.52

The overall interest in this problem is to test whether or not there is an overall difference in brain lipid levels among the six groups of rats. If there are some differences among group means, then within each of the groups, the measurements are assumed to be independently and identically distributed (i.i.d.) having normal distributions and we write $Y_{11}, Y_{12}, \ldots, Y_{1,12}$ i.i.d. $N(\mu_1, \sigma^2)$; $Y_{21}, Y_{22}, \ldots, Y_{2,12}$ i.i.d. $N(\mu_2, \sigma^2)$; $\ldots$; $Y_{61}, Y_{62}, \ldots, Y_{6,12}$ i.i.d. $N(\mu_6, \sigma^2)$. If the measurements all came from the exactly same population, we would write $Y_{11}, Y_{12}, \ldots, Y_{1,12}, Y_{21}, \ldots, Y_{2,12}, \ldots, Y_{61}, \ldots, Y_{6,12}$ i.i.d. $N(\mu, \sigma^2)$, that is, the distributions of *all* measurements across groups would be normally distributed and have the same mean and variance.

A general model for comparing normally distributed outcomes which are divided into g groups is commonly written as

$$\underbrace{y_{ij}}_{\substack{\text{response of } j^{th} \\ \text{subj. in group } i}} = \underbrace{\mu}_{\substack{\text{grand mean over} \\ \text{all groups}}} + \underbrace{\alpha_i}_{\substack{\text{offset due to} \\ \text{group } i}} + \underbrace{\epsilon_{ij}}_{\substack{\text{error for response of} \\ j^{th} \text{ subj. in group } i}} , \tag{5.1}$$

where ϵ_{ij} are independently and identically distributed $N(0, \sigma^2)$, $i = 1, \ldots, g$, $j = 1, 2, \ldots, n_i$ and $\sum_{i=1}^{g} \alpha_i = 0$. The model (5.1) is also commonly written in the following (equivalent) way:

$$\underbrace{y_{ij}}_{\substack{\text{response of } j^{th} \\ \text{subj. in group } i}} = \underbrace{\mu_i}_{\substack{\text{mean of} \\ \text{group } i}} + \underbrace{\epsilon_{ij}}_{\substack{\text{error for response of} \\ j^{th} \text{ subj. in group } i}} \tag{5.2}$$

The version of the model as represented in equation (5.1) is not usually preferable because the model is *overparameterized*, that is, there are $g + 1$ parameters to be estimated but only g means in the experiment. In most cases, the model as written in equation (5.2), called the *cell means model*, is preferable, see Searle [135]. As indicated above, the two subscripts, i and j, designate the group being tested and the subject within group, respectively. In our study, there are $g = 6$ diets and $n_1 = \ldots = n_6 = n = 12$ observations within each diet. This particular study is said to be *balanced* because there are *equal numbers of responses* (observations) in each of the three groups corresponding with diet type. We're interested, first, in testing the hypothesis: $H_0 : \mu_1 = \ldots = \mu_6$ versus H_1: at least two μ_i are different.

Unfortunately, before we test the hypothesis of interest, we run into a technical hitch just as we did in the two-sample case. The hitch, of course, is that we must first test to see whether or not the variances are equal, i.e., test $H_0 : \sigma_1^2 = \ldots = \sigma_6^2$ versus H_1: at least two σ_i^2 are different. Before we embark on this endeavor, we can form an analysis of variance (ANOVA) table. As in Chapter 4, the ANOVA table will consist of the following components: (1) the source of variation (Source); (2) the degrees of freedom (DF); (3) the sums of squares (SS); (4) the mean square (MS); and (5) F-statistics (F). Some books and programs refer to component (4) as "MSE" whereas others use "MS". We start by first writing the deviation from the grand mean for *each* observation into the following form:

$$y_{ij} - \overline{y}_{\bullet\bullet} = (y_{ij} - \overline{y}_{i\bullet}) + (\overline{y}_{i\bullet} - \overline{y}_{\bullet\bullet}), \tag{5.3}$$

where $\overline{y}_{i\bullet} = (\sum_{i=1}^{n_i} y_{ij})/n_i$ and $\overline{y}_{\bullet\bullet} = (\sum_{i=1}^{g} \sum_{i=1}^{n_i} y_{ij})/\sum_{i=1}^{g} n_i$ represent the group means and the mean of all observations, respectively. Squaring each side and summing over all observations yields

$$\sum_{i=1}^{g} \sum_{i=1}^{n_i} (y_{ij} - \overline{y}_{\bullet\bullet})^2 = \sum_{i=1}^{g} \sum_{i=1}^{n_i} (y_{ij} - \overline{y}_{i\bullet})^2 + \sum_{i=1}^{g} \sum_{i=1}^{n_i} (\overline{y}_{i\bullet} - \overline{y}_{\bullet\bullet})^2.$$

Another way of writing this is:

Total Sums of Squares = Within group Sums of Squares + Between group Sums of Squares.

In Example 5.1, $n_1 = \ldots = n_6 = n = 12$ and $g = 6$. We can incorporate these two sources of variation into an Analysis of Variance (ANOVA or AOV) table:

ANOVA table for a One-Way Design

Source	DF	Sums of Squares (SS)	MS	F
Between	$g - 1$	$\sum_{i=1}^{g} \sum_{i=1}^{n_i} (\overline{y}_{i\bullet} - \overline{y}_{\bullet\bullet})^2$	$\frac{SS_B}{g-1}$	$\frac{MS_B}{MS_W}$
Within	$N - g$	$\sum_{i=1}^{g} \sum_{i=1}^{n_i} (y_{ij} - \overline{y}_{i\bullet})^2$	$\frac{SS_W}{N-g}$	
Total	$N - 1$	$\sum_{i=1}^{g} \sum_{i=1}^{n_i} (y_{ij} - \overline{y}_{\bullet\bullet})^2$		

In cases where individual observations are not available, for example, if, say, only the means, $\overline{y}_{i\bullet}$, are given, then the Between SS can be more efficiently computed as

$$\sum_{i=1}^{g} n_i \overline{y}_{i\bullet}^2 - (\sum_{i=1}^{g} n_i \overline{y}_{i\bullet})^2 / N \tag{5.4}$$

where $N = \sum_{i=1}^{g} n_i$. If we also know the values of the group variances, $s_i^2, i = 1, \ldots, g$, then we can compute Within SS $= \sum_{i=1}^{g} (n_i - 1) s_i^2$.

If the group sums, $\overline{y}_{i\bullet}$, are given then

$$\text{Between SS} = \sum_{i=1}^{g} \frac{y_{i\bullet}^2}{n_i} - \frac{y_{\bullet\bullet}^2}{N}, \quad \text{Between MS} = \text{Between SS}/(g - 1) \tag{5.5}$$

where $y_{i\bullet} = \sum_{i=1}^{n_i} y_{ij}$ and $y_{\bullet\bullet} = \sum_{i=1}^{g} \sum_{i=1}^{n_i} y_{ij}$;

$$\text{Total SS} = \sum_{i=1}^{g} \sum_{i=1}^{n_i} y_{ij}^2 - \frac{y_{\bullet\bullet}^2}{N}; \text{ and} \tag{5.6}$$

$$\text{Within SS} = \text{Total SS} - \text{Between SS}, \text{ Within MS} = \text{Within SS}/(N-g). \tag{5.7}$$

The Within MS (MSW) is also known as the *Error MS* (MSE).

Example 5.2. In Example 5.1, the data are already grouped into their summary statistics so that Between SS $= 12[32.57^2 + 36.14^2 + \ldots + 35.56^2] - (12\{32.57 + 36.14 + \ldots + 35.56\})^2/72 = 96646.296 - 95309.78 = 1336.514$. Within MS $= 11(3.71^2) + 11(4.59^2) + \ldots + 11(2.52^2) = 11(3.71^2 + 4.59^2 + \ldots + 2.52^2)/6 = 17.119$. Thus, our ANOVA table is as follows:

ANOVA table for a One-Way Design

Source	DF	SS	MS	F
Between	5	1336.51	267.30	15.614
Within	66	1129.85	17.119	
Total	71	2466.37		

Now, back to testing our "nuisance" parameter, σ^2. To test $H_0 : \sigma_1^2 = \ldots = \sigma_g^2$ versus H_1: at least two σ_i^2 are different, we use Bartlett's test and proceed as follows:

1. Form $X^2 = \frac{\lambda}{c}$ where

$$\lambda = \sum_{i=1}^{g} (n_i - 1) \ln(s^2/s_i^2), \ c = 1 + \frac{1}{3(k-1)} \left[\left(\sum_{i=1}^{g} \frac{1}{n_i - 1} \right) - \frac{1}{N-g} \right],$$

$s_i^2 = $ variance of sample i, and $s^2 = $ within mean squared error (or *pooled variance*).

2. Under H_0, $X \stackrel{\cdot}{\sim} \chi_{g-1}^2$ so the following rule can be used:

   Rule: hypothesis test for equality of variances of g groups

$$\begin{cases} \text{If } X^2 \geq \chi_{g-1,1-\alpha}^2 & \text{then } reject \ H_0 ; \\ \text{Otherwise} & do \ not \text{ reject } H_0. \end{cases}$$

Example 5.3. For the data in Example 5.1, $\lambda = 11\left[\ln \frac{17.11897}{13.7641} + \ln \frac{17.11897}{21.0681} + \ln \frac{17.11897}{24.2064} + \right.$

$\left. \ln \frac{17.11897}{18.6624} + \ln \frac{17.11897}{18.6624} + \ln \frac{17.11897}{6.3504} \right] \implies \lambda \approx 5.798; \ c = 1 + \frac{1}{15}\left[\frac{6}{11} - \frac{1}{66} \right] \approx 1.035$

$\implies X^2 = \frac{\lambda}{c} \approx 5.60.$ $\chi_{5,.95}^2 = 11.07 \implies X^2 < \chi_{5,.95}^2 \implies$ do not reject H_0, that is, we do *not* declare significant differences in variances among the six groups. Hence, we can test $H_0 : \mu_1 = \ldots = \mu_6$ versus H_1: at least two μ_i are different using "standard methodology".

Standard methodology for testing equality of means from normally distributed data refers to the use of an F-test. That is, form

$$F = \frac{\text{Between SS}/(g-1)}{\text{Within SS}/(N-g)}$$

Under H_0, $F \sim F_{g-1,N-g}$.

Hypothesis test for equality of means of g groups
Conditional on not rejecting equality of variances

$$\begin{cases} \text{If } F \geq F_{g-1,N-g,1-\alpha} & \text{then } reject\ H_0\ ; \\ \text{Otherwise} & do\ not\ \text{reject } H_0. \end{cases} \qquad (5.8)$$

Example 5.4. For the data in Example 5.1, $F \approx \frac{267.3}{17.119} \approx 15.614 > F_{5,66,.95} \approx 2.35 \Longrightarrow$ reject $H_0 \Longrightarrow$ at least two means are significantly different. Since the null hypothesis is rejected, we can indeed assume the six groups do not all have a common mean, that is, brain triglyceride levels significantly vary across diets. However, this type of analysis is not sensitive to *where* the differences occur.

5.3 General Contrast

To test, say, whether or not two-group means are equal, that is, $H_0 : \alpha_a = \alpha_b$ versus $H_1 : \alpha_a \neq \alpha_b$, we can form

$$T = \frac{\bar{y}_a - \bar{y}_b}{s\sqrt{\frac{1}{n_a} + \frac{1}{n_b}}} \quad \text{where } s = \sqrt{MS_W}.$$

Under the null hypothesis, H_0, $T \sim t_{N-g}$. Notice here that the degrees of freedom for the test statistic rely on s which is estimated using the data from *all* of the groups in the experiment.

In general, if we want to test $H_0 : \sum_{i=1}^{g} c_i \alpha_i = 0$ versus $H_1 : \sum_{i=1}^{g} c_i \alpha_i \neq 0$ (assuming $\sum_{i=1}^{g} c_i = 0$) we form

$$T = \frac{\sum_{i=1}^{g} c_i \bar{y}_i}{s\sqrt{\sum_{i=1}^{g} \frac{c_i^2}{n_i}}} \quad \sim t_{N-g} \text{ under } H_0. \qquad (5.9)$$

Definition If $\sum_{i=1}^{g} c_i = 0$, then $\sum_{i=1}^{g} c_i \bar{y}_i$ is called a *linear contrast*.

Example 5.5. (A three-group contrast): Consider a three-group case where $y_{ij} = \mu + \alpha_i + \epsilon_{ij}$, $i = 1, 2, 3$ and $j = 1, \ldots, n_i$. Test $H_0 : \alpha_1 = \frac{1}{2}(\alpha_2 + \alpha_3)$ versus $H_1 : \alpha_1 \neq \frac{1}{2}(\alpha_2 + \alpha_3) \Rightarrow H_0 : \alpha_1 - \frac{1}{2}(\alpha_2 + \alpha_3) = 0$ versus $H_1 : \alpha_1 - \frac{1}{2}(\alpha_2 + \alpha_3) \neq 0$. Thus, $c_1 = 1$, $c_2 = -\frac{1}{2}$ and $c_3 = -\frac{1}{2} \Rightarrow c_1^2 = 1$, $c_2^2 = \frac{1}{4}$ and $c_3^2 = \frac{1}{4}$. Thus, one could form the statistic

$$T = \frac{\bar{y}_1 - \frac{1}{2}\bar{y}_2 - \frac{1}{2}\bar{y}_3}{s\sqrt{\frac{1}{n_1} + \frac{1}{4n_2} + \frac{1}{4n_3}}}, \quad \text{which is} \sim t_{N-3} \text{ under } H_0.$$

5.3.1 Trend

When trying to establish a dose response relationship, it is often useful to detect *trends* across dose levels. A *linear* trend is often of particular interest but one may be interested in other types of trends as well (for example, quadratic, cubic, etc.). In analysis of variance, it is mathematically useful to create contrasts in such a manner that they are independent of each other much like the between and within sums of squares are independent. For accomplishing such independence when one is interested in detecting trends across dose levels, one can use *orthogonal polynomial contrasts*. To do this, the *trend contrasts* must be of a special form as given below.

Table 5.1. c_i's Needed for Orthogonal Polynomial Coefficients

g=3		g=4			g=5			
Linear	Quadratic	Linear	Quadratic	Cubic	Linear	Quadratic	Cubic	Quartic
−1	1	−3	1	−1	−2	2	−1	1
0	−2	−1	−1	3	−1	−1	2	−4
1	1	1	−1	−3	0	−2	0	6
		3	1	1	1	−1	−2	−4
					2	2	1	1

g=6					g=7				
Lin.	Quad.	Cub.	Quart.	Quintic	Lin.	Quad.	Cub.	Quart.	Quint.	6th deg.
−5	5	−5	1	−1	−3	5	−1	3	−1	1
−3	−1	7	−3	5	−2	0	1	−7	4	−6
−1	−4	4	2	−10	−1	−3	1	1	−5	15
1	−4	−4	2	10	0	−4	0	6	0	−20
3	−1	−7	−3	−5	1	−3	−1	1	5	15
5	5	5	1	1	2	0	−1	−7	−4	−6
					3	5	1	3	1	1

Example 5.6. (Detecting a linear trend): A simulation based on a study presented in Hassard [67] was conducted. In that study, 24 untrained subjects were classified according to their smoking habits. Each subject exercised for a sustained period of time, then rested for 3 minutes. At this time, their heart rates were taken. The results are recorded below:

Nonsmoker	Light smoker	Moderate Smoker	Heavy Smoker
69	55	66	91
53	61	82	71
70	77	69	81
58	58	77	67
60	62	57	95
64	65	79	84

$$\overline{y}_{1\bullet} = 62\tfrac{1}{3} \qquad \overline{y}_{2\bullet} = 63 \qquad \overline{y}_{3\bullet} = 71\tfrac{2}{3} \qquad \overline{y}_{4\bullet} = 81\tfrac{1}{2}$$
$$s_1 = 6.593 \qquad s_2 = 7.668 \qquad s_3 = 9.416 \qquad s_4 = 10.950$$

For these data, $\overline{y}_{\bullet\bullet} = 69\tfrac{5}{8} = 69.625$ and $s = 8.815$, and $n_1 = n_2 = n_3 = n_4 = 6$. The question of interest here is whether or not there is a linear increase in heart rate across smoking status from nonsmokers to heavy smokers, i.e. test $H_0 : -3\mu_1 - \mu_2 + \mu_3 + 3\mu_4 = 0$ versus $H_1 : -3\mu_1 - \mu_2 + \mu_3 + 3\mu_4 \neq 0$

$$T \approx \frac{-3(62\tfrac{1}{3}) - 63 + 71\tfrac{2}{3} + 3(81\tfrac{1}{2})}{8.815\sqrt{\frac{(-3)^2+(-1)^2+1^2+3^2}{6}}} \approx \frac{66.1667}{16.094} \approx 4.11.$$

Now, $t_{20,.9995} \approx 3.85 \Rightarrow$ reject H_0 $\left[p < .001 \right] \Rightarrow$ 3-minute Heart Rate increases at a significant linear rate as smoking "dose" increases.

We will return to the subject of orthogonal contrasts in Chapter 8 when we discuss how such contrasts are used to detect the mean trend of a population, which is measured longitudinally (that is, repeatedly over time).

5.4 Multiple Comparisons Procedures

Recall that $\alpha = \Pr(\text{Type } I \text{ error}) = \Pr(\text{rejecting } H_0 | H_0 \text{ is true})$. Such an error is called a *comparisonwise error rate*. If, however, p comparisons were made, each at the same α-level, then we define

$$\gamma \equiv \Pr(\text{at least one Type } I \text{ error among the } p \text{ comparisons}) \qquad (5.10)$$

as the *experimentwise error rate* or *familywise error rate* (FWER).

Empirically, we can think of the two error rates as follows:

$$\text{Comparisonwise error rate} = \frac{\#\text{of nonsignificant comparisons falsely declared significant}}{\text{Total } \# \text{ of nonsignificant comparisons}}$$

$$\text{Experimentwise error rate} = \frac{\#\text{of experiments with 1 or more comparisons falsely called sig.}}{\text{Total } \# \text{ of experiments with at least 2 comparisons}}$$

Controlling the experimentwise error rate at, say, $\gamma = .05$ forces an investigator to be more conservative in declaring any two group means to be significantly different. If the p comparisons were *independent* (which is not usually the case) then one would calculate the experimentwise error as $\gamma = 1 - (1 - \alpha)^p \Rightarrow \alpha = 1 - \sqrt[p]{1 - \gamma}$.

Example 5.7. (Experimentwise versus Comparisonwise Error rates): Suppose we have 7 _independent_ comparisons being made in an experiment:

1. If $\alpha = .05$ for each comparison, then $\gamma = 1 - (.95)^7 \approx .302$
2. If we want to hold γ at .05, then each $\alpha = 1 - \sqrt[7]{.95} \approx .0073$.

Unfortunately, in most real situations, the comparisons of interest are _not_ independent nor is the correlation structure among comparisons easily calculated. Also, in some cases, a certain number of comparisons were *planned* prior to the experiment whereas in other cases, none of the post hoc comparisons were planned. In the latter case, since all of the comparisons are *unplanned*, one should conservatively adjust for a large number of comparisons, for example, all pairwise comparisons. Thus, depending on the situation we must use an assortment of multiple comparison procedures to accommodate the given situation and the philosophy of the experimenter (or journal). Listed in Figure 5.1 are some multiple comparisons procedures and where they are on the spectrum with regard to how conservative they are in allowing one to declare significance.

Figure 5.1 *Multiple Comparisons Procedures Listed by Their Relative Propensity to Reject* H_0

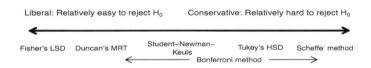

In all of the following multiple comparisons procedures, it is important to *order* the groups according to the value of their means (lowest to highest). So, for example, in the case of four groups, order the groups so that $\bar{x}_A \leq \bar{x}_B \leq \bar{x}_C \leq \bar{x}_D$.

Then, after ordering the groups as outlined above, do the comparisons according to the order given in Figure 5.2. Thus, the most extreme comparison (A versus D) is

Figure 5.2 *Ordering of Pairwise Comparisons For a Multiple Comparisons Procedure*

done first, followed by the two second most extreme comparisons (A versus C and B versus D), etc.

Given this first step of any multiple comparisons procedure, we now compare and contrast several different *multiple comparison* methods. For the purposes of the following discussion, let WMS = within error mean squared = s^2.

5.4.1 Fisher's (Protected) Least Significant Difference [LSD] Method

I. Do an overall F-test

    a. If the F-value is <u>not</u> significant then STOP

    b. If the F-value <u>is</u> significant, then perform pairwise t-tests, $T = \dfrac{\bar{y}_{i\bullet} - \bar{y}_{j\bullet}}{s\sqrt{\frac{1}{n_i} + \frac{1}{n_j}}}$,

    which is compared to $t_{N-g,1-\alpha/2}$ [two-sided hypothesis].

An equivalent way of doing this test is to compare the absolute difference between the mean values of any two groups, $|\bar{y}_{i\bullet} - \bar{y}_{j\bullet}|$ to $t_{N-g,1-\alpha/2}\left(s\sqrt{\frac{1}{n_i} + \frac{1}{n_j}}\right) = t_{N-g,1-\alpha/2}\sqrt{(WMS)(\frac{1}{n_i} + \frac{1}{n_j})}$ (assuming that a two-sided alternative hypothesis).

II. Advantages/Disadvantages

a. Very simple to implement

b. Amenable to construction of confidence intervals

c. Disadvantage in that it *doesn't adequately control experimentwise error rates* especially if the number of groups in a study is > 3, see Einot and Gabriel [37].

5.4.2 Duncan's Multiple Range Test

I. Form $Q = \dfrac{\bar{y}_{i\bullet} - \bar{y}_{j\bullet}}{\sqrt{\frac{s^2}{2}(\frac{1}{n_i} + \frac{1}{n_j})}}$, which is compared $q_{|i-j|+1,N-g,1-\gamma}$ where $q_{|i-j|+1,N-g}$ is the *studentized range statistic* with $|i-j|+1$ and $N-g$ degrees of freedom (Duncan [31]). *An equivalent way* of doing this test is to compare the absolute difference between the mean values of any two groups, $|\bar{y}_{i\bullet} - \bar{y}_{j\bullet}|$ to $q_{|i-j|+1,N-g,1-\gamma}$ $\times \sqrt{\frac{WMS}{2}(\frac{1}{n_i} + \frac{1}{n_j})}$. Note that $q_{|i-j|+1,N-g,1-\alpha/2}$ varies as $|i-j|$ varies with i and j. The critical value, q, increases as the value of $p = |i-j|+1$ increases. The intuitive reason for doing this is that since the groups are ordered according to their (ascending) mean values, one should require more stringent criteria for declaring more extreme means (e.g., lowest versus highest) significant than for declaring less extreme means significant. Another important aspect of Duncan's multiple range test is how we arrive at the value of γ. The test makes use of the fact that for any p groups being compared there are $p-1$ *independent* comparisons that could be made. In Duncan's test, if we set the comparisonwise error rate at α, we would consequently have an experimentwise error rate of $\gamma = 1 - (1-\alpha)^{p-1}$. Thus, for example, $p = 4$ and $\alpha = .05 \implies \gamma = 1 - .05^3 \implies \gamma \approx 0.1426$. Thus, when the most extreme means among four groups are being compared, the studentized range value given in Duncan's tables actually reflect the $(1 - .1426) \times 100^{th} = 85.74^{th}$ percentile of the studentized range statistic.

II. Advantages/Disadvantages

a. Very powerful in most cases

b. Intuitively appealing because it places more stringent criteria for rejecting comparisons as the number of groups increases

c. Not easily amenable to the construction of confidence intervals

d. Like the LSD method, the multiple range test *doesn't adequately control experimentwise error rates*

5.4.3 Student–Newman–Keuls [SNK] Method

I. Form $Q = \dfrac{\bar{y}_{i\bullet} - \bar{y}_{j\bullet}}{\sqrt{\frac{s^2}{2}\left(\frac{1}{n_i} + \frac{1}{n_j}\right)}}$, which is compared to $q_{|i-j|+1, N-g, 1-\alpha}$ where

$q_{|i-j|+1, N-g, 1-\alpha}$ is the $(1-\alpha)^{th}$ percentile of the *studentized range statistic* with $|i-j|+1$ and $N-g$ degrees of freedom. *An equivalent way* of doing this test is to compare the absolute difference between the mean values of any two groups, $|\bar{y}_{i\bullet} - \bar{y}_{j\bullet}|$ to $q_{|i-j|+1, N-g, 1-\alpha}\sqrt{\frac{WMS}{2}\left(\frac{1}{n_i} + \frac{1}{n_j}\right)}$. Note that this method is *exactly the same as Duncan's multiple range test except* that the level for declaring a significant test is changed. In the SNK, the experimentwise error rate is set to some value, say, $\gamma = .05$ and then the α level is calculated, $\alpha = 1 - \sqrt[g]{1-\gamma}$ and thus, depends on how many groups are being considered in each comparison. (see Newman [108] and Keuls [78])

II. Advantages/Disadvantages

    a. Controls well for experimentwise error rate

    b. As in the multiple range test, the rejection "yardstick" *increases* as the number of groups being considered increases

    c. Not easily amenable to the construction of confidence intervals

    d. Somewhat difficult to employ and interpret

5.4.4 Bonferroni's Procedure

I. Form $T = \dfrac{\bar{y}_{i\bullet} - \bar{y}_{j\bullet}}{s\sqrt{\left(\frac{1}{n_i} + \frac{1}{n_j}\right)}}$, which is compared to $t_{N-g, 1-\alpha/(2r)}$ [two-sided test] where r is the number of comparisons of interest. Note that if we were interested in all pairwise comparisons then $r = \binom{g}{2}$. *An equivalent way* of doing this test is to compare the absolute difference between the mean values of any two groups, $|\bar{y}_{i\bullet} - \bar{y}_{j\bullet}|$ to

$$t_{N-g, 1-\alpha/(2r)}\left(s\sqrt{\frac{1}{n_i} + \frac{1}{n_j}}\right) = t_{N-g, 1-\alpha/(2r)}\sqrt{(WMS)\left(\frac{1}{n_i} + \frac{1}{n_j}\right)}$$

(if one were to consider one-sided alternatives, then replace $2r$ with r).

    II. Advantages/Disadvantages

    a. Very simple to implement and interpret

    b. Amenable to construction of confidence intervals

    c. Overly conservative if the number of comparisons of interest is large

5.4.5 Tukey's Honestly Significant Difference [HSD] Method

I. Form $Q = \dfrac{\bar{y}_{i\bullet} - \bar{y}_{j\bullet}}{\sqrt{\frac{s^2}{2}\left(\frac{1}{n_i} + \frac{1}{n_j}\right)}}$, which is compared to $q_{k,N-g,1-\alpha}$ where $q_{k,N-g}$ is the

studentized range statistic with k and $N - g$ degrees of freedom. *An equivalent way* of doing this test is to compare the absolute difference between the mean values of any two groups, $|\bar{y}_{i\bullet} - \bar{y}_{j\bullet}|$ to $q_{g,N-g,1-\alpha}\sqrt{\frac{WMS}{2}\left(\frac{1}{n_i} + \frac{1}{n_j}\right)}$. Note that this method is *exactly the same as the SNK* method *except* that the level for rejection of each comparison does not vary in a given experiment. Thus, the "rejection yardstick," q *stays the same for all comparisons being made in a given experiment.* (See Tukey [152])

II. Advantages/Disadvantages

    a. Controls well for experimentwise error rate

    b. Very easy to employ and interpret

    c. Overly conservative in many cases

5.4.6 Scheffé's Method

I. Form $T_S = \dfrac{\bar{y}_{i\bullet} - \bar{y}_{j\bullet}}{s\sqrt{\left(\sum_{i=1}^{g} \frac{c_i^2}{n_i}\right)}}$ which is compared to $\sqrt{(g-1)F_{g-1,N-g,1-\alpha}}$

where $F_{g-1,N-g,1-\alpha}$ is the $(1-\alpha)^{th}$ percentile of the F distribution with $g-1$ and $N - g$ degrees of freedom. In the case of a *pairwise comparison*, T_S reduces to $T_S = \dfrac{\bar{y}_{i\bullet} - \bar{y}_{j\bullet}}{s\sqrt{\frac{1}{n_i} + \frac{1}{n_j}}}$. *An equivalent way* of doing this test is to compare the absolute difference between the mean values of any two groups, $|\bar{y}_{i\bullet} - \bar{y}_{j\bullet}|$ to

$$\sqrt{(g-1)F_{g-1,N-g,1-\alpha}} \times s \times \sqrt{\sum_{i=1}^{g} \frac{c_i^2}{n_i}} = \sqrt{(g-1)F_{g-1,N-g,1-\alpha}}\sqrt{(WMS)\left(\sum_{i=1}^{g} \frac{c_i^2}{n_i}\right)}$$

(see Scheffé [132, 133])

II. Advantages/Disadvantages

    a. Simple to implement and it controls experimentwise error rate for *any contrast of interest* (not just *pairwise comparisons*)

    b. Amenable to construction of confidence intervals

    c. *Extremely* conservative

Example 5.8. The following example demonstrates how the different multiple comparisons procedures presented so far are implemented.

Consider again the data from Kurban et al. [82] which was presented in Example 5.1. We wish to use this data to demonstrate each of the multiple comparisons procedures presented above. For convenience, we have reproduced the ANOVA table from the earlier example.

Analysis of Variance Table

Source	DF	SS	MS	F
Between	5	1336.51	267.30	15.614
Within	66	1129.85	17.119	
Total	71	2466.37		

$F_{5,66,.95} = 2.35 \implies$ reject $H_0 : \mu_1 = \mu_2 = \mu_3 = \mu_4 = \mu_5 = \mu_6$ at the $\alpha = 0.05$ level.

The comparisons of pairs using each of the multiple comparisons are summarized in the table below.

Multiple comparisons Pairwise Statistics for the data presented in Example 5.1.

Pair	|Diff.|	Fisher's LSD	Duncan	SNK	Tukey's HSD	Scheffé	Bonferroni
M – 0	12.94	3.37	3.81	4.96	4.96	5.79	6.69
C – 0	11.05	3.37	3.75	4.74	4.96	5.79	6.69
M – So	9.05	3.37	3.75	4.74	4.96	5.79	6.69
B – 0	8.06	3.37	3.66	4.45	4.96	5.79	6.69
C – So	7.14	3.37	3.66	4.45	4.96	5.79	6.69
M – Su	5.46	3.37	3.66	4.45	4.96	5.79	6.69
Su – 0	7.48	3.37	3.55	4.05	4.96	5.79	6.69
B – So	4.17	3.37	3.55	4.05	4.96	5.79	6.69
C – Su	3.57	3.37	3.55	4.05	4.96	—	—
M – B	4.88	3.37	3.55	4.05	4.96	—	—
So – 0	3.89	3.37	3.37	3.37	4.96	5.79	6.69
Su – So	3.59	3.37	3.37	3.37	—	—	—
B – Su	0.58	3.37	3.37	3.37	—	—	—
C – B	2.99	3.37	3.37	3.37	—	—	—
M – C	1.89	3.37	3.37	3.37	—	—	—

Pair	Fisher's LSD	Duncan	SNK	Tukey's HSD	Scheffé	Bonferroni
M – 0	sig	sig	sig	sig	sig	sig
C – 0	sig	sig	sig	sig	sig	sig
M – So	sig	sig	sig	sig	sig	sig
B – 0	sig	sig	sig	sig	sig	sig
C – So	sig	sig	sig	sig	sig	sig
M – Su	sig	sig	sig	sig	NS	NS
Su – 0	sig	sig	sig	sig	sig	sig
B – So	sig	sig	sig	NS	NS	NS
C – Su	sig	sig	sig	NS	—	—
M – B	sig	sig	sig	sig	—	—
So – 0	sig	sig	sig	NS	NS	NS
Su – So	sig	sig	sig	—	—	—
B – Su	NS	NS	—	—	—	—
C – B	NS	NS	—	—	—	—
M – C	NS	NS	—	—	—	—

For the two most conservative procedures, only nine of the 15 pairwise comparisons need to be calculated. This is due to the fact that the means are arranged in *increasing order* so that less extreme comparisons nested within more extreme comparisons that have already been declared to be nonsignificant do not have to be recalcuated. For example, the absolute difference between M and B is 4.88 so that in Tukey's HSD method the comparisons between M and C and C and B, which have smaller absolute differences than M and C, necessarily are nonsignificant and thus, do not have to be calculated.

Note that the lines under the labels for the groups indicate that the corresponding members are *not* significantly different according to the associated multiple comparisons group. A somewhat simplistic way of thinking about the visualization of the results is that the lines under

Figure 5.3 *A Visualization of Results by Type of Multiple Comparisons Procedure*

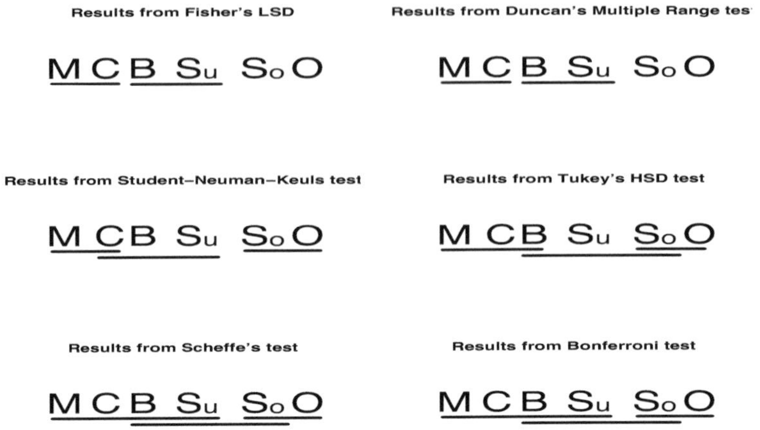

Results from Fisher's LSD

M C B **Su** **So O**

Results from Duncan's Multiple Range tes

M C B **Su** **So O**

Results from Student–Neuman–Keuls test

M C B **Su** **So O**

Results from Tukey's HSD test

M C B **Su** **So O**

Results from Scheffe's test

M C B **Su** **So O**

Results from Bonferroni test

M C B **Su** **So O**

subsets of groups represent clusters of groups with similar mean values. So, for example, the Bonferroni, Scheffé and Tukey methods all identify the groups labelled M, C, B, and Su as one "cluster"; B, Su, and So as another "cluster"; and So and O as a third cluster. Note that a group can be a member of more than one cluster. More groups are clustered or categorized as having similar means if one uses "conservative" methods whereas the opposite is true using "liberal" methods. The methods described above can be employed in SAS with the /lsd, /duncan, /snk, /bon, /tukey, and /scheffe options in the means statement in proc glm.

5.4.7 *Gabriel's Method*

A simple approach proposed by Gabriel [51] avoids the numerous calculations required for the methods given in the previous section. For this method one simply calculates the 95% confidence interval for each of the mean values *adjusted* for multiple comparisons. If the confidence intervals for two means, $\overline{y}_{i\bullet}$ and $\overline{y}_{j\bullet}$, say, (ℓ_i, u_i) and (ℓ_j, u_j) do not overlap then the groups are declared to be significantly different; otherwise, they are not. The confidence intervals formed

$$\overline{y}_{i\bullet} \mp m_{g,N-g,1-\alpha}\left(s/\sqrt{2n_i}\right) \tag{5.11}$$

where $m_{g,N-g,1-\alpha}$ is the $(1-\alpha)^{th}$ percentile of the *studentized maximum modulus distribution*. Tables for that distribution for experimentwise error rates of 1% and 5% can be found in Stoline and Ury [145] and in Rolhf and Sokal [121].

Example 5.9. Consider again the data in the previous example. For $g = 6$ and $N - g = 72 - 6 = 66$, $m_{6,66,.95} \approx 3.03$ and $s/\sqrt{2 \cdot 12} \approx 0.8446 \Longrightarrow m_{6,66,.95}(s/\sqrt{2 \cdot 12}) \approx 2.5598$

so that the Gabriel adjusted confidence intervals are obtained by subtracting and adding 2.56 to each of the means, $\overline{y}_{i\bullet}$. This would be equivalent to a critical difference in the example above of $2 \times 2.56 = 5.12$. Hence, based on this example, the method would lie between that of Tukey and Scheffé with respect to conservativeness. This property is true for most cases. Like several of the other methods, it is useful because it can be employed with unbalanced data. The results for our data are represented graphically in Figure 5.4 below.

Figure 5.4 *Gabriel Adjusted Confidence Intervals*

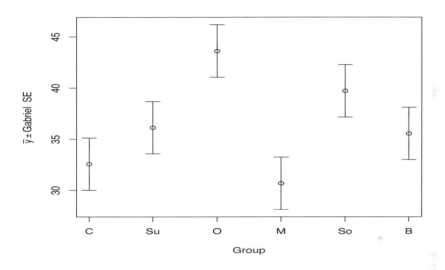

One can see from Figure 5.4 [or calculate from equation (5.11)] that the confidence intervals for the following pairs do *not* cross: M—O, M—So, M—Su, C—O, and C—So. Thus, if indeed all of the pairwise comparisons were unplanned, then we can still declare these pairs to have *significant* differences *adjusting* for the all pairwise comparisons. All other pairwise differences would not achieve statistical significance using this method. A version of Gabriel's method can be obtained in SAS with the /gabriel option in the means statement in proc glm.

5.4.8 Dunnett's Procedure

Dunnett's procedure is one that is applicable to situations where the primary interest is to compare the mean responses of a number of populations to that of a specified population which is often referred to as the *control* population (Dunnett [32, 33]). The assumptions are that the design of the experiment is a "parallel groups" design and that the data are normally distributed with constant variance. So, suppose we have $p = g - 1$ "dosed" groups and one *control* (which is often either *undosed* or placebo) group. For this application, we'll denote the control group using a "0" and the dosed groups with the subscripts $1, 2, \ldots, g - 1$ (sometimes we'll use $p = g - 1$). Assume that we observe $y_{01}, \ldots, y_{0n_0}, \; y_{11}, \ldots, y_{1n_1}, \; \ldots, y_{p1}, \ldots, y_{pn_p}$. Thus, there are a total of g independent groups and we will suppose that the primary interest is to compare the mean of the control group (i.e., Group 0) to *each of the other* $p = g - 1$ groups. Consequently, we interested in testing the following hypotheses:

$H_0 : \mu_0 = \mu_1, \; \mu_0 = \mu_2, \; \ldots, \; \mu_0 = \mu_{g-1}$ versus $H_1 :$ at least one inequality holds.

Then Dunnett's procedures accomplishes two primary aims:

1. Adjusts the experimentwise error rate for the $p = g - 1$ comparisons.
2. Adjusts for the correlation structure induced by the comparisons of each of $p = g - 1$ groups to a single control.

If $n_1 = n_2 = \ldots = n_{g-1}$, that is, if the sample sizes of groups $1, \ldots, g - 1$ are equal (but not necessarily equal to n_0) then Dunnett's procedure is exact $\implies$ the distributional form has been worked out. Approximations are given otherwise (see Dunnett [34]). Under H_0, Dunnett worked out the exact distribution for $n_0 = n_1 = \ldots = n_{g-1}$. To do Dunnett's procedure, form

$$D_i = \frac{\bar{y}_i - \bar{y}_0}{s\sqrt{\frac{1}{n_0} + \frac{1}{n_i}}} = \frac{\bar{y}_i - \bar{y}_0}{s}\sqrt{\frac{n_0 n_i}{n_0 + n_i}}, \text{ where}$$

$s^2 = \sum_{i=0}^{p}(n_i - 1)s_i^2 / \sum_{i=0}^{p}(n_i - 1)$, which is the Within Mean Square (or the Error Mean Square) from the ANOVA table.

Example 5.10. Consider again the data in Examples 5.1 and 5.5. Assume now that the interest is to compare the means of each of the five "active" groups back to the control group. Thus, we would have to control for five comparisons and for the correlation structure, which would be implicit with the fact that each of the five comparisons is compared to a single group.

For these data, $D_1 = \frac{36.14 - 32.57}{4.1375\sqrt{\frac{2}{12}}} \approx 2.11$, $D_2 = \frac{43.62 - 32.57}{4.1375\sqrt{\frac{2}{12}}} \approx 6.54$, $D_3 = \frac{30.68 - 32.57}{4.1375\sqrt{\frac{2}{12}}} \approx -1.12$, $D_4 = \frac{39.73 - 32.57}{4.1375\sqrt{\frac{2}{12}}} \approx 4.24$ and $D_5 = \frac{35.56 - 32.57}{4.1375\sqrt{\frac{2}{12}}} \approx 1.77$.

so that the results could be resummarized as:

Group	n_i	Mean	Std. Dev.	D_i
0 Control [C]	12	32.57	3.71	
1 Sunflower oil [Su]	12	36.14	4.59	2.11
2 Olive oil [O]	12	43.62	4.92	6.54
3 Margarine [M]	12	30.68	4.32	−1.12
4 Soybean oil [So]	12	39.73	4.32	4.24
5 Butter [B]	12	35.56	2.52	1.77

In this example, the error degrees of freedom are $N - g = 66$ and there are $p = 5$ "active" groups being compared to a control. Published versions of Dunnett's tables (which can be obtained by typing "Dunnett's tables" on your favorite browser or refer to, e.g., Dunnett [32] or Fleiss [49]), typically have two-sided Dunnett's values at the 5% α-level for $D_{5,60,.975} =$

2.58 and $D_{5,120,.975} = 2.55$ for the error degrees of freedom 60 and 120, respectively. (The actual critical value of 2.576 can be obtained in SAS with the /dunnett option in the means statement in proc glm. Thus, only $|D_2|$ and $|D_4|$ are greater than the critical value and so only the olive oil and soybean oil groups had significantly greater mean brain triglyceride levels than the control group.

Dunnett's procedure is a special type of contrast that is useful in many laboratory or clinical situations where the comparisons of interest may involve several different types of compounds as compared to a control. Some general comments about Dunnett's procedure are as follows:

1. It is *not* necessary to do an F-test before doing Dunnett's procedure *if* the primary interest a priori of the study is to compare a number of group means to a control mean.

2. Dunnett's procedure is _not_ useful for establishing a dose response relationship for a single drug. Use a linear trend contrast (as was presented earlier in this chapter) or isotonic regression instead.

3. The optimal (most powerful) sample size allocation for Dunnett's procedure is to assign $n_0 = \sqrt{p}\, n_i$ observations to the control group and keep n_i fixed for $i = 1, \ldots, p$. Thus, for example, if $p = 4$, then the optimal sample size allocation would be to assign twice as many observations to the control group as to each of the four experimental groups.

For more discussion about this procedure, see Dunnett [32, 33] and Fleiss [49].

5.4.9 False Discovery Rate and the Benjamini–Hochberg Procedure

The multiple comparison procedures developed so far are applicable to classical applications of ANOVA where the number of comparisons may be as high as 50. However, in some modern applications such as analyses of microarray, proteomic and neuroscience data, the number of comparisons being made often range into 1000s. For example, DNA microarrays are used to measure changes in gene expression levels which may result in over 20, 000 tests being performed. Furthermore, the purpose of the comparisons may be to *discover* differential patterns in the data which may be used to generate hypotheses to be tested in follow-up experiments. Thus, the usual standard of controlling the error rate so that Pr(no Type I errors) is less than, say, 0.05 may be too stringent as one would be apt to miss interesting differences or patterns characteristic of complex data. Consequently, in situations with large numbers of comparisons where one wishes to perform exploratory analyses, the number of Type I errors may be allowed to be some small number > 0. Such considerations lead to the principle of *False Discovery Rate (FDR)* (Benjamini and Hochberg [10]). Consider Table 5.2 which summarizes the results of testing m null hypotheses, of which, m_0 are true.

Table 5.2. *Number of Errors Committed when Performing* m *Tests*

	Test Declared Not Significant	Test Declared Significant	Total
H_0 is true	U	V	m_0
H_0 is not true	T	S	$m - m_0$
	$m - R$	R	m

Of interest is the proportion of tests which are falsely declared to be significant, that is, $Q = V/(V + S) = V/R$. The variables denoted with upper case letters are all random variables; however, only R is observable. The formal definition of FDR is given as

$$\text{FDR} = E[Q)] = E[V/R] \tag{5.12}$$

From equation (5.12), we see that FDR is the *expected* proportion of significant tests that are falsely significant. If all of the null hypotheses in Table 5.2 were true then the FDR would have the same value as the experimentwise or *familywise* error rate [FWER] given in Section 5.2. Otherwise, the value of the FDR is larger. The following algorithm has been shown by Benjamini and Hochberg [10] to control the FDR so that its value is $\leq \alpha$ for some prechosen α.

- For m tests, sort the p-values from the smallest to the largest, $p_{(1)} \leq p_{(2)} \leq \cdots \leq p_{(m)}$, and
- Pick k to be the largest i where $p_{(i)} \leq \frac{i}{m}\alpha$, where $p_{(i)}$ is i^{th} smallest p-value.

Example 5.11. Consider the following simulation where $10,000$ two-sample t-tests are considered. Each dataset is simulated from two normal distributions of size $n = 20$ which both have a standard deviation of 5. The mean values for the two samples are $\mu_1 = 18$ and $\mu_2 = 20$. The number of comparisons declared to be significant using the Benjamini–Hochberg (B–H) and Bonferroni criteria, respectively, are calculated. A program listing to do this comparison is given below.

```
N.sim <- 10000
n.1<- 20; n.2<- 20
mn.1<- 18; mn.2<- 20
Delta<- mn.2-mn.1
std.1<- 5; std.2 <- 5
p.value<- rep(NA,N.sim)
for (j in 1:N.sim){
    x<- rnorm(n.1,mn.1,std.1)
    y<- rnorm(n.2,mn.2,std.2)
    t.xy <- t.test(x,y)
    p.value[j]<- t.xy$p.value
}
rank.pval<-rank(p.value)
sort.pval<-sort(p.value)
alpha<-.05
Bon.alpha<-alpha/N.sim
n.sig.Bon<-ifelse(length(which(sort.pval<Bon.alpha))==0,
```

```
                                  0,max(which(sort.pval<Bon.alpha)) )
BH.comp.alpha <- (1:N.sim)*alpha/N.sim
n.sig.BH<-ifelse(length(which(sort.pval<BH.comp.alpha))==0,
                                  0,max(which(sort.pval<BH.comp.alpha)) )
BH.alpha<-ifelse(length(which(sort.pval<BH.comp.alpha))==0,
                                  Bon.alpha,BH.comp.alpha[n.sig.BH])
cbind(Delta,n.sig.BH,BH.alpha,n.sig.Bon,Bon.alpha)
```

Using the program above, one can construct a table comparing the B–H procedure to the the Bonferroni procedure with respect to how many significant results can be found if 10,000 t-tests were performed.

Number of significant tests for 10,000 simulated datasets

Δ	# Sig. Tests B–H criterion	B–H α	# Sig. Test Bonforroni crit.	Bonferroni α α
0.0	0	0.000005	0	0.000005
0.5	1	0.000005	1	0.000005
1.0	0	0.000005	0	0.000005
1.5	0	0.000005	0	0.000005
2.0	128	0.000640	4	0.000005
2.5	871	0.004355	5	0.000005
3.0	2402	0.012010	20	0.000005

The example shows that using the B–H criterion, one can achieve the same value of the FWER as is achieved by using the Bonferroni procedure but the B–H procedure will result in higher numbers of significant tests if a true difference is present. This means that the B–H procedure has superior *power*. The topic of power is further explored in Chapter 11 of this book.

5.5 Two-Way Analysis of Variance: Factorial Design

5.5.1 Introduction

Factorial designs are useful when an investigator wants to assess the effect of combining *every* level of one variable (or *factor*) with *every* level of another variable (or *factor*). The simplest such design is that of a 2×2 factorial design. An $R \times C$ factorial design has R levels of factor A and C levels of factor B. A special case, of course, is the 2×2 factorial for which $R = 2$ and $C = 2$. Examples of the data structure for factorial designs are as follows:

Equal replications per cell	Unequal replications per cell	No replications per cell
xxx | xxx | xxx | xxx	xx | xx | xxx | xxx	x | x | x | x
xxx | xxx | xxx | xxx	xx | xx | xxx | xxx	x | x | x | x
xxx | xxx | xxx | xxx	xx | x | xxx | xxx	x | x | x | x

The factorial design is often called a "parallel groups" design because the observations in the cells are randomized in parallel, i.e., observations are independent both within *and across* all cells. For our present purposes, we'll assume that all observations are from normal population(s) with a common variance and are independent from each other. We'll also assume that the factors have fixed effects (i.e., we'll limit inference to *only* the given levels of each factor). The classical way of writing this model is

$$
\underbrace{y_{ijk}}_{\substack{\text{response for } k^{th} \\ \text{subject in cell } ij}} = \underbrace{\mu}_{\substack{\text{grand} \\ \text{mean}}} + \underbrace{\alpha_i}_{\substack{\text{effect due to } i^{th} \text{ level} \\ \text{of Factor A}}} + \underbrace{\beta_j}_{\substack{\text{effect due to } j^{th} \text{ level} \\ \text{of Factor B}}} + \underbrace{(\alpha\beta)_{ij}}_{\text{interaction}} + \underbrace{\epsilon_{ijk}}_{\substack{\text{error term for } k^{th} \\ \text{subject in cell } ij}} \qquad (5.13)
$$

$i = 1, \ldots, r;\ j = 1, \ldots, c;$ and $k = 1, \ldots, n_{ij}$. As with the one-way AOV design, (5.13) can be written as a cell means model:

$$
\underbrace{y_{ijk}}_{\substack{\text{response for } k^{th} \\ \text{subject in cell ij}}} = \underbrace{\mu_{ij}}_{\substack{\text{mean reponse} \\ \text{in cell } ij}} + \underbrace{\epsilon_{ijk}}_{\substack{\text{error term for } k^{th} \\ \text{subject in cell } ij}}, \qquad (5.14)
$$

$i = 1, \ldots, r; j = 1, \ldots, c;$ and $k = 1, \ldots, n_{ij}$.

Equations (5.13) and (5.14) allow one to represent the factorial model as a general linear model, $\mathbf{Y} = \mathbf{X}\boldsymbol{\mu} + \boldsymbol{\epsilon}$. The latter (cell means) model, however, allows one to obtain a unique solution, $\hat{\boldsymbol{\mu}} = (\mathbf{X}'\mathbf{X})^{-1}\mathbf{X}'\,\mathbf{Y}$. For example, in the two by two case,

$$
\mathbf{X} = \begin{pmatrix} 1 & 0 & 0 & 0 \\ \vdots & \vdots & \vdots & \vdots \\ 1 & 0 & 0 & 0 \\ 0 & 1 & 0 & 0 \\ \vdots & \vdots & \vdots & \vdots \\ 0 & 1 & 0 & 0 \\ 0 & 0 & 1 & 0 \\ \vdots & \vdots & \vdots & \vdots \\ 0 & 0 & 1 & 0 \\ 0 & 0 & 0 & 1 \\ \vdots & \vdots & \vdots & \vdots \\ 0 & 0 & 0 & 1 \end{pmatrix}, \quad \mathbf{Y} = \begin{pmatrix} Y_{111} \\ \vdots \\ Y_{11n_{11}} \\ Y_{121} \\ \vdots \\ Y_{12n_{12}} \\ Y_{211} \\ \vdots \\ Y_{21n_{21}} \\ Y_{221} \\ \vdots \\ Y_{22n_{22}} \end{pmatrix}, \quad \text{and } \boldsymbol{\mu} = \begin{pmatrix} \mu_{11} \\ \mu_{12} \\ \mu_{21} \\ \mu_{22} \end{pmatrix},
$$

so that

$$
\hat{\boldsymbol{\mu}} = \begin{pmatrix} n_{11} & 0 & 0 & 0 \\ 0 & n_{12} & 0 & 0 \\ 0 & 0 & n_{21} & 0 \\ 0 & 0 & 0 & n_{22} \end{pmatrix}^{-1} \begin{pmatrix} \sum_{k=1}^{n_{11}} y_{11k} \\ \sum_{k=1}^{n_{12}} y_{12k} \\ \sum_{k=1}^{n_{21}} y_{21k} \\ \sum_{k=1}^{n_{22}} y_{22k} \end{pmatrix} = \begin{pmatrix} \bar{y}_{11\bullet} \\ \bar{y}_{12\bullet} \\ \bar{y}_{21\bullet} \\ \bar{y}_{22\bullet} \end{pmatrix}.
$$

Thus, the least squares solution for $\boldsymbol{\mu}$ is nothing more than the means of the four cells. From equations (5.13) and (5.14), we can also see that the parallel groups factorial design can be also formulated in terms of a one-way design where, for example, in the two by two case, comparing the difference of the average of μ_{11} and μ_{12} to that of μ_{21} and μ_{22} is equivalent to examining whether or not Factor A is effective. Likewise, comparing the difference of the average of μ_{11} and μ_{21} to that of μ_{12} and

μ_{22} is equivalent to examining whether or not Factor B is effective. By examining the *difference of the differences*, one can examine whether or not an *interaction* is present.

5.5.2 Interaction

Factors A and B have an interaction (denoted $A \times B$) when the difference(s) between (among) levels of A *change over the levels* of B. Interaction is sometimes called *effect modification*. In contrast, if there is *not* an interaction (effect modification) then the model is said *additive*. Pictorial representations of qualitative, quantitative, and no interactions are given on the next page of this chapter. Interactions *can only be detected if there is more than one observation per cell*. Therefore, each cell must have replications for an interaction term to be included in an analysis of variance table.

Figure 5.5 *A Visualization of Types of Interactions Between Two Factors*

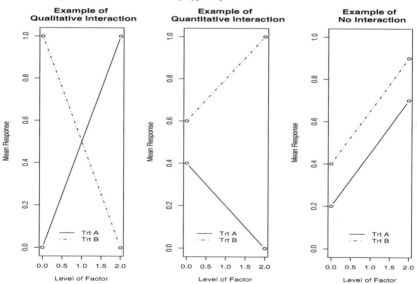

5.5.3 Calculations When the Numbers of Observations in All Cells Are Equal.

If there are equal numbers of observations in each of the cells, then the calculations for constructing an Analysis of Variance (ANOVA) table are quite straight forward. To start, suppose that there are r levels of factor A and c levels of factor B and that there are n observations for *each combination* of the $r \times c$ levels of the two factors.

Thus, there are a total of $N = rcn$ total observations. Also define a *correction term for the mean* as

$$C_M \equiv \left(\sum_{i=1}^{r} \sum_{j=1}^{c} \sum_{k=1}^{n} y_{ijk} \right)^2 \bigg/ N$$

Then, the analysis of variance table for a *balanced* two-way factorial design is as follows:

ANALYSIS OF VARIANCE TABLE (All cells with equal observations)

SOURCE	DF	SS	MS
Row (Factor A)	$r-1$	$\sum_{i=1}^{r} \left(\sum_{j=1}^{c} \sum_{k=1}^{n} y_{ijk} \right)^2 \bigg/ (cn) - C_M$	$\frac{SS_A}{DF}$
Column (Factor B)	$c-1$	$\sum_{j=1}^{c} \left(\sum_{i=1}^{r} \sum_{k=1}^{n} y_{ijk} \right)^2 \bigg/ (rn) - C_M$	$\frac{SS_B}{DF}$
$A \times B$	$(r-1)(c-1)$	$\sum_{i=1}^{r} \sum_{j=1}^{c} \left(\sum_{k=1}^{n} y_{ijk} \right)^2 \bigg/ n - SS_A - SS_B - C_M$	$\frac{SS_I}{DF}$
Error	$N-rc$	$\sum_{i=1}^{r} \sum_{j=1}^{c} \left(\sum_{k=1}^{n} y_{ijk} \right)^2 \bigg/ n - C_M$	—
Total	$N-1$	$\sum_{i=1}^{r} \sum_{j=1}^{c} \sum_{k=1}^{n} y_{ijk}^2 - C_M$	—

Example 5.12. The data below are from Zar, *Biostatistical Analysis*, 3rd edition, 1996 [161]. The numbers represent values of plasma concentrations in mg/ml for of birds of both sexes, half of the birds of each sex being treated with a hormone and the other half not treated with the hormone. Let i be the level of Treatment (Trt), $i = 1, 2$ (1 = No Hormones, 2 = Hormones); j be the level of Sex $j = 1, 2$ (1 = Female, 2 = Male); and k be the observation within Treatment (Trt) and Sex ($k = 1, \ldots, 5$).

Sex

Treatment	FEMALE	MALE
NONE	16.5, 18.4, 12.7, 14.0, 12.8	14.5, 11.0, 10.8, 14.3, 10.0
	$\sum_{k=1}^{5} y_{11k} = 74.4$	$\sum_{k=1}^{5} y_{12k} = 60.6$
HORMONE	39.1, 26.2, 21.3, 35.8, 40.2	32.0, 23.8, 28.8, 25.0, 29.3
	$\sum_{k=1}^{5} y_{21k} = 162.6$	$\sum_{k=1}^{5} y_{22k} = 138.9$
Totals by Sex	$\sum_{i=1}^{2} \sum_{k=1}^{5} y_{i1k} = 237.0$	$\sum_{i=1}^{2} \sum_{k=1}^{5} y_{i2k} = 199.5$
Totals by Trt	$\sum_{j=1}^{2} \sum_{k=1}^{5} y_{1jk} = 135.0$	$\sum_{j=1}^{2} \sum_{k=1}^{5} y_{2jk} = 301.5$

Grand totals

$$\sum_{i=1}^{2}\sum_{j=1}^{2}\sum_{k=1}^{5} y_{ijk} = 436.5, \quad \sum_{i=1}^{2}\sum_{j=1}^{2}\sum_{k=1}^{5} y_{ijk}^2 = 11354.31, \quad C = \frac{(436.5)^2}{20} = 9526.6125$$

$$SS_A = \frac{135^2 + 301.5^2}{10} - C = 1386.1125, \quad SS_B = \frac{237^2 + 199.5^2}{10} - C = 70.3125$$

$$SS_I = \frac{74.4^2 + 60.6^2 + 162.6^2 + 138.9^2}{5} - C - SS_A - SS_B = 4.9005,$$

$$SS_T = 11354.31 - 9526.6125 = 1827.6975,$$

$$SS_E = 1827.6975 - SS_A - SS_B - SS_I = 366.3720 .$$

An analysis of variance table for the model is summarized below:

ANALYSIS OF VARIANCE (ANOVA) TABLE

SOURCE	DF	SS	MS	F	p-value
Hormone (Row)	1	1386.1125	1386.1125	60.53	< 0.0001
Sex (Column)	1	70.3125	70.3125	3.07	0.099
Hormone × Sex	1	4.9005	4.9005	0.214	0.65
Error	16	366.3720	22.8982	—	
Total	19	1827.6975	—	—	

Since the interaction term from this model is not even close to being significant at the $\alpha = .05$ level, it is customary to fold that term into the error term yielding the following table associated with an *additive* model (no interaction term):

ANALYSIS OF VARIANCE TABLE (Additive Model)

SOURCE	DF	SS	MS	F	p-value
Hormone (Row)	1	1386.1125	1386.1125	63.47	< 0.0001
Sex (Column)	1	70.3125	70.3125	3.22	0.091
Error	17	371.2725	21.8396	—	
Total	19	1827.6975	—	—	

In either case, the hormone treatment level has a much greater effect on the plasma concentration level than does the gender of the animal.

5.5.4 Calculations When the Numbers of Observations in the Cells Are Not All Equal

In the event that there are differing numbers of observations per cell, the mathematics for testing the significance of the factors in a two-way design becomes a lot more complex. Most authors and statistics packages recommend the use of an *unweighted means* analysis. In statistics, "unweighted" often means *equally* weighted. In many cases, the weights are all equal to 1. The idea behind unweighted means analysis in multiway ANOVA is to multiply the sums of squares of each factor by a weight related to the mean of the sample sizes of each "cell."

Algorithm for Computing an unweighted means analysis

Let n_{ij} = the number of observations in cell ij and let $\bar{y}_{ij\bullet}$ be the mean of the ij^{th} cell.

1. Compute the weighting factor, n_w, where $\dfrac{1}{n_w} = \dfrac{\sum_{i=1}^{r}\sum_{j=1}^{c}\frac{1}{n_{ij}}}{rc}$

2. Compute the Row (Factor A) Sums of Squares as

$$n_w\left[\frac{\sum_{i=1}^{r} y_{i\bullet\bullet}^{*2}}{c} - \frac{y_{\bullet\bullet\bullet}^{*2}}{rc}\right] \text{ where } y_{i\bullet\bullet}^{*} = \sum_{j=1}^{c}\bar{y}_{ij\bullet}, \text{ and } y_{\bullet\bullet\bullet}^{*} = \sum_{i=1}^{r}\sum_{j=1}^{c}\bar{y}_{ij\bullet}$$

Then, compute the Row (Factor A) Mean Square as (Row SS) / $(r-1)$.

3. Compute the column (Factor B) Sums of Squares as

$$n_w\left[\frac{\sum_{j=1}^{c} y_{\bullet j\bullet}^{*2}}{r} - \frac{y_{\bullet\bullet\bullet}^{*2}}{rc}\right] \text{ where } y_{\bullet j\bullet}^{*} = \sum_{i=1}^{r}\bar{y}_{ij\bullet}$$

Then, compute the column (Factor B) Mean Square as (Column SS) / $(c-1)$.

4. Compute the interaction Sums of Squares as

$$n_w\left[\sum_{i=1}^{r}\sum_{j=1}^{c}\bar{y}_{ij\bullet}^{2} - \frac{y_{\bullet\bullet\bullet}^{*2}}{rc}\right] - \text{ Row SS } - \text{ Column SS}$$

Then, compute the interaction Mean Square as (Interaction SS) / $[(r-1)(c-1)]$.

5. Compute the Error Mean Square as

$$\frac{\sum_{i=1}^{r}\sum_{j=1}^{c}(n_{ij}-1)s_{ij}^{2}}{N-rc} = \frac{\left[\left(\sum_{i=1}^{r}\sum_{j=1}^{c}\sum_{k=1}^{n_{ij}} y_{ijk}^{2}\right) - \sum_{i=1}^{r}\sum_{j=1}^{c}\frac{y_{ij\bullet}^{*2}}{n_{ij}}\right]}{N-rc}$$

$$\text{where } \quad y_{ij\bullet}^{*} = \sum_{k=1}^{n_{ij}} y_{ijk}, \text{ and } N = \sum_{i=1}^{r}\sum_{j=1}^{c} n_{ij}.$$

From the five steps given above, one can form an ANOVA table (see example below).

Example 5.13. Source: Sokal and Rohlf, *Biometry*, 3rd edition, 1995, pp. 358–359 [141]

The numbers below represent values of seven-week weights in grams for male and female untreated chicks versus male and female chicks that have been injected with thyroxin:

Thryroxin

Sex	NO	YES
MALE	560, 500, 350, 520, 540, 620, 600, 560, 450, 340, 440, 300 $\sum_{k=1}^{12} y_{11k} = 5780$	410, 540, 340, 580, 470, 550, 480, 400, 600, 450, 420, 550 $\sum_{k=1}^{12} y_{12k} = 5790$
FEMALE	530, 580, 520, 460, 340, 640, 520, 560 $\sum_{k=1}^{8} y_{21k} = 4150$	550, 420, 370, 600, 440, 560, 540, 520 $\sum_{k=1}^{8} y_{22k} = 4000$

Summary statistics

$N = 40$, $n_{11} = 12, \bar{y}_{11\bullet} \approx 481.6667$, $n_{12} = 12, \bar{y}_{12\bullet} = 482.50$,

$\bar{y}_{\bullet\bullet\bullet} = 493$ $s_{11} \approx 106.073$ $s_{12} \approx 81.5893$

$s \approx 88.6133$ $n_{21} = 8, \bar{y}_{21\bullet} = 518.75$, $n_{22} = 8, \bar{y}_{22\bullet} = 500.0$,

   $s_{21} \approx 89.1928$ $s_{22} \approx 80.1784$

Unweighted Means Analysis

1. $\frac{1}{n_w} = (\frac{1}{12} + \frac{1}{12} + \frac{1}{8} + \frac{1}{8})/4 \approx .1042 \Longrightarrow n_w \approx 9.6$

2. $y_{1\bullet\bullet}^* \approx 964.1667$, $y_{2\bullet\bullet}^* = 1018.75$, $y_{\bullet\bullet\bullet}^* \approx 1982.9167$, $\frac{(y_{\bullet\bullet\bullet}^*)^2}{rc} \approx 982,989.6267$, Row SS $= (9.6)744.8351 \approx 7150.42 =$ Row MS (because $r - 1 = 1$)

3. $y_{\bullet1\bullet}^* \approx 1000.4167$, $y_{\bullet2\bullet}^* = 982.5$, Column SS $= (9.6)80.2517 \approx 770.4163 =$ Column MS (because $c - 1 = 1$)

4. $(481.6667^2 + 482.5^2 + 518.75^2 + 500^2) - \frac{(y_{\bullet\bullet\bullet}^*)^2}{rc} -$ Row SS - Column SS ≈ 920.42

5. $11(106.073)^2 + 11(81.5893)^2 + 7(89.1928)^2 + 7(80.1784)^2 = 297,679.1667$, $\Longrightarrow$ Error MS $\approx 297,679.1667/36 \approx 8268.87$

The resulting analysis of variance table for the model is summarized below:

ANALYSIS OF VARIANCE TABLE (Unweighted Means)

SOURCE	DF	SS	MS	F	p-value
Row (Sex)	1	7150.4	7150.4	0.8647	0.36
Column (Thyroxin)	1	770.42	770.42	0.0932	0.76
Interaction	1	920.42	920.42	0.1113	0.74
Error	36	297679	8268.9	—	

Assuming that the variables are called chickwt, sex and chickwt, the calculations could be performed in SAS with the following code:

```
proc glm;
class sex thyrox;
model chickwt=sex thyrox sex*thyrox;
run; quit;
```

In R or S-plus, the code would look like this:

```
results<- lm(chick.wt~sex*thyrox)
anova(results)
```

The interaction plot below indicates that no obvious interaction exists that is consistent with the results from the ANOVA table. What the plot lacks, however, are standard error bars. The pooled standard error for these data is approximately 90.9, indicating that neither main effects or interactions show any differences that are near significance.

Figure 5.6 *An Interaction Plot for Sex × Thyroxin level*

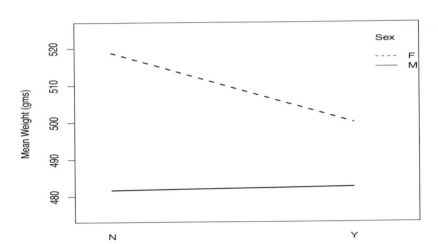

The unweighted means analysis for row effects in our above example is testing

$$H_0 : \frac{1}{2}(\mu_{11} + \mu_{12}) = \frac{1}{2}(\mu_{21} + \mu_{22}) \text{ versus } H_1 : \frac{1}{2}(\mu_{11} + \mu_{12}) \neq \frac{1}{2}(\mu_{12} + \mu_{22})$$

[μ_{ij} here denotes the population mean for the ith row and the jth column.] Similar results would be obtained if the test of interest were column (or Thyroxin in our case) effects. The *unweighted* means analysis obviously gives equal weight to each cell and therefore, is usually the preferred method of analysis in cases where there are unequal numbers of observations in each cell. In cases where there are equal numbers of observations in each cell, any type of analysis (weighted or unweighted) will give the same results.

5.6 Two-Way Analysis of Variance: Randomized Complete Blocks

5.6.1 Introduction

Blocking is a technique used by investigators to control for factors *other* than the one that is of primary interest. The idea is to match the members within each block so that they closely resemble each other with respect to a set of factors considered to confound the question of true interest. For example, one may want to use sex, age and smoking status as factors to block on when studying the effects of a set of dietary regimens on cholesterol levels. Consequently, one would hope to reduce the variability within each "block" due to sex, age and smoking status so that the we can determine the influence of the factor of interest (diet) on the outcome (cholesterol level). The simplest type of blocked design is where only two treatments are being tested. This design is associated with a paired t-test.

A typical data structure for a randomized complete blocks experiment is shown below. Our discussion in this section will be limited to the case where there is one observation for each combination of Block i and Treatment j. Also, for the purposes of this discussion, we'll assume that the outcome of interest comes from a population that is normally distributed. This can obviously be extended to cases where there are multiple observations for each block and also to cases where the outcomes are from populations that are not normally distributed.

5.6.2 Structure of Data From a Completely Randomized Blocks Experiment

A typical stucture for a completely randomized block design is outlined in the table below.

	Treatment					
Block	1	...	j	...	k	**Mean**
1	Y_{11}	...	Y_{1j}	...	Y_{1k}	$\overline{Y}_{1\bullet}$
$\vdots$	$\vdots$	$\vdots$	$\vdots$	$\vdots$	$\vdots$	$\vdots$
i	Y_{i1}	...	Y_{ij}	...	Y_{ik}	$\overline{Y}_{i\bullet}$
$\vdots$	$\vdots$	$\vdots$	$\vdots$	$\vdots$	$\vdots$	$\vdots$
b	Y_{b1}	...	Y_{bj}	...	Y_{bk}	$\overline{Y}_{b\bullet}$
Mean	$\overline{Y}_{\bullet 1}$	...	$\overline{Y}_{\bullet j}$	...	$\overline{Y}_{\bullet k}$	$\overline{Y}_{\bullet\bullet}$

The model for an experiment using a randomized complete blocks design can be written as

$$\underbrace{Y_{ij}}_{\substack{\text{response for } i^{th} \text{ block and } j^{th} \\ \text{treatment}}} = \underbrace{\mu}_{\substack{\text{grand mean}}} + \underbrace{\beta_i}_{\substack{\text{offset from mean} \\ \text{due to block } i}} + \underbrace{\tau_j}_{\substack{\text{offset from mean} \\ \text{due to treatment } j}} + \underbrace{\epsilon_{ij}}_{\substack{\text{error term for } i^{th} \\ \text{block in treatment } j}} \quad (5.15)$$

where $\epsilon_{ij} \sim N(0, \sigma^2)$, $i = 1, \ldots b$, $j = 1, \ldots k$.

Notice that there is *no interaction term* in the above model. This type of a model is called an <u>*additive model*</u>. In cases where there are no replications in each block, an interaction cannot be estimated because there would be no degrees of freedom left over for the error. An analysis of variance table for the model is given below:

ANALYSIS OF VARIANCE (ANOVA) TABLE

SOURCE	DF	SS	MS	F
Treatments	$k-1$	$b\left(\sum_{j=1}^{k}\left(\overline{Y}_{\bullet j}-\overline{Y}_{\bullet\bullet}\right)^2\right)$	$\frac{TSS}{k-1}$	$\frac{TMS}{EMS}$
Blocks	$b-1$	$k\left(\sum_{i=1}^{b}\left(\overline{Y}_{i\bullet}-\overline{Y}_{\bullet\bullet}\right)^2\right)$	$\frac{BSS}{b-1}$	$\frac{BMS}{EMS}$
Error	$(k-1)(b-1)$	$\sum_{i=1}^{b}\sum_{j=1}^{k}\{Y_{ij}-\overline{Y}_{i\bullet}-\overline{Y}_{\bullet j}+\overline{Y}_{\bullet\bullet}\}^2$		
Total	$kb-1$	$\sum_{i=1}^{b}\sum_{j=1}^{k}\left(Y_{ij}-\overline{Y}_{\bullet\bullet}\right)^2$		

An ANOVA table is intuitively easy to understand because the sums of squares for the blocks or treatments are simply the sums of the squared variations of block or treatment means about the grand mean. However, the formulas in the above table are not typically used to calculate an ANOVA. Alternatively, the formulas (which are equivalent to those in the table above) in the ANOVA table below are used.

ANALYSIS OF VARIANCE TABLE (alternative form for easier calculation)

SOURCE	DF	SS	MS	F
Treatments	$k-1$	$b(\sum_{j=1}^{k}\overline{Y}_{\bullet j}^2 - k\overline{Y}_{\bullet\bullet}^2)$	$\frac{TSS}{k-1}$	$\frac{TMS}{EMS}$
Blocks	$b-1$	$k(\sum_{i=1}^{b}\overline{Y}_{i\bullet}^2 - b\overline{Y}_{\bullet\bullet}^2)$	$\frac{BSS}{b-1}$	$\frac{BMS}{EMS}$
Error	$(k-1)(b-1)$	$\sum_{i=1}^{b}\sum_{j=1}^{k}\{Y_{ij}-\overline{Y}_{i\bullet}-\overline{Y}_{\bullet j}+\overline{Y}_{\bullet\bullet}\}^2$		
Total	$kb-1$	$\sum_{i=1}^{b}\sum_{j=1}^{k}Y_{ij}^2 - ng\overline{Y}_{\bullet\bullet}^2$		

Example 5.14. The data below were taken from Fleiss, *The Design and Analysis of Clinical Experiments*, Wiley & Sons, New York, pp. 127 [†] [49]. The outcome of interest was clotting times in minutes for comparing four treatments. Each subject in the experiment was used as his or her own control. Thus, a block here is really a subject. The components of a block are the repeated clotting time measurements made on a subject.

**Clotting Times (in Minutes) for 4 Treatments
in a Randomized Complete Blocks Design[†]**

Subject	1	2	3	4	Mean
1	8.4	9.4	9.8	12.2	9.950
2	12.8	15.2	12.9	14.4	13.825
3	9.6	9.1	11.2	9.8	9.925
4	9.8	8.8	9.9	12.0	10.125
5	8.4	8.2	8.5	8.5	8.400
6	8.6	9.9	9.8	10.9	9.800
7	8.9	9.0	9.2	10.4	9.375
8	7.9	8.1	8.2	10.0	8.550
Mean	9.300	9.7125	9.9375	11.025	9.9938
S.D.	1.550	2.294	1.514	1.815	

[†] Reproduced with permission from John Wiley & Sons

The results of the analysis are given below.

$$\overline{Y}_{\bullet\bullet} = \frac{319.8}{32} \approx 9.9938, \ \text{Treatment SS} = 8\left[9.3^2+9.7125^2+9.9375^2+11.025^2-4(9.99375^2)\right] \approx 13.0163$$

$$\text{Subject SS} = 4\left[9.95^2+13.825^2+9.925^2+10.125^2+8.4^2+9.8^2+9.375^2+8.55^2-8(9.99375^2)\right] \approx 78.9888$$

$$\text{Total SS} = \left[8.4^2+12.8^2+9.6^2+9.8^2+8.4^2+8.6^2+8.9^2+7.9^2+9.4^2+15.2^2+9.1^2+8.8^2+8.2^2+9.9^2\right.$$

$$+9^2+8.1^2+9.8^2+12.9^2+11.2^2+9.9^2+8.5^2+9.8^2+9.2^2+8.2^2+12.2^2+14.4^2+9.8^2+12^2+8.5^2+10.9^2$$

$$\left. +10.4^2 + 10^2\right] - 32(9.99375^2) \approx 105.7788,$$

$$\text{Error SS} = \text{Total SS} - \text{Treatment SS} - \text{Subject SS} \approx 105.7788 - 13.0163 - 78.9888 = 13.7737$$

$$\text{Treatment MS} \approx \frac{13.0163}{3} \approx 4.3388, \ \text{Subject MS} \approx \frac{79.9888}{7} \approx 11.2841, \ \text{Error MS} \approx \frac{13.7737}{21} \approx 0.6559.$$

ANOVA table for clotting time data[†]

Source	DF	SS	MS	F ratio
Treatment	3	13.0163	4.3388	6.62
Subject (Block)	7	78.9888	11.2841	
Error	21	13.7737	0.6559	
Total	31	105.7788		

[†] Reproduced with permission from John Wiley & Sons

One can easily write a SAS program to analyze the above data. For example, if the data are contained in a computer file called 'C:\fleiss-block-data.dat' and are arranged so that each record contains the clotting time, the subject number and the treatment number in that order, then a program would look as follows:

```
* SAS program to analyze clotting data from Fleiss, The Design and
  Analysis of Clinical Experiments, 1986, Wiley and Sons, p. 125-129 ;
options ls=70 pageno=1;
data block0;
infile 'C:\fleiss-block-data.dat';
input clottime subject treatment;
run;

proc glm;
class treatment subject;
model clottime=treatment subject;
title 'Analysis of a randomized block design on clotting times';
means treatment subject; *<- Get marginal mean clotting times;
run;
```

Since $F \approx 6.62 > F_{3,21,.95} \approx 3.07$, we reject the null hypothesis that the four treatments are equal ($p \approx 0.003$). At this point, it could be of interest to do pairwise comparisons or use some other set of linear contrasts to test various hypotheses of interest. The usual multiple comparisons procedures can be used if one is interested in testing more than one contrast.

If the blocking procedure employed is indeed effective, then we would expect that the variation *across* blocks would be large in comparison to the variation *within* blocks. If, on the other hand, the blocking was *ineffective*, then it would have been a lot easier to simply employ a "parallel groups" (completely randomized) design instead of trying to match (or "block") individuals or experimental units on a number of confounding factors. Of course, once someone has already done a particular study, (s)he cannot change the design. However, in many cases, one would like to assess a previous study for purposes of designing *future* studies. One way to test whether or not blocking was effective is to calculate a measure of *Relative Efficiency* (RE). The relative efficiency measures whether or not the blocking was helpful or hurtful to the design of the experiment. Essentially, relative efficiency is a positive number that measures the ratio of sample size per group needed for the two designs (randomized blocks and "parallel groups") to be equivalent. Given a dataset, one can estimate the RE as

$$\widehat{RE} = c + (1 - c)\left(\frac{MS_{Blocks}}{MS_{Error}}\right), \tag{5.16}$$

where $c = b(k - 1)/(bk - 1)$. If $\widehat{RE} = 1$, then the two designs would require equal sample sizes per group. If $\widehat{RE} < 1$, then the "parallel groups" design would require less subjects per group whereas if $\widehat{RE} > 1$, then the randomized block design would require less subjects per group. In most cases, we would require that $\widehat{RE}$ to be at least 2 because it is much more difficult to employ a block design than a "parallel groups" design.

Example 5.15. In our previous example, $b = 8$ and $k = 4$ so that $c = (8 \times 3)/31 \approx 0.7742$ and $\widehat{RE} \approx 0.7742 + (1 - 0.7742) \times \left(\frac{11.2841}{0.6559}\right) \approx 4.66$. Hence, in our example, the blocking was quite effective.

5.6.3 Missing Data

The strength of the randomized complete block design is that it can be used to reduce the variability of factors not of primary interest to the investigator. The *weakness of the randomized complete blocks design in the health sciences* is that the power of the resulting analysis can be reduced substantially if there are even a modest number of missing observations. Missing observations are, of course, somewhat more common in the health sciences than in other sciences because an investigator may not have complete control over the experimental situation due to patient/subject dropout, measurement error, etc. The attenuation of power is especially noticeable if the pattern of missingness is such that a *large proportion of subjects* have, say, one or two missing values. The attenuation of power is smaller if a *small proportion of subjects* have a large number of missing values.

The usual way that most statistical packages handle one or more missing observations within a block is to drop the block out of the analysis. More recent work, however, uses techniques such as maximum likelihood estimation to make use of all of the available data. However, one must be cautioned that most of these techniques rely on an assumption that the pattern of missingness is not related to the unmeasured values of the outcome variable. An example of where this assumption might not hold is in a case when blocks with missing values contain sicker subjects/patients than blocks without missing values. The outcome values that haven't been measured might be lower (or higher) because of the poorer health of the associated subject or patient. Therefore, the available information may be biased towards values of healthier subjects or patients. The situation is further exasperated when the pattern of missingness is different in different treatment groups.

5.7 Analysis of Covariance [ANCOVA]

Analysis of covariance models involve using both discrete and continuous variables to predict a continuous outcome. Suppose, now, that we're interested in comparing continuous outcomes (which are assumed to be normally distributed) across some grouping variable such as treatment and suppose that some confounding variable(s) exist(s). For example, when comparing two treatments for blood pressure, one might use age and sex as covariates. When comparing the mean weights among rats randomized into several diet regimens, one might use the initial weight of each rat as a covariate.

The analysis of covariance *combines analysis of variance models with regression models*. A typical analysis of covariance model may be written as

$$
\underbrace{Y_{ij}}_{\substack{\text{response for } j^{th} \\ \text{subject in } i^{th} \text{ group}}} = \underbrace{\mu_i}_{\substack{\text{mean} \\ \text{of } i^{th} \\ \text{group}}} + \underbrace{\beta}_{\substack{\text{slope associated with} \\ \text{covariates}}} \underbrace{X_{ij}}_{\substack{\text{covariate value for } j^{th} \\ \text{subject in } i^{th} \text{ group}}} + \underbrace{\epsilon_{ij}}_{\substack{\text{error for } j^{th} \text{ subject in} \\ i^{th} \text{ group}}} , \qquad (5.17)
$$

where the errors, $\epsilon_{ij} \sim N(0, \sigma^2)$, $i = 1, \ldots, k$, $j = 1, \ldots, n_i$. For some applications, we wish to adjust for the mean of the $X's$ so that the model is written as

$$
Y_{ij} = \mu_i + \beta(X_{ij} - \overline{X}_{..}) + \epsilon_{ij}, \qquad (5.18)
$$

$i = 1, \ldots, k$, $j = 1, \ldots, n_i$.

5.7.1 Modeling Considerations for ANCOVA Models

When one goes through the steps of assessing a "best" model in analysis of covariance, a number of cases naturally arise depending on the significance or lack of significance of either the discrete or continuous covariates. Five such cases for models with one discrete predictor and one continuous predictor are given below. We'll assume, for simplicity, that the discrete predictor has only two levels although, in practice, it could have more.

<u>Case 1</u>: No Significant Difference Between Groups and No Significant Covariate Effect. (One Mean Line Fits All Observations.)

$$\underline{\text{Model:}} \ Y_{ij} = \mu + \epsilon_{ij} \tag{5.19}$$

<u>Case 2</u>: Groups are Significantly Different But No Significant Covariate Effect. (Equivalent to a One-Way ANOVA Model With Unequal Means.)

$$\underline{\text{Model:}} \ Y_{ij} = \mu_i + \epsilon_{ij} \tag{5.20}$$

Figure 5.7 *Cases 1 and 2: No Trend Exists*

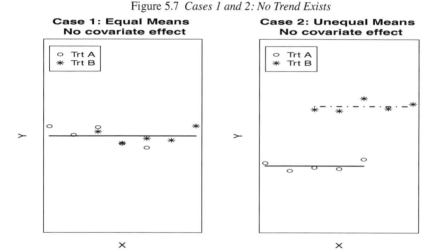

<u>Case 3</u>: Covariate Effect is Significant But *No* Significant Difference Between Group Slopes or Intercepts. (One Regression Line Fits All Groups.)

$$\underline{\text{Model:}} \ Y_{ij} = \mu + \beta X_{ij} + \epsilon_{ij} \tag{5.21}$$

<u>Case 4</u>: Covariate Effect is Significant and Group Effect is Significant. Slopes are Not Significantly Different Among Groups.

$$\underline{\text{Model:}} \ Y_{ij} = \mu_i + \beta X_{ij} + \epsilon_{ij} \tag{5.22}$$

Figure 5.8 *Cases 3 and 4: Significant Trend Exists*

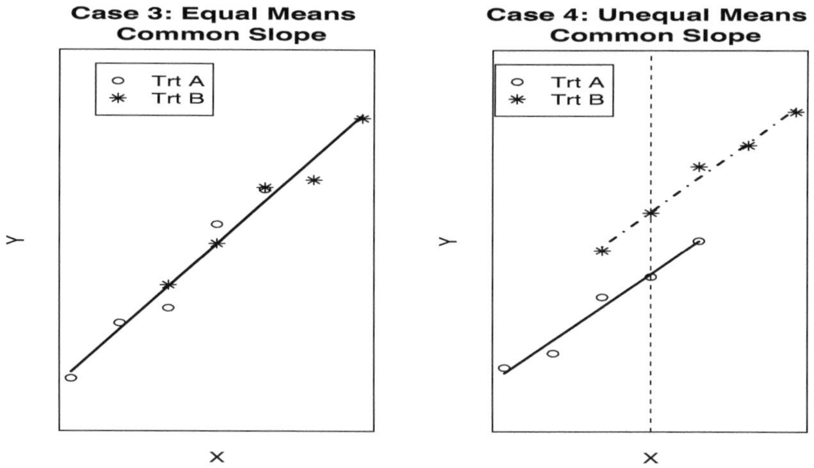

Case 3: Equal Means
Common Slope
Trt A
Trt B
Y
X
Case 4: Unequal Means
Common Slope
Trt A
Trt B
Y
X

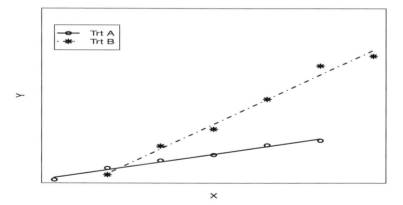

Case 5: Significant Group and Covariate Effects. Both Slopes and Intercepts are Different for the Groups.

$$\underline{\text{Model:}}\ Y_{ij} = \mu_i + \beta_i X_{ij} + \epsilon_{ij}\ . \tag{5.23}$$

Figure 5.9 *Case 5: Significant But Different Trends Exist*

Case 5: Unequal Means and Unequal Slopes
Trt A
Trt B
Y
X

Example 5.16. Example of a Program to Fit an Analysis of Covariance Model

The following data were taken from Hassard's book [67], *Understanding Biostatistics*, chapter 12 and involve blood pressures for subjects taking two different types of medication. A SAS program to identify a reasonable model for these data is given below.

```
options linesize=70 nodate;
data example;
input bp age TRT;
datalines;
  120  26  1
  114  37  1
  132  31  1
  130  48  1
  146  55  1
  122  35  1
  136  40  1
  118  29  1
  109  33  2
  145  62  2
  131  54  2
  129  44  2
  101  31  2
  115  39  2
  133  60  2
  105  38  2
; run;

data ex2; set example;
label trt='Treatment' bp='Systolic Blood Pressure';
run;

proc glm;
 class TRT;
 model bp = TRT age(TRT);*<-Model bp with effects due to trt&ages within trt;
 estimate 'Age 1' age(TRT) 1 0;*<-- estimate age effect within trt 1;
 estimate 'Age 2' age(TRT) 0 1;*<-- estimate age effect within trt 2;
 estimate 'Age 1 vs 2' age(TRT) 1 -1;*<-- estimate DIFFERENCE of age effects;
TITLE 'Analysis of Blood Pressure by Treatment adjusting for Age';
title2 'Model using Separate Slopes for each Treatment';
run;

proc glm;
 class TRT;
 model bp = TRT age;*<-Model bp with effects due to trt&age pooled across trt;
 estimate 'age' age 1; *<- Estimate POOLED age effect;
 lsmeans trt; *<-- Give adjusted bp's for each trt;
TITLE 'Example of analysis BP by Treatment adjusting for Age';
title2 'Model using a Common Slope for both Treatments';
run; quit;
```

Selected Output from a Program to Fit an Analysis of Covariance Model

```
        Analysis of Blood Pressure by Treatment adjusting for Age        2
             Model using Separate Slopes for each Treatment

                         The GLM Procedure
Dependent Variable: bp   Systolic Blood Pressure
                                 Sum of
Source                     DF    Squares      Mean Square   F Value
Model                       3   2082.488831   694.162944     14.04
Error                      12    593.261169    49.438431
Corrected Total            15   2675.750000
```

```
                       Source                Pr > F
                       Model                 0.0003
                       Error

                       Corrected Total

           R-Square     Coeff Var       Root MSE        bp Mean
           0.778282      5.664650       7.031247       124.1250


Source                        DF     Type I SS    Mean Square   F Value
TRT                            1    156.250000    156.250000       3.16
age(TRT)                       2   1926.238831    963.119416      19.48

                       Source                Pr > F
                       TRT                   0.1008
                       age(TRT)              0.0002


Source                        DF    Type III SS   Mean Square   F Value
TRT                            1    230.984810    230.984810       4.67
age(TRT)                       2   1926.238831    963.119416      19.48

                       Source                Pr > F
                       TRT                   0.0516
                       age(TRT)              0.0002
```

```
    Analysis of Blood Pressure by Treatment adjusting for Age      3
           Model using Separate Slopes for each Treatment

                       The GLM Procedure
Dependent Variable: bp    Systolic Blood Pressure

                                  Standard
Parameter                  Estimate         Error  t Value  Pr > |t|
Age 1                    0.78379878    0.27045757     2.90    0.0134
Age 2                    1.21660340    0.22006251     5.53    0.0001
Age 1 vs 2              -0.43280462    0.34867579    -1.24    0.2382
```

```
    Example of analysis BP by Treatment adjusting for Age          5
           Model using a Common Slope for both Treatments

                       The GLM Procedure

Dependent Variable: bp    Systolic Blood Pressure
                                   Sum of
Source                        DF     Squares    Mean Square   F Value
Model                          2   2006.315161  1003.157581     19.48
Error                         13    669.434839    51.494988
Corrected Total               15   2675.750000

                       Source                Pr > F
                       Model                 0.0001
                       Error
                       Corrected Total


           R-Square     Coeff Var       Root MSE        bp Mean
           0.749814      5.781270       7.176001       124.1250


Source                        DF     Type I SS    Mean Square   F Value
TRT                            1    156.250000    156.250000       3.03
age                            1   1850.065161   1850.065161      35.93
```

```
              Source                      Pr > F
              TRT                         0.1051
              age                         <.0001
```

Source	DF	Type III SS	Mean Square	F Value
TRT	1	700.293014	700.293014	13.60
age	1	1850.065161	1850.065161	35.93

```
              Source                      Pr > F
              TRT                         0.0027
              age                         <.0001
```

```
        Example of analysis BP by Treatment adjusting for Age       6
             Model using a Common Slope for both Treatments

                         The GLM Procedure
Dependent Variable: bp   Systolic Blood Pressure
                                          Standard
    Parameter                 Estimate       Error   t Value  Pr > |t|
    age                     1.04420215   0.17421019     5.99    <.0001


        Example of analysis BP by Treatment adjusting for Age       7
             Model using a Common Slope for both Treatments

                         The GLM Procedure
                        Least Squares Means

                     TRT        bp LSMEAN
                      1        131.165758
                      2        117.084242
```

From the above analyses, one can see that there is a significant treatment effect and a significant covariate (age) effect and that there is an overall model effect. However, when separate slopes for age are fit to the two treatments, a test of equality of the slopes was nonsignificant ($p \approx 0.24$). Thus, the "best" model among the ones presented above appears to be that given in Case 4 [equation (5.22)].

One interesting feature of the output from SAS is that is produces both Type I and Type III sums of squares. The Type I values give the values that are obtained when the model is fitted *sequentially* so that the statements, model bp = TRT age; and model bp = TRT; would produce the same Type I between treatment sums of squares (that is, 156.25). The Type III sums of squares are obtained from the *simultaneous* fit of the parameters. Hence, the *order* that the factors appear in the model statement will affect the Type I sums of squares but will not affect the Type III sums of squares. For a thorough discussion of Types $I - IV$ sums of square, see Searle [135].

5.7.2 *Calculation of Adjusted Means*

Once an ANCOVA model is fit, one can *adjust* the means of the groups of interest based on the model parameters. For instance, in Example 5.16, the mean value of systolic blood pressure for treatments 1 and 2 are 127.25 and 121, respectively. The mean ages are 37.625 and 45.125, respectively. The mean age over both groups is 41.375. Assuming that the model described equation (5.22) is the correct one, we can calculate *age adjusted means* for persons in each of the treatment groups. From equation (5.18), we can see that

$$\hat{\mu}_i = \overline{y}_{i.} - \hat{\beta}(\overline{X}_{i.} - \overline{X}_{..}).$$

For treatment 1,

$$\hat{\mu}_1 = \overline{y}_{1,\text{adjusted}} \approx 127.25 - 1.0442(37.625 - 41.375) \approx 127.25 + 3.92 = 131.17.$$

For treatment 2,

$$\hat{\mu}_2 = \overline{y}_{2,\text{adjusted}} \approx 121 - 1.0442(45.125 - 41.375) \approx 121 - 3.92 = 117.09.$$

Notice that, in this particular case, the age adjusted mean systolic blood pressures by group are further apart than the raw mean systolic blood pressures by group.

5.8 Exercises

1. For the data in Example 5.3,

(a) Plot the means and standard errors of the four groups.

(b) Test whether or not a quadratic trend exists.

(c) Based on the results given in the example itself and on those obtained in parts **(a)** and **(b)**, what are your conclusions?

test whether or not a quadratic trend exists.

2. In R, download the library called ISwR (Dalgaard [26]) and consider the data set called zelazo. That dataset, originally published by Zelazo, et al. [162] and later reproduced by Altman [4] and Dalgaard [26] compares four groups of infants with respect to age (in months) when they first walked. The four groups were a control group (C) measured at 8 weeks but not monitored; a group that got no training (N) but were monitored weekly from weeks 2 – 8, a "passive" training group (P), which received motor and social stimulation and monitored weekly from week 2 - 8; and an active training group (A), which received 4 3-minute walking and reflex training sessions daily from weeks 2 - 8.

(a) Calculate an ANOVA table and an F-test to test the hypothesis, $H_0 : \mu_C = \mu_N = \mu_P = \mu_A$ versus H_1: at least one μ_i is different.

(b) Perform each of the multiple comparisons procedures covered in sections 5.4.1 – 5.4.7. Display your results similar to those displayed in Figure 5.3.

(c) Perform Dunnett's test comparing the control group to each of the other groups. What is your conclusion? [Hint: For this problem, you can either use R directly or paste the data into a file and analyze the data in SAS. However, it may not be possible to perform all of the procedures in one package. Alternatively, by using the commands lapply zelazo 1 mean and lapply zelazo 1 sd, one can obtain all of the information needed and easily perform the calculations by hand (and the use of, say, R functions as pf and ptukey.)]

3. In the study by Kurban et al. [82] referred to in Example 5.1, brain cholesterol was also measured. The means $\mp$ standard deviations for that endpoint for each of the six groups are as follows: 22.58 ∓ 2.29 [C]; 30.51 ∓ 2.30 [Su]; 31.80 ∓ 2.90 [O]; 24.49 ∓ 2.00 [M]; 27.81 ∓ 1.71 [So]; and 24.43 ∓ 1.47 [B].

(a) Perform the same analyses done in Examples 5.2 – 5.4 for these data.

(b) Do the analyses presented in example **5.8** for the brain cholesterol endpoint.

(c) Perform Gabriel's method for the brain cholesterol data.

(d) Do Dunnett's test for the brain cholesterol data.

(e) State your conclusions. Are the conclusions the same as for the brain triglyceride endpoint?

4. Consider an investigation of Low Density Lipid (LDL) cholesterol in subjects with high cholestorol levels, half of whom are < 50 years of age and half of whom are $50+$ years of age. Their LDL measurements are taken when the study begins and then they are assigned to either a low-cholesterol diet or a normal diet. Their measurements are taken again after 6 weeks on the respective diets. The data are as follows:

Normal diet Age < 50		Normal diet Age $50+$		Low chol. diet Age < 50		Low chol. diet Age $50+$	
Pre	Post	Pre	Post	Pre	Post	Pre	Post
132	183	191	199	169	135	164	183
192	144	150	142	170	135	162	145
152	172	163	161	158	134	183	157
162	158	189	135	152	141	168	148
170	166	148	147	174	151	155	160
160	169	191	153	179	143	157	161
161	174	176	170	173	153	179	143
172	184	156	189	163	128	160	164

(a) Calculate the change in LDL (ΔLDL = Post LDL - Pre LDL) for each subject.

(b) Create an interact plot to visualize the interaction between diet in age for the ΔLDL endpoint (see Figure 5.6).

(c) Create an ANOVA table for the two-way factorial design for the outcome, ΔLDL (factors: diet, age, diet $\times$ age).

(d) Do a formal test of diet $\times$ age interaction. *If this term is not significant* at the $\alpha = 0.05$ level, then refit using an additive model.

(e) With the most appropriate model, test whether or not there are age and diet main effects.

(f) State your conclusions.

5. Consider the Anorexia data from *Small Data Sets*, data set # 285 in Hand et al. [61]. That data, originally provided by Brian Everitt, consists of pre- and postbody weights, in lbs., of 72 young girls receiving three different therapeutic approaches for treating anorexia. In the study, 29 girls received cognitive behavorial training [CBT], 26 received a control [C] treatment, and 17 received family therapy [FT]. The data can be obtained from the MASS library in R (Venables and Ripley [153]) or from http://www.stat.ncsu.edu/sas/sicl/data/.

(a) Overlay mean and standard error bars for the pre- and postweights for each of the three treatments. Does there appear to be any different patterns of pre and or postweights across treatment groups?

(b) Fit analysis of covariance models using the postbody weight as the outcome, the treatment as a class variable of interest and the prebody weight as a continuous covariate.

   (I) For the first model, test whether or not a significant pre body weight × treatment interaction exists. Do the test at the $\alpha = 0.05$ level. (Hint: Use treatment and prebody weight as main effects and also include an interaction term.)

   (II) If the overall interaction in (I) is *not* significant then drop that term and refit the model with just the main effects. If the main effects are not significant are not significant, refit again until only significant effects (at the $\alpha = 0.05$ level) are in the model.

   (III) For your "best" model, calculate the adjusted mean post body weight by treatment arm.

   (IV) State your conclusions.

(c) Overlay mean plots of the post- and pre-*differences* (i.e., postweight − preweight) by treatment arm.

(d) Fit an ANOVA model comparing the differences across treatment arms. Do a Dunnett's test comparing the CBT and FT groups to the C group. State your conclusions.

(e) Based on your results in parts (b) and (d), discuss the advantages and disadvantages of the ANCOVA approach (using post- and preweights *separately* in the model) versus the ANOVA approach, which uses the difference of the post- and prebody weights as the outcome.

CHAPTER 6

Discrete Measures of Risk

6.1 Introduction

In sections 1.8 and 1.11 of this book, we briefly introduced methods for comparing proportions of "discrete outcomes" or "events." In this chapter, we will expand on that discussion by introducing odds ratios and relative risk. We will limit the discussion to cases where the outcome is binary. We will also show how one can compare proportions (or odds ratios) of events between two or more groups while adjusting for a set of strata (discrete variables) and/or continuous variables which, if ignored, could influence or bias the inference made about the event proportions or odds ratios. This will be accomplished by the Mantel–Haenszel procedure and logistic regression. Finally, we'll look at paired binary measurements and discuss methods for quantifying and testing discordance and concordance among those pairs.

6.2 Odds Ratio (OR) and Relative Risk (RR)

Let p = the probability of a success in a Bernoulli trial and let $q = 1 - p$. The *odds in favor of success* are defined as Odds $\equiv \frac{p}{1-p} = \frac{p}{q}$.

Now, let D = the event of disease, $\overline{D}$ = the event of no disease, E = the event of exposure, $\overline{E}$ = the event of no exposure. We can then arrange a contingency table of the number of events and nonevents by exposure level:

	E	$\overline{E}$	Total
D	a	b	$n_{1\bullet} = a + b$
$\overline{D}$	c	d	$n_{2\bullet} = c + d$
Total	$n_{\bullet 1} = a + c$	$n_{\bullet 2} = b + d$	$N = n_{\bullet\bullet} = a + b + c + d$

Suppose, now, that p_1 and p_2 are the proportions of "success" (say, disease) in two groups of, say, exposure levels (e.g., Yes / No). Then, the odds *ratio* is defined as $OR \equiv \frac{p_1/q_1}{p_2/q_2} = \frac{p_1 q_2}{p_2 q_1}$. The *estimated* odds ratio is $\widehat{OR} = \frac{\hat{p}_1 \hat{q}_2}{\hat{p}_2 \hat{q}_1} = \frac{a/c}{b/d} = \frac{ad}{bc}$. This is also known as the estimate of the *crude* odds ratio.

The relative risk (RR) or risk ratio is defined to be $RR = \frac{\Pr(D|E)}{\Pr(D|\bar{E})}$. Its estimate is $\widehat{RR} \equiv \frac{a/(a+c)}{b/(b+d)} = \frac{a(b+d)}{b(a+c)}$.

Example 6.1. Evans County (Georgia) Heart Disease Study (1960 – 1969). This study was presented earlier in Kleinbaum, Kupper, Morgenstern [79]. In the study, 609 males in age range of 40–76 years old were followed. The objective was to assess the putative association of endogenous catecholamines (CAT) with coronary heart disease (CHD) incidence in a seven-year period.

		CAT	
CHD	High	Low	Total
Yes	27	44	71
No	95	443	538
	122	487	609

$\widehat{OR} = \frac{27(443)}{44(95)} \approx 2.861, \quad \widehat{RR} = \frac{27/122}{44/487} \approx 2.450.$

6.2.1 Confidence Intervals for Odds Ratio – Woolf's Method

The two most common methods for determining the $(1 - \alpha)\%$ confidence intervals for odds ratios are Woolf's method and Miettenen's test-based method.

Woolf's method is based on transforming the estimated odds ratio using a natural logarithm (denoted $\ln(\widehat{OR})$). The transformation is asymptotically normally distributed with $\mathrm{var}[\ln(\widehat{OR})] \approx \frac{1}{a} + \frac{1}{b} + \frac{1}{c} + \frac{1}{d}$. Then the $(1 - \alpha)\%$ confidence interval is given as

$$\left(e^{\ln(\widehat{OR}) - Z_{1-\frac{\alpha}{2}} \sqrt{\frac{1}{a} + \frac{1}{b} + \frac{1}{c} + \frac{1}{d}}}, e^{\ln(\widehat{OR}) + Z_{1-\frac{\alpha}{2}} \sqrt{\frac{1}{a} + \frac{1}{b} + \frac{1}{c} + \frac{1}{d}}} \right) \tag{6.1}$$

Woolf's method has the property that for the calculation of confidence intervals for odds ratios, it generally yields confidence intervals with closer approximations to nominal levels than does the test-based method. Simple modifications of equation (6.1) will yield confidence intervals for $\widehat{RR}$.

6.2.2 Confidence Intervals for Odds Ratio – Miettenen's Method

Two-sided $100 \times (1 - \alpha)\%$ confidence intervals for the test-based method are given by

$$\begin{cases} \left(\widehat{OR}^{1-\sqrt{\chi^2_{1,1-\alpha}/X^2}}, \widehat{OR}^{1+\sqrt{\chi^2_{1,1-\alpha}/X^2}} \right) & \text{if } \widehat{OR} \geq 1; \\ \left(\widehat{OR}^{1+\sqrt{\chi^2_{1,1-\alpha}/X^2}}, \widehat{OR}^{1-\sqrt{\chi^2_{1,1-\alpha}/X^2}} \right) & \text{if } \widehat{OR} < 1, \end{cases} \tag{6.2}$$

where $X^2 = N\left[|ad - bc| - \frac{N}{2}\right]^2 / \left[(a+b)(c+d)(a+c)(b+d)\right]$.

The advantage of the test-based method is that $(1 - \alpha)\%$ confidence intervals are consistent with the 1 results of a χ^2 test. That is, if a hypothesis testing the equality of two proportions is rejected at some α level, then the $(1 - \alpha)\%$ confidence interval will not include 1.0. Conversely, if the hypothesis does not reject the equality of the proportions at level α, then the $(1 - \alpha)\%$ confidence will include 1.0. Kleinbaum, Kupper, and Morgenstern [79] point out that test-based confidence intervals can be used for any type of ratio of ratios. Thus, in equation (6.2), we could substitute, e.g., $\widehat{RR}$ for $\widehat{OR}$.

Example 6.2. Using Woolf's method, from Example 6.1, we get

$$\widehat{OR} \approx 2.861 \implies \ln(\widehat{OR}) \approx 1.051 \text{ and } var[\ln(\widehat{OR})] \approx \frac{1}{27} + \frac{1}{44} + \frac{1}{95} + \frac{1}{443} \approx .0725$$

$$\implies 95\% \text{ C.I. } \approx \left(e^{1.051 - 1.96\sqrt{.0725}}, e^{1.051 + 1.96\sqrt{.0725}}\right) \approx \left(e^{0.5234}, e^{1.5793}\right) \approx \left(1.688, 4.851\right)$$

Using Miettenen's test-based method, $X^2 = 609\left[|27(443) - 95(44)| - \frac{609}{2}\right]^2 / \left[71 \cdot 538 \cdot 122 \cdot 487\right] \approx 15.000.$

$$\implies 95\% \text{ C.I. } \approx \left(2.8615^{1 - \sqrt{\frac{3.84}{15.000}}}, 2.8615^{1 + \sqrt{\frac{3.84}{15.000}}}\right) \approx \left(2.8615^{0.494}, 2.8615^{1.506}\right)$$

$$\approx \left(1.681, 4.871\right).$$

6.3 Calculating Risk in the Presence of Confounding

A problem with measures such as the crude odds ratio is that they do not adjust for any *confounding* variables. In general, a confounding variable is defined as one which, along with the variable(s) of primary interest, is associated with the outcome (say, disease) and which can change the nature of the relationship between the outcome and the variable(s) of interest. Confounding variables are not of primary interest to the investigator (rather, they *confound* the investigator).

- Example #1: When associating cigarette smoking with cancer or heart disease, age and sex are confounders.
- Example #2: When associating passive smoking with, for example, cancer risk, one's own smoking status is a confounder.

Three ways of controlling for confounding are:

1. *Matching* (pairing or blocking) individuals on the confounding factors.
2. *Stratifying* (either at randomization time or post hoc) by the confounding factors.

3. Modeling confounding factors post hoc using such techniques as the *Mantel–Haenszel procedure* or *logistic regression*.

Matching involves ensuring that for each individual in one risk set, there is an (set of) individual(s) in the other risk set(s) who has (have) roughly the same value(s) with respect to the confounding variable(s). Stratification involves ensuring that the proportions of individuals in each risk set are roughly equal with respect to having similar values of the confounding variable(s). Using logistic regression, we attempt to compare the two risk groups while *adjusting for* the confounding variable(s). *It is always preferable to ensure that the confounding factors are identified and adjusted for prior to the beginning of a study.* If possible, one should have control over the design of the experiment to guarantee that known confounders are properly controlled for.

The Mantel–Haenszel procedure is used to test $H_0 : p_1 = p_2$ versus, say, $H_1 : p_1 \neq p_2$ *adjusting for* k levels of a set of confounding variables. If there are a reasonably small number of discrete levels associated with the confounding factors, then the Mantel–Haenszel adjustment is often preferred over logistic regression. Logistic regression (introduced in section 6.4) is preferable when there are many levels of the discrete confounders or if the confounders are continuous variables.

<u>Algorithm for Performing the Mantel–Haenszel Procedure</u>

1. Form k strata from the confounding variable(s).
2. Compute the total number of observed units a particular cell, say, the $(1, 1)^{th}$ cell, that is, the cell in the upper left-hand corner of a contingency table. Make sure to total *over all of the k strata*.

a_1	b_1
c_1	d_1

Stratum 1

a_2	b_2
c_2	d_2

Stratum 2

$\cdots$

a_i	b_i
c_i	d_i

Stratum i

$\cdots$

a_k	b_k
c_k	d_k

Stratum k

3. Compute the observed and expected number of units in the k strata:

$$O = \sum_{i=1}^{k} O_i = \sum_{i=1}^{k} a_i, \quad E = \sum_{i=1}^{k} E_i = \sum_{i=1}^{k} \frac{(a_i + b_i)(a_i + c_i)}{N_i}. \qquad (6.3)$$

4. Compute $V = \mathrm{Var}(O - E)$:

$$V = \sum_{i=1}^{k} V_i = \sum_{i=1}^{k} \left[\frac{(a_i + b_i)(c_i + d_i)(a_i + c_i)(b_i + d_i)}{N_i^2(N_i - 1)} \right] \qquad (6.4)$$

5. Form $Z_{MH}^2 = \frac{\left(|O - E| - \frac{1}{2} \right)^2}{V} \dot{\sim} \chi_1^2$ under H_0. For a hypothesis test, reject H_0 if $Z_{MH}^2 \geq \chi_1^2$; otherwise, do not reject. A p-value is calculated as $p = \Pr(\chi_1^2 > Z_{MH}^2)$.

6. Use the large sample approximation if $V \geq 5$.

Notice that the requirement that $V \geq 5$ does *not* preclude situations where there is relatively little information in a specific stratum. If the total number of observations in a given cell is relatively large across all strata, then the large sample approximation used in the M–H procedure is quite good.

Example 6.3. Evans county heart disease study (continued)

The data relating CAT level to incidence of CHD were further broken down as follows:

Observed

Men < 55 years old

CAT

CHD	High	Low	
Yes	4	24	28
No	21	309	330
	25	333	358

Men > 55 years old

CAT

CHD	High	Low	
Yes	23	20	43
No	74	134	208
	97	154	251

Expected

Men < 55 years old

CAT

CHD	High	Low	
Yes	1.955	26.045	28
No	23.045	306.955	330
	25	333	358

Men > 55 years old

CAT

CHD	High	Low	
Yes	16.618	26.382	43
No	80.382	127.618	208
	97	154	251

We want to test:

$H_0 : \Pr(\text{CHD}|\text{CAT}) = \Pr(\text{CHD}|\overline{\text{CAT}})$ versus $H_1 : \Pr(\text{CHD}|\text{CAT}) \neq \Pr(\text{CHD}|\overline{\text{CAT}})$ *adjusted for* age.

From the $(1, 1)^{th}$ cell, $O = 4 + 23 = 27$, $E \approx 1.955 + 16.618 = 18.573$, $V \approx 1.681 + 8.483 = 10.164 \geq 5 \Longrightarrow$ use the large sample approximation.

$$Z^2_{MH} = \frac{\left(|O - E| - \frac{1}{2}\right)^2}{V} = \frac{\left(7.927\right)^2}{10.164} \approx 6.183$$

$\Longrightarrow$ we reject H_0 at the $\alpha = .05$ $(p = 0.013)$ and, therefore, even after adjusting for age, we declare a significant difference in the amount of endogenous catecholemines in those men having coronary heart disease and those not having coronary heart disease. If no continuity correction is implemented then the

$$Z^2_{MH} = \frac{\left(|O - E|\right)^2}{V} = \frac{\left(8.427\right)^2}{10.164} \approx 6.987 \Longrightarrow p \approx 0.008$$

The continuity corrected Mantel–Haenszel procedure can be implemented in R with the following code:

```
CHD.risk <- as.table(array(c(4, 21, 24, 309, 23, 20, 74, 134), dim = c(2, 2, 2),
            dimnames = list("CHD" = c("Yes", "No"), "Cat" = c("High", "Low"),
            Age = c("Young", "Old")))))
CHD.risk
mantelhaen.test(CHD.risk,correct=TRUE)
```

In SAS, the Mantel–Haenszel procedure *without* the continuity correction can be implemented with the following code:

```
data chap6_2;
input agecat $ chd $ cat $ count@@;
datalines;
young yes high 4   young yes low 24   young no high 21   young no low 309
old yes high 23   old yes low 20   old no high 74   old no low 134
; run;

proc freq data=chap6_2 order=data;
table agecat*chd*cat/binomialc cmh1;
weight count; run;
```

6.3.1 Mantel–Haenszel Adjusted Odds Ratio (Test-based Method)

Another way to quantify the risk of a given factor while adjusting for confounders is to use the Mantel–Haenszel odds ratio and its $100 \times (1 - \alpha)\%$ confidence intervals:

$$\widehat{OR}_{MH} = \frac{\sum_{i=1}^{k} a_i d_i / N_i}{\sum_{i=1}^{k} b_i c_i / N_i} \tag{6.5}$$

The two-sided $100 \times (1 - \alpha)\%$ confidence intervals are given by

$$\begin{cases} \left(\widehat{OR}_{MH}^{1-\sqrt{\chi^2_{1,1-\alpha}/Z^2_{MH}}}, \widehat{OR}_{MH}^{1+\sqrt{\chi^2_{1,1-\alpha}/Z^2_{MH}}} \right) & \text{if } \widehat{OR}_{MH} \geq 1; \\ \left(\widehat{OR}_{MH}^{1+\sqrt{\chi^2_{1,1-\alpha}/Z^2_{MH}}}, \widehat{OR}_{MH}^{1-\sqrt{\chi^2_{1,1-\alpha}/Z^2_{MH}}} \right) & \text{if } \widehat{OR}_{MH} < 1. \end{cases} \tag{6.6}$$

Woolf's method for calculated confidence intervals for adjusted odds ratio is much more involved and is not presented here. One can get a detailed description of this method in Hosmer and Lemeshow [70].

Example 6.4. Evans county heart disease study (continued)

The adjusted odds ratio in our example is

$$\frac{\frac{4(309)}{358} + \frac{23(134)}{251}}{\frac{24(21)}{358} + \frac{20(74)}{251}} \approx \frac{15.731}{7.304} \approx 2.154$$

The 95% confidence intervals are given by

$$\left(2.154^{1-\sqrt{3.84/6.183}}, 2.154^{1+\sqrt{3.84/6.183}} \right) \approx \left(1.177, 3.943 \right).$$

Thus, the summary of the 95% confidence intervals for the crude odds ratio versus the odds ratio *adjusted for age* is as follows:

$$\text{Crude } \widehat{OR} : 2.861, \quad 95\% \text{ C.I.} : \left(1.681, 4.871\right)$$

$$\text{AGE Ajusted } \widehat{OR} : 2.154, \quad 95\% \text{ C.I.} : \left(1.177, 3.943\right)$$

6.3.2 Simpson's Paradox: A Fallacy in Interpreting Risk

Before going further, we caution the reader that the proper *interpretation* of risk, especially in complex datasets, can sometimes be elusive. One situation where risk can easily be misinterpreted involves assuming that inferences about subsets of individuals can be combined to make inferences about the complete population. This particular pittfall is known as *Simpson's paradox* and is demonstrated in the following example.

Example 6.5. (Simpson's paradox) Consider the following example where a new treatment appears to be better than a standard treatment in both men and women but not when the data are combined.

	Men				Women		
Result	New trt	Std trt	Total	Result	New trt	Std trt	Total
Effective	24	6	30	Effective	16	42	58
Not effective	36	14	50	Not effective	4	18	22
Total	60	20	80	Total	20	60	80

Thus, for men, the new treatment was effective for 40% and the standard treatment was effective for 30% leading to the conclusion that the new treatment is better for men. Likewise, for women, using the new treatment, the new treatment was effective for 80% whereas the standard treatment was effective for 70% leading to the conclusion that the new treatment is better for women as well. Hence, one might infer at this point that the new treatment was better. However, *combining the two subsets* yields:

	All patients		
Result	New trt	Std trt	Total
Effective	40	48	88
Not effective	40	32	72
Total	80	80	160

The results indicate that for those using the new treatment was effective for 50% whereas standard treatment was effective for 60% from which one would conclude that in the complete population, the standard treatment is better. But, why does one come to these opposite conclusions?

One can start to unravel this paradox by observing that the sampling scheme for the standard treatment is weighted towards the group that benefits most, that is, the women.

6.4 Logistic Regression

Another way to adjust for confounding factors is to include them as covariates in a regression model. For example, if $p =$ the probability of an outcome (e.g., occurrence of a disease) then a possible model could be of the form

$$E(p) = \beta_0 + \beta_1 X_1 + \ldots + \beta_{k-1} X_{k-1} = \underbrace{\mathbf{x}}_{1 \times k} \underbrace{\boldsymbol{\beta}}_{k \times 1}, \tag{6.7}$$

where "E" stands for expected value. However, the above model has the property that sometimes the right-hand side [RHS] of 6.7 can be < 0 or > 1, which *is meaningless* when we were speaking of probability. (Recall that probabilties must be between 0 and 1, inclusively.) Since the values on the RHS of equation (6.7) can potentially take on values between $-\infty$ and ∞, we must somehow transform both sides of the equation so that a very large (or small) value on the RHS of (6.7) maps onto a value between 0 and 1, inclusive.

One convenient transformation is that called the *logistic* or *logit* transformation of p. If we first note that the odds measure, $\frac{p}{q} = \frac{p}{1-p}$, has the property that $0 \leq \frac{p}{1-p} < \infty$, then we can see that if we take the natural log of the odds we have the property that $-\infty < \ln\left(\frac{p}{1-p}\right) < \infty$. Thus, instead of using the model represented in (6.7), we use

$$\text{logit}(p) = \ln\left(\frac{p}{1-p}\right) = \ln\left(\frac{p}{q}\right) = \ln(\text{odds}) = \beta_0 + \beta_1 X_1 + \ldots + \beta_{k-1} X_{k-1} \tag{6.8}$$

In order to translate the logistic model to a value that is associated with a probability, we can solve for p in equation (6.8). For example, equation (6.8)

$$\Longrightarrow \frac{p}{1-p} = \exp\left(\beta_0 + \beta_1 X_1 + \ldots + \beta_{k-1} X_{k-1}\right) = \exp(\mathbf{x}\boldsymbol{\beta})$$

$$\Longrightarrow p = (1-p)\exp(\mathbf{x}\boldsymbol{\beta}) \Longrightarrow p + p\exp(\mathbf{x}\boldsymbol{\beta}) = \exp(\mathbf{x}\boldsymbol{\beta})$$

$$\Longrightarrow p = \frac{\exp(\mathbf{x}\boldsymbol{\beta})}{1 + \exp(\mathbf{x}\boldsymbol{\beta})} = \frac{\exp\left(\beta_0 + \beta_1 X_1 + \ldots + \beta_{k-1} X_{k-1}\right)}{1 + \exp\left(\beta_0 + \beta_1 X_1 + \ldots + \beta_{k-1} X_{k-1}\right)} \tag{6.9}$$

Example 6.6. Evans county heart disease study (continued)

Define the variables ECG and CAT as

$$ECG = \begin{cases} 1, & \text{if the electrocardiogram is abnormal} \\ 0, & \text{otherwise} \end{cases}$$

and

$$CAT = \begin{cases} 1, & \text{if the endogenous catecholemine level is high} \\ 0, & \text{otherwise} \end{cases}$$

A logistic model was fit to ascertain the joint effects of catecholemine concentration, age, and ECG status on the *probability* of having coronary heart disease (CHD) in a seven-year period. AGE here is modeled as a continuous quantity.

$$\text{logit}\left[\Pr(CHD|CAT, AGE, ECG)\right] = \beta_0 + \beta_1(CAT) + \beta_2(AGE) + \beta_3(ECG)$$

The results of the model were as follows:

Source	$\hat{\beta}_i$	$se(\hat{\beta}_i)$	$\hat{\beta}_i/se(\hat{\beta}_i)$	χ^2	p-value
Intercept	-3.9110	0.8004	-4.89	23.88	$< .001$
CAT	0.6516	0.3193	2.04	4.16	0.041
AGE	0.0290	0.0146	1.99	3.94	0.047
ECG	0.3423	0.2909	1.18	1.38	0.239

Hence, men with increased catecholemine levels and at greater ages had significantly higher risks of having CHD over a seven-year period.

Suppose, now, that we want to obtain an odds ratio (odds *in favor of success* of individual 1 with a covariate "vector," $\mathbf{x}_1$, versus individual 2 with a covariate "vector," $\mathbf{x}_2$. Suppose, further that the vectors differ only in the jth component. Thus,

$$\ln\left[\frac{p_1/q_1}{p_2/q_2}\right] = \ln\left(\frac{p_1}{q_1}\right) - \ln\left(\frac{p_2}{q_2}\right) = \mathbf{x}_1\boldsymbol{\beta} - \mathbf{x}_2\boldsymbol{\beta} = \left(\mathbf{x}_1 - \mathbf{x}_2\right)\boldsymbol{\beta}$$

$$= \left[(1 \quad x_1 \quad \cdots \quad x_j + \Delta \quad \cdots \quad x_{k-1}) - (1 \quad x_1 \quad \cdots \quad x_j \quad \cdots \quad x_{k-1})\right]\begin{pmatrix} \beta_0 \\ \beta_1 \\ \vdots \\ \beta_j \\ \vdots \\ \beta_{k-1} \end{pmatrix}$$

$$= \underbrace{(0 \quad \cdots \quad 0 \quad \Delta \quad 0 \quad \cdots \quad 0)}_{1 \times k}\underbrace{\begin{pmatrix} \beta_0 \\ \beta_1 \\ \vdots \\ \beta_j \\ \vdots \\ \beta_{k-1} \end{pmatrix}}_{k \times 1} = \underbrace{\Delta\beta_j}_{1 \times 1} \text{ (a scalar quantity).}$$

Calculating $\frac{p_1/q_1}{p_2/q_2} = e^{\Delta\beta_j}$ gives the odds ratio in favor of success for individual 1 having the jth covariate value $x_j + \Delta$ versus individual 2 having the jth covariate value x_j. This calculation *adjusts for all other* covariates.

Example 6.7. Evans county heart disease study (continued)

In our example, the vector of covariate values include catecholamine status, age and ECG status. Suppose individual 1 has *high CAT* values and individual 2 has *low CAT* values while all other covariate values are equal in the two individuals. Then

$$\widehat{OR} = e^{\Delta \beta_1} = e^{(1-0)(0.6516)} \approx 1.919$$

The approximate two-sided 95% confidence interval for the odds ratio is (assuming $\Delta \geq 0$) is given by

$$\left(e^{\left[\hat{\beta}_i - z_{1-\alpha/2} se(\hat{\beta}_i) \right] \Delta}, e^{\left[\hat{\beta}_i + z_{1-\alpha/2} se(\hat{\beta}_i) \right] \Delta} \right).$$

Thus, for the data given in the Evans county heart disease study, the 95% confidence interval is given by

$$\left(e^{0.6516 - 1.96(0.3193)}, e^{0.6516 + 1.96(0.3193)} \right) \approx \left(e^{0.0258}, e^{1.2774} \right) \approx \left(1.026, 3.587 \right).$$

In this particular example, $\Delta = 1$. However, Δ could take on any value. If we compare the point estimate and 95% confidence intervals of the odds ratio for high CAT versus low CAT (adjusted for AGE and ECG status) to the crude values and the values adjusted for AGE alone, we get the following summary:

$$\text{Crude } \widehat{OR} : 2.861, \quad 95\% \text{ C.I.} : \left(1.681, 4.871 \right)$$

$$\text{AGE Ajusted } \widehat{OR} : 2.154, \quad 95\% \text{ C.I.} : \left(1.177, 3.943 \right)$$

$$\text{AGE}^* \text{ and ECG Ajusted } \widehat{OR} : 1.919, \quad 95\% \text{ C.I.} : \left(1.026, 3.587 \right)$$

* AGE is a continuous variable in the Logistic Model

In all cases, the two-sided 95% confidence intervals for the odds ratio do *not* include 1. Thus, catecholamine status has a significant effect on the probability of having coronary heart disease after adjusting for other risk factors. However, we also see that when the other risk factors are adjusted for, the effect of CAT status on the probability of having coronary heart disease (CHD) is attenuated quite a bit.

Finally, using the logistic model one can calculate the estimated probability of having CHD in the seven-year period. For example, if the estimated probability of a 51-year old man with high CAT and normal ECG having CHD in the next seven years would be

$$\hat{p} = \frac{e^{-3.9110 + 0.6516(1) + 0.029(51)}}{1 + e^{-3.9110 + 0.6516(1) + 0.029(51)}} \approx \frac{e^{-1.7804}}{1 + e^{-1.7804}} \approx \frac{0.1686}{1.1686} \approx .144.$$

Thus, according to our model, such an individual would have about a 14.4% chance of developing CHD in the next seven years. A complete description of the all aspects of this analysis is given in Kleinbaum, Kupper, and Morgenstern [79].

6.5 Using SAS and *R* for Logistic Regression

Both SAS and *R* provide very useful algorithms for conducting logistic regression. We only provide a skeleton presentation here but for interested readers, books by Allison [3] and Everitt and Hothorn [39] for fitting and interpreting logistic regression models in SAS and *R*, respectively, are recommended. One caveat for implementation of logistic regression and other types of models with outcomes that are discrete in *R* via the function glm is that one should be somewhat familiar with *generalized linear models*. A very short overview of such models are given by Everitt and Hothorn [39]. A more theoretical exposition of this topic is given in McCullagh and Nelder [94]. For the purposes of fitting binary logistic regression via glm, we note that because we are interested in modeling proportions of *binary events* (e.g., 0's and 1's), the family of distributions, which is apropos is the binomial as is indicated in the example below.

Example 6.8. Consider the data called malaria in the library in *R* called ISwR. To do this, download ISwR from the web, load it by typing library(ISwR) at the *R* console, then type ?malaria at the *R* console. The dataset consist of measurements on a random sample of 100 children aged 315 years from a village in Ghana. Each child's age and the level of an antibody was measured at the beginning of the study. After 8 months, each child was assessed for symptoms of malaria.

The goal of this analysis is to assess whether or not age and antibody levels are related to the 8-month incidence of malaria symptoms. An *R* session to do this task is given below.

```
> library(ISwR)
> attach(malaria)
> names(malaria)
[1] "subject" "age"     "ab"      "mal"
> apply(cbind(age,ab),2,summary) #<- summary info for the 2 designated columns
           age      ab
Min.       3.00     2.0
1st Qu.    5.75    29.0
Median     9.00   111.0
Mean       8.86   311.5
3rd Qu.   12.00   373.8
Max.      15.00  2066.0
> table(mal)
mal
 0  1
73 27

> fit.malaria<-glm(mal~age+ab,family=binomial)
> summary(fit.malaria)

Call:
glm(formula = mal ~ age + ab, family = binomial)

Deviance Residuals:
    Min       1Q   Median       3Q      Max
-1.1185  -0.8353  -0.6416   1.2335   2.9604

Coefficients:
              Estimate Std. Error z value Pr(>|z|)
(Intercept)  0.152899   0.587756   0.260   0.7948
age         -0.074615   0.065438  -1.140   0.2542
ab          -0.002472   0.001210  -2.042   0.0412 *
```

```
---
Signif. codes:   0 *** 0.001 ** 0.01 * 0.05 . 0.1   1

(Dispersion parameter for binomial family taken to be 1)

    Null deviance: 116.65  on 99  degrees of freedom
Residual deviance: 105.95  on 97  degrees of freedom
AIC: 111.95

Number of Fisher Scoring iterations: 6
```

Note that antibody level was significantly related to the probability of having symptoms of malaria but age was not. One could also do this is SAS by exporting the data into a file external to R and then reading that file into SAS. To do easily do this, one can use the library foreign written by Thomas Lumley and Stephen Weigand. The R session would look similar to the session below.

```
> library(foreign)
> ?write.foreign
> setwd("c:\\programs")
> getwd()
[1]  "c:/programs"
> write.foreign(malaria,"malaria.dat","malaria.sas",package="SAS")
```

A data file is created and the following SAS code is created (in a different display) to read in the data from the created file.

```
* Written by R;
*  write.foreign(malaria, "malaria.dat", "malaria.sas", package = "SAS") ;
DATA  rdata ;
INFILE  "c:\book\programs\malaria.dat"  DSD  LRECL= 16 ;
INPUT subject age ab mal; RUN;
```

One can then use proc logistic in SAS to fit the logistic regression.

```
proc logistic data=malaria;
model mal (events='1')=ab age;
run;
```

Note that in the model statement, one needs to specify '(events='1') to ensure that the risk of malaria *events* are being modeled. The default for SAS is to use the level, 0, as the event instead of 1. This is usually the opposite of how health data are coded.

6.6 Comparison of Proportions for Paired Data

6.6.1 McNemar's Test

Up to this point, we've only considered count or frequency data for independent groups similar to a parallel groups design described earlier. Consider now a design where data are paired; either by matching individuals with similar characteristics or by using the same individual twice. Each member of a pair may receive, for example, a different treatment. The outcome of interest may be an event (e.g., illness, death) and we are interested in testing which treatment is more effective with respect to that event. Thus, our interest may be in testing the efficacy in the pairs for which *one treatment is effective while the other one is not and vice versa* (called *discordant pairs*). The pairs for which both or neither treatment is effective (called *concordant pairs*) do not give any information about the differential effectiveness between the two treatments.

Since the observation (cell) counts described represent *pairs* of observations, the typical χ^2 or Fisher's exact analyses are not apropos. Instead, we'll use *McNemar's* test. The formal hypothesis of interest for employing McNemar's test is $H_0 : p = \frac{1}{2}$ versus, say, $H_1 : p \neq \frac{1}{2}$, where p represents the proportion of *discordant* pairs for which treatment 1 is effective but treatment 2 is not. The expected number of pairs under H_0 for which treatment 1 is more effective than treatment 2 is $\frac{n_D}{2}$ where n_D is total number of discordant pairs. If $X = $ the number of pairs for which treatment 1 is effective and 2 is not, then $X \sim b(n_D, \frac{1}{2})$ under the null hypothesis, H_0. Thus, as stated above, if H_0 is true, then $E(X) = \frac{n_D}{2}$ and $Var(X) = n_D(\frac{1}{2})(\frac{1}{2}) = \frac{n_D}{4}$. If $\frac{n_D}{4} \geq 5$, or, equivalently, $n_D \geq 20$, then a normal approximation to the binomial distribution is reasonably accurate and so, can be used to test H_0 versus H_1 whereas if $\frac{n_D}{4} < 5$ (or $n_D < 20$) then an exact method is usually recommended.

For the normal approximation method, we form

$$Z = \frac{|n_1 - \frac{n_D}{2}| - \frac{1}{2}}{\frac{1}{2}\sqrt{n_D}} = \frac{2(|n_1 - \frac{n_D}{2}| - \frac{1}{2})}{\sqrt{n_D}}, \tag{6.10}$$

which, if H_0 is true, then $Z \stackrel{\cdot}{\sim} N(0,1)$. Another way to do this is to form

$$Z^2 = \frac{\left(|n_1 - \frac{n_D}{2}| - \frac{1}{2}\right)^2}{\frac{n_D}{4}} = \frac{4(|n_1 - \frac{n_D}{2}| - \frac{1}{2})^2}{n_D}, \tag{6.11}$$

which, if H_0 is true, then $Z^2 \stackrel{\cdot}{\sim} \chi_1^2$. It should be noted that some versions of McNemar's test do not include the continuity correction so that the $-1/2$ in the numerators of (6.10) and (6.11) are excluded in the calculation of the test statistic.

If the alternative hypothesis is *one*-sided then the rule for rejecting H_0 (or not) is as follows:

$$\begin{cases} \text{If } |Z| \geq Z_{1-\alpha} & \text{then } \textit{reject } H_0 \text{ ;} \\ \text{Otherwise} & \textit{do not} \text{ reject } H_0. \end{cases}$$

For a two-sided alternative the rule can be written either as

$$\begin{cases} \text{If } |Z| \geq Z_{1-\alpha/2} & \text{then } \textit{reject } H_0 \text{ ;} \\ \text{Otherwise} & \textit{do not} \text{ reject } H_0. \end{cases}$$

or

$$\begin{cases} \text{If } Z^2 \geq \chi_{1,1-\alpha}^2 & \text{then } \textit{reject } H_0 \text{ ;} \\ \text{Otherwise} & \textit{do not} \text{ reject } H_0. \end{cases}$$

As mentioned above, in cases where the data are too sparse for the normal approximation to be accurate, an *exact* method is employed. Using this method for testing $H_0 : p = \frac{1}{2}$ vs $H_1 : p \neq \frac{1}{2}$, one calculates the p-value as

$$2 \times Pr(X \leq n_1) = 2 \sum_{k=0}^{n_1} \binom{n_D}{k} \left(\frac{1}{2}\right)^{n_D} \quad \underline{\text{if }} n_1 < n_D/2. \tag{6.12}$$

or as

$$2 \times Pr(X \geq n_1) = 2 \sum_{k=n_A}^{n_D} \binom{n_D}{k} \left(\frac{1}{2}\right)^{n_D} \quad \underline{\text{if}} \; n_1 \geq n_D/2. \qquad (6.13)$$

If the alternative is *one-sided*, then drop the factor 2 from the calculations in equations (6.12) and (6.13).

Then the rejection rule that follows is given as

$$\begin{cases} \text{If } p-\text{value} \leq \alpha & \text{then } \textit{reject } H_0 \; ; \\ \text{Otherwise} & \textit{do not } \text{reject } H_0. \end{cases}$$

Example 6.9. Suppose that 300 men are carefully matched into pairs based on similar ages, demographic and clinical conditions to assess the effects of two drug regimens, A and B, on the treatment of AIDS. Thus, there are a total of 150 pairs. In 55 pairs, both regimens are effective; in 75 pairs, neither drug is effective; in 8 pairs, regimen A is effective whereas regimen B is not; and in 22 pairs drug B is effective whereas regimen A is not.

A contingency table for the pairs of observations is presented below.

		Regimen B	
		Effective	Ineffective
Regimen A	Effective	55	8
	Ineffective	22	75

For the hypothetical AIDS data given above, $n_D = 30 \Longrightarrow$ a normal approximation is a reasonably good approximation. We're testing a two-sided alternative hypothesis, so we must calculate $Z = Z = \dfrac{2(|8 - 15| - 1/2)}{\sqrt{30}} \approx \dfrac{13}{5.4772} \Longrightarrow Z \approx 2.37346 > 1.96 = Z_{.975}$ (or $p \approx 0.018$). If the χ^2 version of the test were used then we would have $Z^2 = (2.37346)^2 \approx 5.6333 > 3.84 = \chi^2_{1,.95}$ and would arrive at the exact same conclusion. Either way, we reject H_0 at the $\alpha = .05$ level and we conclude that there regimen B is significantly superior to regimen B. An R session to do this test is given below.

```
> x<-matrix(c(55, 22, 8, 75),nrow=2)
> mcnemar.test(x)
        McNemar's Chi-squared test with continuity correction
data:   x
McNemar's chi-squared = 5.6333, df = 1, p-value = 0.01762
```

As can be seen from Example 6.9, one particular disadvantage of McNemar's test is that it only uses part of the information (that is, the discordant pairs) to draw an inference. One argument for the strategy employed by McNemar's test, however, is that the information regarding the *differences* in regimens is contained in the discordant pairs. However, the bigger picture (in this hypothetical situation) is that even though regimen B has a statistically significant advantage over regimen A, neither regimen is effective in half of the pairs of men considered.

6.6.2 Cohen's Kappa Statistic

In the previous section, we were interested in examining paired data for analyzing differences in discordant pairs of events. However, it is often the case where one wishes to analyze *concordant* pairs. Such applications may arise in determining the reliability of repeated administrations of the same survey or the reproducibility of a clinical or laboratory test. One method for determining the degree of agreement was proposed by Cohen [19]. His formulation takes into account the *agreement by chance*. For example, if two people who have no pathologic training are reading the same set of slides then, by *chance*, they will agree on the diagnosis for a certain proportion of the slides.

The kappa statistic can be written as follows:

$$\kappa = \frac{p_C - p_E}{1 - p_E} \tag{6.14}$$

where p_C is the proportion of pairs that are concordant and p_E is the proportion of agreement between two pairs expected by chance. In some cases, the κ statistic is expressed as a percentage so that the right-hand side of equation (6.14) would be multiplied by 100%. (We will not do that in this section.) A demonstration of how a calculation of the κ statistic is performed is given in the example below.

Example 6.10. One prognostic pathologic marker for treatment of breast cancer is known hormone receptor status. One particular hormone marker is called *estrogen receptor* (ER) status which is typically diagnosed as being positive or negative. Two large studies in stage I breast cancer patients, Protocols B-14 and B-23, from the National Surgical Adjuvant Breast and Bowel Project (NSABP), indicate that the drug, tamoxifen, had large disease-free survival and survival effects as compared to a placebo, in patients whose tumors were ER positive (Fisher et al. [44]) whereas the effects were small or neglible in those whose tumors were ER negative (Fisher et al. [45]). In an unpublished study, pathologists at two laboratories affiliated with the NSABP, were asked to re-evaluate 181 specimens where patients had been originally diagnosed as being ER positive. For 34 specimens, both labs diagnosed the tumor as being ER negative; for 9 specimens, the first lab diagnosed the specimen as being ER negative while the second lab diagnosed the specimen as being ER positive; for 1 specimen, the first lab diagnosed the specimen as being ER positive while the second lab diagnosed the specimen as being ER negative; and for 137 specimens, both labs diagnosed the tumor as being ER positive.

A contingency table for the pairs of observations is presented below.

		Lab #2		
		ER negative	ER positive	Total
Lab #1	ER negative	34	9	43
	ER positive	1	137	138
	Total	35	146	181

For this example, $p_C = \dfrac{34 + 137}{181} = \dfrac{171}{181} \approx 0.945$. To calculate p_E, one first notes that in laboratory #1, 43 out of the 181 (23.76%) were diagnosed as being ER negative, and the rest (76.24%) were diagnosed as being ER positive. Thus, if the assignment of ER status was done based on no set criteria, then one would expect that $35 \times 0.2376 \approx 8.315$ cases from laboratory #2 would be diagnosed as being ER negative. Furthermore, one would expect that $146 \times 0.7624 \approx 111.315$ of the laboratory #2 cases would be diagnosed as being ER positive. The expected counts are summarized in the table below.

Expected Counts by Chance

		Lab #2		
		ER negative	ER positive	Total
Lab #1	ER negative	8.315	34.685	43
	ER positive	26.685	111.315	138
	Total	35	146	181

Thus, the proportion of cases *expected to be in agreement* is $p_E \approx \frac{8.315+119.63}{181} \approx 0.6609$. Hence, we can calculate the kappa statistic as $\kappa = \frac{.94475-.66093}{1-.66093} \approx \frac{.2838}{.3391} \approx .837$.

The SAS program below will produce the desired results.

```
data Ckappa;
input diag1 $ diag2 $  count;
label diag1='Diagnosis lab 1' diag2='Diagnosis lab 2';
datalines;
N N   34
N P    9
P N    1
P P 137
;   run;

proc freq data=Ckappa;
table diag1*diag2/agree;
weight count;
run;
```

One question that arises from viewing the results of an analysis using Cohen's Kappa is: What level of κ assures an investigator that the agreement associated with a survey or diagnostic test is good? The answer to this question, of course, depends largely on the application at hand. However, most investigators in the health sciences (Landis and Koch [84] and Gordis [55]) assert that $\kappa < .40$ indicates poor agreement, κ between 0.40 and 0.75, inclusively, indicates intermediate to good agreement; and $\kappa > 0.75$ indicates excellent agreement. Thus, in the example above, the results of the analysis indicate that the two labs had excellent agreement with respect to diagnosis of ER status.

6.7 Exercises

1. Consider the data in Example 6.2.

(a) Do a simple χ^2 test of proportions for the combined data.

(b) Calculate the odds ratio of getting better versus not for the new treatment versus the standard treatment. Give approximate 95% confidence intervals for your results.

(c) Do a Mantel–Haenzsel procedure for the data using the men's and women's data.

(d) Calculate the odds ratio of getting better versus not for the new treatment versus the standard treatment *adjusting for sex*. Give approximate 95% confidence intervals for your results.

(e) Do the calculations in parts **(a)** – **(d)** provide any resolution to Simpson's paradox? Why or why not?

2. Consider the birthweight data in the R library called MASS. (Type library (MASS) then ?birthwt . Then, under "Examples," one can copy all of the commands and paste them into the R console.) The data, which were presented earlier by Hosmer and Lemeshow [70] and Venables and Ripley [153], involve characteristics of the mother that may affect whether or not her child has a low birthweight. The dataset includes an outcome variable, low, which is an indicator of whether the birthweight is < 2.5 kg and several characteristics associated with the mother: her age in years (age), weight (in *lbs*) at her last menstrual period (lwt), race (1 = white, 2 = black, 3 = other), smoking status during pregnancy (smoke), number of previous premature labors (pt1), her prior history of hypertension (ht), the presence of uterine irritability(ui), and the number of physician visits during the first trimester (ftv). The initial model fitted in the R example uses all of the predictor variables. One can refit the model with the code,

```
logist2.low<-glm(low ~ lwt+race+smoke+ptd+ht, binomial, bwt)
summary(logist2.low)
```

to include only variables that are significant at the $p = 0.05$ level. The output is as follows:

```
            Estimate Std. Error z value Pr(>|z|)
(Intercept)  0.09462    0.95704   0.099  0.92124
lwt         -0.01673    0.00695  -2.407  0.01608 *
raceblack    1.26372    0.52933   2.387  0.01697 *
raceother    0.86418    0.43509   1.986  0.04701 *
smokeTRUE    0.87611    0.40071   2.186  0.02879 *
ptdTRUE      1.23144    0.44625   2.760  0.00579 **
htTRUE       1.76744    0.70841   2.495  0.01260 *
```

The variables summarized in the table above include weight at last menstrual period (continuous variable), race (black and other versus white), smoking status (smoker versus nonsmoker), previous premature labors (one or more versus none), and history of hypertension (yes [TRUE] versus no).

(a) Provide the odds ratio estimates of having a low weight baby versus not for each of the levels given in the table above (ignore the intercept).

(b) Give approximate 95% confidence intervals for your estimates in part **(a)**.

(c) Give the odds ratio of having a low birth baby versus not for a woman who is 20 lbs heavier than another woman with all other factors being equal.

(d) From the model, calculate the probability of having a low birth child for a 150-lb white woman who smokes, has had no previous premature labors but has had a history of hypertension.

## CHAPTER 7

Multivariate Analysis

7.1 Introduction

In Chapters 1–6, we investigated statistical methods assuming that each outcome or observation of interest had a single value. In this chapter, we will explore methods that accommodate situations where each outcome has several values which may be correlated. We first introduce the multivariate normal or multinormal distribution followed by multivariate analogs to t-tests, analysis of variance (ANOVA), and regression. We also explore how multivariate methods can be used to discriminate or classify observations to different populations.

7.2 The Multivariate Normal Distribution

In Chapter 1, we introduced the normal or Gaussian distribution, which has a probability density function of the form

$$f(y) = \frac{1}{\sqrt{2\pi\sigma^2}} \exp\left\{-\frac{(y-\mu)^2}{2\sigma^2}\right\} \tag{7.1}$$

When a random variable, Y, is normally distributed, then we write this as $Y \sim N(\mu, \sigma^2)$. Suppose now that we have r normal random variables that are correlated. The joint distribution of these variables is called a *multivariate normal* or a *multinormal* distribution.

If $\mathbf{Y}$ is multinormal, we denote $\mathbf{Y}$ as $\mathbf{Y} \sim MVN(\boldsymbol{\mu}, \boldsymbol{\Sigma})$

$$\text{where } \mathbf{Y} = \begin{pmatrix} Y_1 \\ Y_2 \\ \vdots \\ Y_r \end{pmatrix}, \boldsymbol{\mu} = \begin{pmatrix} \mu_1 \\ \mu_2 \\ \vdots \\ \mu_r \end{pmatrix} \text{ and } \boldsymbol{\Sigma} = \begin{pmatrix} \sigma_1^2 & \sigma_{12} & \cdots & \sigma_{1r} \\ \sigma_{12} & \sigma_2^2 & \cdots & \sigma_{2r} \\ \vdots & \vdots & \ddots & \vdots \\ \sigma_{1r} & \sigma_{2r} & \cdots & \sigma_r^2 \end{pmatrix}.$$

and has a probability density function with the form

$$f(\mathbf{y}) = \frac{1}{\sqrt{(2\pi)^r |\boldsymbol{\Sigma}|}} \exp\left\{-\frac{1}{2}(\mathbf{y} - \boldsymbol{\mu})' \boldsymbol{\Sigma}^{-1} (\mathbf{y} - \boldsymbol{\mu})\right\} \tag{7.2}$$

where μ is called the *mean vector*, Σ is called the *covariance matrix* and $|\Sigma|$ is the determinant of the covariance matrix. The covariance matrix has the property that the diagonal elements, σ_i^2 represent the variances of the random variables, Y_i, and the off-diagonal elements, σ_{ij}, are the covariance between the variables Y_i and Y_j. As can be seen from equations (7.1) and (7.2), the multivariate distribution is a clear generalization of its univariate counterpart. A list of some other properties of the multinormal distribution is given below. More complete discussions of the properties of the multinormal distribution can be found in Anderson [6] and Mardia et al. [95]. A three-dimension plot of a multinormal distribution and a contour plot of equal probabilities are given in Figure 7.1.

Figure 7.1 *A Three-Dimensional and a Contour Plot of a Bivariate Normal Distribution*

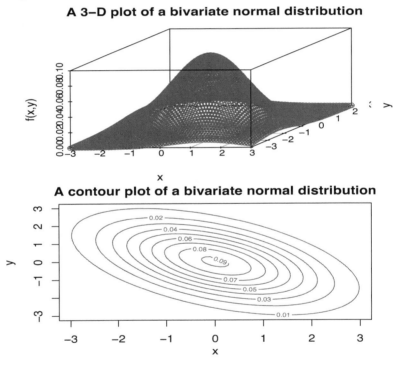

A 3–D plot of a bivariate normal distribution

A contour plot of a bivariate normal distribution

Other Properties of the Multinormal Distribution

1. A r-variate multinormal distribution can be specified by identifying the r parameters of the mean vector and the $r(r + 1)/2$ parameters of the covariance matrix so that a total of $r(r + 3)/2$ parameters must be specified.

2. The multinormal distribution has the property in that if there is zero correlation between two of the components, they are independent. This property is <u>not</u> true for many multivariate distributions. Conversely, however, if two component distributions are independent then they *always* are uncorrelated.

3. If $\mathbf{Y}$ is a multinormal vector, then each component, Y_i has a normal distribution and all linear combinations, $\mathbf{c}'\mathbf{Y}$ ($\mathbf{c} \neq \mathbf{0}$) of the multinormal vector are themselves normally distributed.

7.3 One- and Two-Sample Multivariate Inference

7.3.1 One-Sample Multivariate Data

The multivariate analog to the t-distribution was first introduced by Harold Hotelling in 1931 [71]. Suppose we have a single sample of r-variate observations, $\mathbf{Y}_1, \mathbf{Y}_2, \ldots, \mathbf{Y}_n$ and suppose we wish to test whether or not that sample has a particular mean value (which would be a vector consisting of the r component means). To test $H_0 : \boldsymbol{\mu} = \boldsymbol{\mu}_0$ versus $H_1 : \boldsymbol{\mu} \neq \boldsymbol{\mu}_0$, form

$$T^2 = \frac{(n-r)n}{(n-1)r}\left[(\overline{\mathbf{Y}} - \boldsymbol{\mu})'\mathbf{S}^{-1}(\overline{\mathbf{Y}} - \boldsymbol{\mu})\right] \tag{7.3}$$

which, under the null hypothesis, H_0 is distributed as $F_{r,n-r}$. In the equation above, the $\mathbf{S}$ is given by $\underbrace{\mathbf{S}}_{r \times r} = \frac{1}{n-1}\sum_{i=1}^{n}(\mathbf{Y}_i - \overline{\mathbf{Y}})(\mathbf{Y}_i - \overline{\mathbf{Y}})'$

$$= \frac{1}{n-1}\begin{pmatrix} \sum_{i=1}^{n}(Y_{i1} - \overline{Y}_1)^2 & \cdots & \sum_{i=1}^{n}(Y_{i1} - \overline{Y}_1)(Y_{ir} - \overline{Y}_r) \\ \sum_{i=1}^{n}(Y_{i2} - \overline{Y}_2)(Y_{i1} - \overline{Y}_1) & \cdots & \sum_{i=1}^{n}(Y_{i2} - \overline{Y}_2)(Y_{ir} - \overline{Y}_r) \\ \vdots & \vdots & \vdots \\ \sum_{i=1}^{n}(Y_{ir} - \overline{Y}_r)(Y_{i1} - \overline{Y}_1) & \cdots & \sum_{i=1}^{n}(Y_{ir} - \overline{Y}_r)^2 \end{pmatrix} \tag{7.4}$$

As can be seen from the above formulas, the calculations needed to perform a multivariate test are indeed daunting. However, the computer allows us, as is demonstrated below, to easily perform this type of a calculation.

Example 7.1. In a dataset presented by Verzani [154], the heights in inches of 1078 father–son pairs are given ($\Longrightarrow n = 1078$ and $r = 2$). Box plots and a scatterplot of that dataset are given in Figure 7.2.

Suppose that we want to test $H_0 : \begin{pmatrix} \mu_1 \\ \mu_2 \end{pmatrix} = \begin{pmatrix} 68 \\ 68 \end{pmatrix}$ versus $H_1 : \begin{pmatrix} \mu_1 \\ \mu_2 \end{pmatrix} \neq \begin{pmatrix} 68 \\ 68 \end{pmatrix}$. The mean height vector and its variance–covariance matrix are

$$\begin{pmatrix} \overline{\mathbf{Y}}_1 \\ \overline{\mathbf{Y}}_2 \end{pmatrix} \approx \begin{pmatrix} 67.6871 \\ 68.6841 \end{pmatrix}, \text{ and } \mathbf{S} \approx \begin{pmatrix} 7.5343 & 3.8733 \\ 3.8733 & 7.9225 \end{pmatrix}$$

so that $(\mathbf{Y} - \boldsymbol{\mu})'\mathbf{S}^{-1}(\mathbf{Y} - \boldsymbol{\mu})$ is approximately

$$\begin{pmatrix} -0.3129 & 0.6841 \end{pmatrix}\begin{pmatrix} 0.1773 & -0.08667 \\ -0.08667 & 0.1686 \end{pmatrix}\begin{pmatrix} -0.3129 \\ 0.6841 \end{pmatrix}$$

Figure 7.2 *Plots of Heights from 1078 Father–Son Pairs*

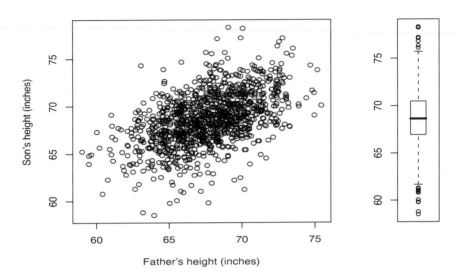

Father's height (inches)

$$\approx 0.13336 \implies F = \frac{1078 \times 1076}{2 \times 1077} \, 0.13336 \approx 71.813 \; .$$

The critical F-value at the $\alpha = 0.00001$ level is $F_{2,1076,0.99999} \approx 11.64$ so that $p < 0.00001$. This result is highly significant. The observed differences from the values in this arbitrary null hypothesis are not large but since the sample size here (1078) is very large, even small deviations from the null can lead to significant results. Thus, large sample sizes enable us to have high *power* to detect small differences from the null hypothesis.

7.3.2 Two-Sample Multivariate Test

Suppose $\mathbf{Y}_1, \mathbf{Y}_2, \ldots, \mathbf{Y}_{n_1}$ and $\mathbf{Z}_1, \mathbf{Z}_2, \ldots, \mathbf{Z}_{n_2}$ are two independent random samples of independent r-variate outcomes that are multinormally distributed as $MVN(\boldsymbol{\mu}_1, \boldsymbol{\Sigma})$ and $MVN(\boldsymbol{\mu}_2, \boldsymbol{\Sigma})$, respectively. To test $H_0 : \boldsymbol{\mu}_1 = \boldsymbol{\mu}_2$ versus $H_1 : \boldsymbol{\mu}_1 \neq \boldsymbol{\mu}_2$, form

$$T^2 = \frac{n_1 n_2}{n_1 + n_2} (\overline{\mathbf{Y}} - \overline{\mathbf{Z}})' \mathbf{S}_p^{-1} (\overline{\mathbf{Y}} - \overline{\mathbf{Z}}) \tag{7.5}$$

where $\overline{\mathbf{Y}} = \left(\sum_{i=1}^{n_1} \mathbf{Y}_i \right) / n_1, \overline{\mathbf{Z}} = \left(\sum_{i=1}^{n_2} \mathbf{Z}_i \right) / n_2$ are the group mean vectors,

$$S_p = \frac{\left[(n_1 - 1)\mathbf{S}_1 + (n_2 - 1)\mathbf{S}_2\right]}{(n_1 + n_2 - 2)}$$ is the pooled estimated covariance matrix and $\mathbf{S}_1$

and $\mathbf{S}_2$ are group covariance matrix estimates, each similar to that given in equation (7.4). The null hypothesis is rejected if $\frac{f-r+1}{fr} T^2 \geq F_{r, f-r+1}$ where $f = n_1 + n_2 - 2$. The r here is the dimension of each of the vectors.

Example 7.2. A dataset consisting of 32 skull measurements for Tibetan soldiers was collected by Colonel L.A. Waddell from a battlefield and different grave sites in neighboring areas of Tibet. The data consisted of 45 measurements per individual. Five of these measurements were analyzed by by Morant [102] and later reproduced in *Small Data Sets* [61], dataset #144. We consider here the first three measurements (in mm): greatest length of skull, greatest horizontal breadth of skull and the height of skull. The "Type A" skulls from Khams were originally thought to be from a different human type than the "Type B" skulls, which were from the predominant Mongolian and Indian groups typically found in Tibet. Box plots for each of the three measurements considered here are given side by side by "Type" in Figure 7.3.

The mean coefficient vectors for the 17 "type A" soldiers and 15 "type B," respectively, are

$$\hat{\boldsymbol{\mu}}_A = \overline{\mathbf{Y}} \approx \begin{pmatrix} 174.82 \\ 139.35 \\ 132.00 \end{pmatrix} \text{ and } \hat{\boldsymbol{\mu}}_B = \overline{\mathbf{Z}} \approx \begin{pmatrix} 185.73 \\ 138.73 \\ 134.77 \end{pmatrix} \Longrightarrow \overline{\mathbf{Y}} - \overline{\mathbf{Z}} \approx \begin{pmatrix} -10.910 \\ 0.61961 \\ -2.7667 \end{pmatrix}$$

and their covariance matrices are

$$\mathbf{S}_A = \begin{pmatrix} 45.529 & 25.222 & 12.391 \\ 25.222 & 57.805 & 11.875 \\ 12.391 & 11.875 & 36.094 \end{pmatrix} \text{ and } \mathbf{S}_B = \begin{pmatrix} 74.423810 & -9.522 & 22.737 \\ -9.5226 & 37.352 & -11.263 \\ 22.737 & -11.263 & 36.317 \end{pmatrix}$$

and the pooled covariance matrix is

$$\mathbf{S}_p = \frac{16\mathbf{S}_A + 14\mathbf{S}_B}{30} \approx \begin{pmatrix} 59.013 & 9.0081 & 17.219 \\ 9.0081 & 48.261 & 1.0772 \\ 17.219 & 1.0772 & 36.198 \end{pmatrix}$$

$$T^2 = \frac{15(17)}{15 + 17} (-10.910 \quad 0.61961 \quad -2.7667) \begin{pmatrix} 59.013 & 9.0081 & 17.219 \\ 9.0081 & 48.261 & 1.0772 \\ 17.219 & 1.0772 & 36.198 \end{pmatrix} \begin{pmatrix} -10.910 \\ 0.61961 \\ -2.7667 \end{pmatrix}$$

$$\approx 17.021 \Longrightarrow F = \frac{f - r + 1}{fr} T^2 = \frac{30 - 3 + 1}{30(3)} T^2 \approx 5.2956 \sim F_{3,30-3+1} = F_{3,28}$$

under $H_0 : \boldsymbol{\mu}_A = \boldsymbol{\mu}_B$. If this test was performed at the 5% level then $F > F_{3,28,.95} = 2.947$ so that we reject the null hypothesis ($p \approx 0.0051$) and conclude that the populations had significant differences with respect to these three skull dimensions.

Figure 7.3 *Skull Lengths, Breadths, and Heights of Tibetan Type A and B Soldiers*

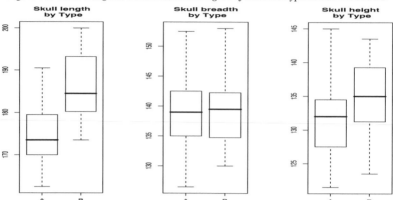

7.4 Multivariate Analysis of Variance (MANOVA)

A more general approach to the T^2 method introduced in the previous section involves the comparison of $g \geq 2$ groups of independent r-variate multinormal observations. Consider a set of $N = \sum_{i=1}^{g} n_i$ observations, each of which are r-variate and which are also divided up into g subpopulations such that $2 \leq g < N$. We can write such a model as

$$
\underbrace{\begin{pmatrix} \mathbf{Y}_{11} \\ \vdots \\ \mathbf{Y}_{1n_1} \\ \mathbf{Y}_{21} \\ \vdots \\ \mathbf{Y}_{2n_2} \\ \vdots \\ \mathbf{Y}_{g1} \\ \vdots \\ \mathbf{Y}_{gn_g} \end{pmatrix}}_{N \times r} = \underbrace{\begin{pmatrix} 1 & 0 & \cdots & 0 \\ \vdots & \vdots & \vdots & \vdots \\ 1 & 0 & \cdots & 0 \\ 0 & 1 & \vdots & 0 \\ \vdots & \vdots & \vdots & \vdots \\ 0 & 1 & \cdots & 0 \\ \vdots & \vdots & \vdots & \vdots \\ 0 & 0 & \cdots & 1 \\ \vdots & \vdots & \vdots & \vdots \\ 0 & 0 & \cdots & 1 \end{pmatrix}}_{N \times g} \underbrace{\begin{pmatrix} \boldsymbol{\mu}_1 \\ \boldsymbol{\mu}_2 \\ \vdots \\ \boldsymbol{\mu}_g \end{pmatrix}}_{g \times r} + \underbrace{\begin{pmatrix} \mathbf{E}_{11} \\ \vdots \\ \mathbf{E}_{1n_1} \\ \mathbf{E}_{21} \\ \vdots \\ \mathbf{E}_{2n_2} \\ \vdots \\ \mathbf{E}_{g1} \\ \vdots \\ \mathbf{E}_{gn_g} \end{pmatrix}}_{N \times r} \tag{7.6}
$$

where $\mathbf{Y}_{11}, \mathbf{Y}_{11}, \ldots, \mathbf{Y}_{gn_g}$ are r-variate random variables, $\boldsymbol{\mu}_1, \boldsymbol{\mu}_2, \ldots, \boldsymbol{\mu}_g$ are r-variate mean vectors and $\mathbf{E}_{11}, \mathbf{E}_{11}, \ldots, \mathbf{E}_{gn_g}$ are errors that are independently and identically distributed as r-variate $MVN(\mathbf{0}, \boldsymbol{\Sigma})$. This model can be written more succinctly as

$$
\mathbf{Y} = \mathbf{X}\boldsymbol{\mu} + \mathbf{E}. \tag{7.7}
$$

Suppose we wish to test $H_0 : \boldsymbol{\mu}_1 = \boldsymbol{\mu}_2 = \ldots = \boldsymbol{\mu}_g$ versus H_1: at least one $\boldsymbol{\mu}_i$ is different from one of the other $g - 1$ vectors. The multivariate analog to ANOVA is

MANOVA which stands for *Multivariate* Analysis of Variance. Similar to the univariate case, we can form between, within and total sums of squares for a MANOVA as

$$\mathbf{B} = \sum_{i=1}^{g} n_i (\overline{\mathbf{Y}}_{i\bullet} - \overline{\mathbf{Y}}_{\bullet\bullet})'(\overline{\mathbf{Y}}_{i\bullet} - \overline{\mathbf{Y}}_{\bullet\bullet}) \;,\;\; \mathbf{W} = \sum_{i=1}^{g}\sum_{j=1}^{n_i}(\mathbf{Y}_{ij} - \overline{\mathbf{Y}}_{i\bullet})'(\mathbf{Y}_{ij} - \overline{\mathbf{Y}}_{i\bullet})$$

$$\text{and } \mathbf{T} = \sum_{i=1}^{g}\sum_{j=1}^{n_i}(\mathbf{Y}_{ij} - \overline{\mathbf{Y}}_{\bullet\bullet})'(\mathbf{Y}_{ij} - \overline{\mathbf{Y}}_{\bullet\bullet}) \;.$$

The difference in the multivariate case is that the corresponding "sums of squares", sometimes denoted as SSP, are actually $r \times r$ *matrices* that have sums of squares as each element whereas in the univariate case, there are only single values.

There are several different multivariate tests that are used to determine whether or not there is overall heterogeneity across the g groups. The most widely used of these tests is the *Wilk's* Λ, which is simply

$$\Lambda = |\mathbf{W}| \Big/ |\mathbf{T}| \tag{7.8}$$

where "$|\;|$" denotes a determinant. Hence, Λ is a single value or a *scalar*.

Example 7.3. In this example we present an analysis of the Egyptian skull data available in *A Handbook of Small Data Sets* [61]. This dataset involves four different skull measurements in mm for 30 skulls taken in each of five periods of human development: Early Predynastic, Late Predynastic, the 12th & 13th Dynasties, the Ptolemaic Period, and the Roman Period so that $N = 150$. The periods are ordered chronologically and designated 1 to 5, respectively, in this analysis. A complete analysis of these data are given in Manley [90]. For our analysis, we will focus on two measurements, basibregmatic height and basibregmatic length.

One question arises as to whether or not there are any differences among these periods with respect to the mean multivariate observions, that is, we wish to first test $H_0 : \boldsymbol{\mu}_1 = \ldots = \boldsymbol{\mu}_5$ versus H_1 : at least one $\boldsymbol{\mu}_i$ is different from the others, $i = 1, \ldots, 5$. The mean basibregmatic values in each of the five respective periods are

$$\overline{\mathbf{Y}}_1 = \begin{pmatrix} 133.6 \\ 99.17 \end{pmatrix},\; \overline{\mathbf{Y}}_2 = \begin{pmatrix} 132.7 \\ 99.07 \end{pmatrix},\; \overline{\mathbf{Y}}_3 = \begin{pmatrix} 133.8 \\ 96.03 \end{pmatrix},\; \overline{\mathbf{Y}}_4 = \begin{pmatrix} 132.3 \\ 94.53 \end{pmatrix} \text{ and } \overline{\mathbf{Y}}_5 = \begin{pmatrix} 130.3 \\ 93.50 \end{pmatrix},$$

respectively.

MANOVA table for Egyptian Skull Data

Source	DF	SSP
Between	8	$\begin{pmatrix} 229.907 & 292.28 \\ 292.28 & 803.293 \end{pmatrix}$
Within (Error)	288	$\begin{pmatrix} 3405.27 & 754 \\ 754 & 3505.97 \end{pmatrix}$
Total	296	$\begin{pmatrix} 3635.173 & 1046.28 \\ 1046.28 & 4309.26 \end{pmatrix}$

$$\Lambda = \left| \begin{pmatrix} 3405.27 & 754 \\ 754 & 3505.97 \end{pmatrix} \right| \Big/ \left| \begin{pmatrix} 3635.173 & 1046.28 \\ 1046.28 & 4309.26 \end{pmatrix} \right|$$

$$\approx 11370235 / 14570205 \approx 0.7804 \;.$$

This particular Λ-value is associated with an F-value of about 4.75, which is $\sim F_{8,288}$ under H_0 and thus, $p < 0.0001$. A SAS program to perform the analysis is given below.

```
/* Data Source: Small datasets #361 and
Manley, BFJ. Multivariate Statistical Methods: A primer, Chapman/Hall, 2004 */

options pageno=1 ls=75 nodate;
* First read the tab deliminated file;
proc import datafile="C:\book\data\Small Data Sets\skulls.dat" out=skulls
            dbms=tab replace;
   getnames=no;
run;

data skull1(rename=(var1=maxbr var2=basiht var3=basilgth var4=nasalht));
set skulls; Group=1; keep Group var1--var4; run;
data skull2(rename=(var5=maxbr var6=basiht var7=basilgth var8=nasalht));
set skulls; Group=2; keep Group var5--var8; run;
data skull3(rename=(var9=maxbr var10=basiht var11=basilgth var12=nasalht));
set skulls; Group=3; keep Group var9--var12; run;
data skull4(rename=(var13=maxbr var14=basiht var15=basilgth var16=nasalht));
set skulls; Group=4; keep Group var13--var16; run;
data skull5(rename=(var17=maxbr var18=basiht var19=basilgth var20=nasalht));
set skulls; Group=5; keep Group var17--var20; run;

data skull; set skull1 skull2 skull3 skull4 skull5;
  label maxbr='Maximum breadth' basiht='Basibregmatic height'
        basilgth='Basibregmatic length' nasalht='Nasal height';
run;

proc format;
value gp 1='Early Predynastic' 2='Late Predynastic' 3='12th & 13th Dynasties'
     4='Ptolemaic Period' 5='Roman Period';
run;
proc print label noobs; format Group gp.; run;
title1 'Example of a balanced One-Way MANOVA: Egyptian Skull Measurements (mm)';
title2 'from Manley, ''Multivariate Statistical Methods: A primer,'' 3rd ed.';

proc glm data = skull;
  class Group;
  model basiht basilgth = Group/nouni;
  manova h = Group/printe printh ;
title3 'General MANOVA model: Basibregmatic Measurements Only';
run;

proc glm data = skull;
```

```
      class Group ;
      model  basiht basilgth= Group/nouni;
      contrast 'Test: Compare to oldest' Group 1 -1 0 0 0,
                                         Group 1 0 -1 0 0,
                                         Group 1 0 0 -1 0,
                                         Group 1 0 0 0 -1;
      manova/printe printh;
title3 'Testing Basibregmatic measurements for all gps vs Early Predynastic';
run;
```

As a practical matter, in an analysis of data such as this, one would typically employ one of the multiple comparisons procedures introduced in sections 5.4 – 5.6 to account for the fact that several comparisons are being made which, without any adjustment, would lead to inflated type I error rates.

7.5 Multivariate Regression Analysis

Earlier, when characterizing n single outcomes, $y_1, y_2, \ldots, y_n$, by an intercept and, say, $p-1$ covariates, $X_1, \ldots, X_{p-1}$, we introduced the model, $\underset{n \times 1}{y} = \underset{n \times p}{X} \underset{p \times 1}{\beta} + \underset{n \times 1}{\epsilon}$.
If we now have outcomes that are r-variate, we can write an analog to the earlier model as

$$\underset{n \times r}{Y} = \underset{n \times p}{X} \underset{p \times r}{B} + \underset{n \times r}{E} \tag{7.9}$$

The least squares solution to this model is

$$\underset{p \times r}{\widehat{B}} = (\underset{p \times n}{X}' \underset{n \times p}{X})^{-1} \underset{p \times n}{X}' \underset{n \times r}{Y} . \tag{7.10}$$

Thus, instead of a $p \times 1$ column *vector* of parameter estimates, we now have r such column vectors that are represented as an $p \times r$ *matrix* of parameters estimates.

7.6 Classification Methods

Many applications in the biological and health sciences require one to *classify* individuals or experimental units into different populations or groups according to several measurements or characteristics of that individual. Such classification may be necessary to identify, e.g., subclasses of species, genetic signatures of patients or prognostic groups that may require different treatment strategies.

Three popular classification methods are

- Discriminant analysis;
- Logistic regression; and
- Classification trees.

Discriminant analysis requires the multivariate observations made on individuals to be multinormally distributed. For the two other methods, the observations are *not* required to be multinormally distributed. Both of these latter methods have an advantage over discriminant analysis in that they both can classify individuals according to both discrete and continuous measurements. However, if the observations are truly multinormally distributed then discriminant analysis is preferred.

7.6.1 Linear Discriminant Analysis

It is often of interest in medical, education, and psychology applications to classify individuals into distinct populations based on observing multiple characteristics of the individuals. The classical method to do this is known as linear *discriminant analysis*. In discriminant analysis, the underlying assumption is that the populations to be classified have multinormal distributions with a common covariance structure but with different means. The statistical problem is to find a boundary that is used to best discriminate between two or more populations. In the classical case, that boundary is calculated by defining a rule which classifies each observation into a particular population. The rule used in classical analysis is equivalent to minimizing the squared "multivariate distance" between each observation, y_i and the mean of the population into which it is allocated. The distance used is typically the *Mahalonobis distance*, defined as

$$\Delta = (\mathbf{y} - \boldsymbol{\mu}_k)' \boldsymbol{\Sigma}^{-1} (\mathbf{y} - \boldsymbol{\mu}_k), \tag{7.11}$$

where $\mathbf{y}$ is a vector of observations, $\boldsymbol{\mu}_k$ is the population mean vector of group k, and $\boldsymbol{\Sigma}$ is the covariance matrix associated with the multivariate observations in a sample. The Mahalonobis distance represents the distance between a r-dimensional point and the center of an ellipsoid defined by $\boldsymbol{\mu}_k' \boldsymbol{\Sigma}^{-1} \boldsymbol{\mu}_k$.

It turns out that the solution to this problem, which allows one to best classify observations into two or more populations (groups), $\pi_1, \ldots, \pi_g$, can be accomplished by maximizing the ratio of the between-population variance to the within-population variance. An equivalent method (although philosophically different) is to allocate observations into a population based on obtaining a maximum value of a likelihood equation in the allocated population as compared to other potential populations. So, assuming that observations are distributed as $MVN(\boldsymbol{\mu}_i, \boldsymbol{\Sigma})$, $i = 1, \ldots, g$, it can be shown that the discriminant rule leading to a maximum ratio between the allocated population, π_j and other populations is that value of i, which minimizes $\Delta = (\mathbf{y} - \boldsymbol{\mu}_i)' \boldsymbol{\Sigma}^{-1} (\mathbf{y} - \boldsymbol{\mu}_i)$. In the special case that we wish to classify an observation, $\mathbf{x}$ into $g = 2$ populations, then $\mathbf{x}$ is allocated to π_1 if

$$\left(\boldsymbol{\mu}_1 - \boldsymbol{\mu}_2 \right)' \boldsymbol{\Sigma}^{-1} \left(\mathbf{x} - \frac{1}{2} (\boldsymbol{\mu}_1 + \boldsymbol{\mu}_2) \right) > 0 \tag{7.12}$$

and to π_2 otherwise. Hence, an r-dimensional plane provides the boundary for classifying a given observation into one of the two populations of interest.

Example 7.4. The following data were obtained from the statistical dataset library at the University of Massachusetts at Amherst (http://www.umass.edu/statdata/statdata/data/). The data consist of hospital stay data for 500 patients. The outcome of interest for this example is the vital status (dead/alive) at discharge. (The length of stay is provided but is not used for this example.) Two variables related to the vital status are age and systolic blood pressure. Listed below is a SAS program using PROC DISCRIM and some of the output produced by SAS. Also given is an R program (lda from the MASS library). Figure 7.4 depicts the data (circles are those alive, and +'s are deaths) overlayed by the discriminant function.

SAS code for linear discriminant analysis example

```
data whas500;
infile 'C:\book\data and data sources\whas500.txt';
input
id age gender hr sysbp diasbp bmi cvd afb sho chf av3 miord mitype year
 @72 admitdate mmddyy10. @72 disdate mmddyy10. @96 fdate mmddyy10.
los dstat lenfol fstat; run;

proc discrim data=whas500;
class dstat;
var age sysbp; run;
--------------------------- Partial Output ---------------------------
                    Class Level Information
```

	Variable				Prior
dstat	Name	Frequency	Weight	Proportion	Probability
0	_0	461	461.0000	0.922000	0.500000
1	_1	39	39.0000	0.078000	0.500000

```
              Pooled Covariance Matrix Information

                            Natural Log of the
                 Covariance   Determinant of the
                Matrix Rank   Covariance Matrix
                     2            12.23338

                    The DISCRIM Procedure
       Pairwise Generalized Squared Distances Between Groups
```

$$D^2(i|j) = (\bar{X}_i - \bar{X}_j)' \, COV^{-1} \, (\bar{X}_i - \bar{X}_j)$$

```
            Generalized Squared Distance to dstat
            From dstat          0            1
                  0             0         0.96444
                  1          0.96444         0

                Linear Discriminant Function
```

$$Constant = -.5 \, \bar{X}_j' \, COV^{-1} \, \bar{X}_j \qquad Coefficient\ Vector = COV^{-1} \, \bar{X}_j$$

```
              Linear Discriminant Function for dstat
              Variable          0            1
              Constant      -21.92700    -22.77297
              age             0.33438      0.38495
              sysbp           0.14186      0.12048

                    The DISCRIM Procedure
```

```
Classification Summary for Calibration Data: WORK.WHAS500
Resubstitution Summary using Linear Discriminant Function

            Generalized Squared Distance Function
```

$$D_j^2(X) = (X-\bar{X}_j)' \; COV^{-1} \; (X-\bar{X}_j)$$

```
    Posterior Probability of Membership in Each dstat
```

$$Pr(j|X) = \exp(-.5 \; D_j^2(X)) \; / \; SUM_k \; \exp(-.5 \; D_k^2(X))$$

```
    Number of Observations and Percent Classified into dstat
        From dstat          0          1       Total
             0             313        148         461
                         67.90      32.10      100.00
             1              12         27          39
                         30.77      69.23      100.00
          Total           325        175         500
                         65.00      35.00      100.00
        Priors            0.5        0.5

              Error Count Estimates for dstat
                            0          1       Total
        Rate            0.3210     0.3077      0.3144
        Priors          0.5000     0.5000
```

_R_ code for linear discriminant analysis example

```
library(MASS)
model.dstat <- lda(dstat ~ age+sysbp) #<-- the "response" is grouping factor
p.dstat<-predict(model.dstat)
```

Figure 7.4 *Data Overlayed by the Discriminant Function for Hospital Stay Data*

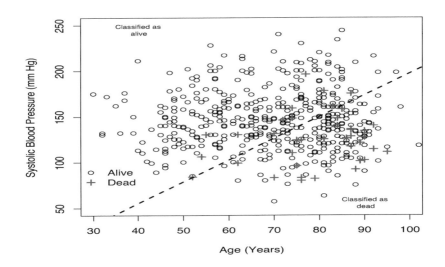

As can be seen from the output above and Figure 7.4, 30.8% of the individuals who died were classified as being alive and 32.1% of those alive were classified as being dead. Such high misclassification rates would be unacceptable and hence, one would want to use a lot more information to properly classify the discharge deaths and nondeaths.

7.6.2 Logistic Regression as a Tool for Classification

The logistic regression method is implemented by fitting a set of multivariate items $\mathbf{X}_j$ to a logistic model,

$$Pr(y_j \in \pi_i | \mathbf{X}_j) = \frac{\exp\{\mathbf{X}_j \boldsymbol{\beta}\}}{1 + \exp\{\mathbf{X}_j \boldsymbol{\beta}\}}$$

and then establishing a cutoff probability (often prespecified in statistical packages as 0.5) in order to assign the observations of say, the jth individual into a particular population, i. The predictor vector, $\mathbf{X}_j$, for individual j can consist of both discrete and continuous variables.

Example 7.5. The following data were obtained from the library DAAG in R. They involve characteristics and mortality outcomes in 1295 women who have previously had myocardial infarctions. In this particular example, we set up two *dummy* variables, which characterize the variable associated with whether or not a woman has had angina. There are three levels of that variable: yes (y), no (n), and not known (nk). For this example, two separate indicator variables are created that indicate a "y" or an "nk," that is, a 1 is given for the specified level and a 0 is given for all other levels. Using both dummy variables together, one can then determine whether or not angina is a significant predictor of subsequent mortality. The predictor will be associated with two degrees of freedom (see the label called t_angina at the end of the output below). A similar strategy is used for the variable "smstat" but there are three dummy variables instead of two (see the label smoking below).

```
options nonumber nodate;
data mi;
infile 'c:\mifem.txt';
input outcome $ age yronset premi $ smstat $ diabetes $ highbp $
      hichol $ angina $ stroke $;
run;
data mi2; set mi;
   angin_ny=(angina eq 'y'); angin_nu=(angina eq 'nk'); *<- "dummy variables";
   smoke_nc = (smstat eq 'c'); smoke_nx = (smstat eq 'x');
   smoke_nu = (smstat eq 'nk');
run;

proc logistic data=mi2;* descending;
model outcome=age yronset angin_ny angin_nu
      smoke_nc smoke_nx smoke_nu /outroc=roc1;* rsquare;* /selection=b;
t_angina: test angin_ny,angin_nu;
smoking: test smoke_nc,smoke_nx,smoke_nu;
output out = pred_dth p = prdeath xbeta = logit ;
title 'Logistic Regression model for Women''s MI data';
title2 'Data from DAAG';
run;

proc print data=pred_dth (obs=10);
```

```
var outcome age yronset angina smstat prdeath;
format prdeath f6.3;
title2 '10 observations and model predictions';
run;
```

```
------------------------- PARTIAL OUTPUT -------------------------
           Logistic Regression model for Women's MI data
                          Data from DAAG

                       The LOGISTIC Procedure
                        Model Information

          Data Set                    WORK.MI2
          Response Variable           outcome
          Number of Response Levels   2
          Number of Observations      1295
          Model                       binary logit
          Optimization Technique      Fisher's scoring

                       Response Profile
                  Ordered                        Total
                  Value      outcome          Frequency
                     1       dead                   321
                     2       live                   974

        Probability modeled is outcome='dead'.

                       The LOGISTIC Procedure
            Analysis of Maximum Likelihood Estimates

                               Standard        Wald
     Parameter   DF   Estimate     Error  Chi-Square   Pr > ChiSq
     Intercept    1     3.1568    2.6411      1.4287       0.2320
     age          1     0.0466    0.0122     14.5731       0.0001
     yronset      1    -0.0846    0.0287      8.6746       0.0032
     angin_ny     1     0.3888    0.1515      6.5861       0.0103
     angin_nu     1     2.6041    0.3191     66.6069       <.0001
     smoke_nc     1    -0.3298    0.1833      3.2349       0.0721
     smoke_nx     1    -0.2613    0.1893      1.9060       0.1674
     smoke_nu     1     1.2375    0.2959     17.4930       <.0001

                      Odds Ratio Estimates

                          Point           95% Wald
            Effect      Estimate     Confidence Limits
            age            1.048     1.023      1.073
            yronset        0.919     0.869      0.972
            angin_ny       1.475     1.096      1.985
            angin_nu      13.519     7.234     25.267
            smoke_nc       0.719     0.502      1.030
            smoke_nx       0.770     0.531      1.116
            smoke_nu       3.447     1.930      6.156

   Association of Predicted Probabilities and Observed Responses
         Percent Concordant     72.7    Somers' D    0.459
         Percent Discordant     26.8    Gamma        0.461
         Percent Tied            0.5    Tau-a        0.171
         Pairs                312654    c            0.729

               Linear Hypotheses Testing Results
                          Wald
          Label      Chi-Square    DF    Pr > ChiSq
          t_angina     67.3250      2       <.0001
          smoking      28.2305      3       <.0001
```

```
          Logistic Regression model for Women's MI data
               10 observations and model predictions

Obs     outcome     age     yronset     angina     smstat     prdeath
 1       live        63       85          n          x         0.205
 2       live        55       85          n          c         0.142
 3       live        68       85          y          nk        0.682
 4       live        64       85          y          x         0.284
 5       dead        67       85          nk         nk        0.949
 6       live        66       85          nk         x         0.800
 7       live        63       85          n          n         0.250
 8       dead        68       85          y          n         0.384
 9       dead        46       85          nk         c         0.595
10       dead        66       85          n          c         0.216
```

The results indicate that all of the covariates considered in this analysis, that is, age, year of onset, angina, and smoking status were significant predictors of mortality. For the two discrete variables, angina and smoking status, "no" or "non" were used as the baseline values for the other levels to be compared to. Also noted is that 72.9% of the outcomes were concordant with that predicted by the model. This is calculated with the following strategy. If the concordance between predicted and observed was perfect, that is, 100%, then one would expect that the predicted probability of death for each woman who actually died would be greater than that of each woman who remained alive. In reality, this is not true. To test this condition in a dataset, one must compare predicted mortality probabilities for each possible pair of women, one of whom died and one who was alive. In our dataset, 321 women died and 974 were alive so that $974 \times 321 = 312654$ pairs were checked. Among these pairs, 72.9% had the predicted probability of death higher for the woman who died compared to the one who was alive (concordant pairs); 26.8% had the predicted probability of death higher for the woman who was alive (discordant pairs); and there were tied predicted probabilities for 0.5% of the women. A printout of the outcome, the covariate values, and the predicted mortality is given above for 10 of the women in the database.

7.6.3 Classification Trees

Another type of classification involves discriminating populations in a heirarchical manner. One such classification model is known as a *binary tree-structured model* (Gordon [56]). The most well known and popular of these was developed in 1984 by Breiman, et al. [13] and is referred to as **C**lassification **A**nd **R**egression **T**rees (CART). In these models, the factors or covariates most highly related to outcome are located at the top of the "tree" and those less related form the "branches" of the tree. The models are created via the use of "recursive" binary partitions in the covariate space to create subgroups that are assumed to be approximately homogeneous. Each split point is referred to as a "node." At each node, a program "runs through" all possible binary splits of each of the candidate variables and selects the "best split" based on minimizing the "impurity" within each of the two "daughter nodes" or equivalently, maximizing the heterogeneity between the two "daughter nodes." The criterion for splitting is based on defining a measure of homogeneity with respect to the outcome of interest. One can run through a tree until the data has been divided into groups so small as to make the resulting classifications meaningless or incomprehensible. Alternatively, one can "prune" a tree from the bottom up based on throwing away splits that do not greatly increase the heterogeneity across the nodes.

Other than the CART algorithm developed by the authors themselves, another convenient algorithm for creating tree-structured models was developed in R and S-plus by Therneau and Atkinson [149]. For information on this R procedure, download the rpart library, then at your R console, type `libary(rpart)` followed by `?rpart`.

Example 7.6. Consider the myocardial infarction (MI) data from the previous example. Suppose we wish to classify patients into whether or not they have a high probability of dying (or remaining alive) based on their clinical and demographic characteristics. Among the characteristics considered are age, whether or not they had a previous MI event (yes, no, not known), blood pressure status (high, low, not known), cholesterol status (high, low, unknown), smoking status (current, non, not known), angina status (yes, no, unknown) and stroke status (yes, no, not known). The parameters `cp` and `maxdepth` are set so that only a moderate amount of pruning takes place. The results are that the root node is determined by angina status which has the strange property that the "not known" group has the worst mortality status and is therefore split from the yes/no categories. The yes/no group formed a terminal node classified as "live" as 957 in that cohort were alive while 239 died. The "angina not known" group was further split into the not known group (as a terminal node) among whom 65 died and only 3 remained alive and the "stroke yes/no" group that was further split into two terminal nodes, classified as "live" if the smoking status were current or ex (10 alive, 6 dead) and "dead" if the smoking status were non or not known (4 alive, 11 dead). This additional structure is questionable based on the sparse numbers in the right-hand branches of the tree and indicates that possibly more pruning would be warranted.

```
library(rpart)
library(DAAG)
attach(mifem)
fit<-rpart(outcome~age+premi+highbp+hichol+smstat+angina+stroke,
    method="class",data=mifem, control=rpart.control(maxdepth=3,cp=.001))
par(mfrow=c(1,2), xpd=NA, mar=c(4,4,4,0)+.1)
plot(fit,compress=T)
text(fit,use.n=T, pretty=0,all=T,cex=.6)
#summary(fit)
```

Figure 7.5 *An Example of a Tree-Structured Model*

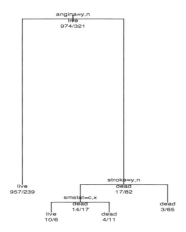

7.7 Exercises

1. A famous study originally published by Frets (1921) compared 25 sets of brothers with respect to different head measurements. Two of the measurements taken were length and breadth of the head. The data can be obtained by downloading `boot` library from R. (See "Install Package(s)" on the "Packages" tab in the R menu.) Once `boot` is downloaded, type the following at the R command line: `library boot; data(frets); attach(frets); frets.`

(a) Make scatterplots (second eldest versus eldest in each) of (1) the head length and (2) the head breadth measurements.

(b) Create side by side box plots for the eldest and second eldest brothers of (1) the head length and (2) the head breadth measurements.

(c) Find sample means, standard errors for the measurements and correlation between measurements.

(d) Test $H_0 : \mu_1 = \mu_2$ versus $H_1 : \mu_1 \neq \mu_2$ where μ_1 and μ_2 represent the true means of the head lengths and breadths for the eldest versus the second eldest sons, respectively. (Hint: First, take the differences between the each set of brothers with respect to length and breadth and then, using methods outlined in section 7.3.1 [one-sample multivariate methods], test whether or not the mean (bivariate) vector is different from 0. The analyses can be performed either in R or SAS or other packages. For those using packages other than R, data can be pasted into the packages from R as input.)

(e) Does the result of the hypothesis test seem to confirm the visual summarization you did in part **(b)**?

2. Consider again the Anorexia data from *Small Data Sets*, dataset # 285 [61] which was referred to in Exercise 5.5. The data consists of pre- and postbody weights, in *lbs*, of 72 young girls receiving three different therapeutic approaches for treating anorexia; cognitive behavorial training [CBT], control [C] treatment, and family therapy [FT].

(a) Consider the pre- and postweights to be bivariate outcomes and conduct a MANOVA to determine if the treatments significantly differ. If so, perform pairwise comparisons to determine where the differences occur.

(b) State your conclusions.

(c) Compare this approach to that in Exercise 5.5 where the posttreatment weight was analyzed as a univariate outcome using the pretreatment weight as a covariate. State the pros and cons of both approaches.

3. Consider the birthweight data in the R library called MASS presented in Exercise 6.2. Refer back to that exercise to see the full description of the data. (Do not forget to type library(MASS) then?birthwt. Then, under "Examples," one can copy all of the commands and paste them into the R console.)

(a) Refit the logistic model as was done in Exercise 6.2 (or the data could be pasted into SAS and proc logistic could be run). Based on the predicted model, for each birth, calculate the probability of that baby having a low birthweight. (If you use R, type ?predict.glm at the console. You can also use proc logistic to do this problem [see Example 7.5].)

   (I) If the probability you calculate from your model is < 0.5, then classify the predicted birthweight as being normal whereas as if is ≥ 0.5, then classify the predicted birthweight as low. Compare your predictions to the actual outcomes (low). If the predicted outcome and observed outcome are both the same, then you'll label the prediction as a "correct classification," whereas if the predicted outcome and observed outcome are not the same you'll label the prediction as a "misclassification." After this has been done for all of the births, calculate the misclassification proportion.

   (II) Repeat the process in part **(I)**, but use the probability cutoff values as < 0.6 and ≥ 0.6 to classify the birth weights as low or normal. You can use either R or SAS to do this.

(b) From the results obtained in part **(a)**, state your conclusions.

4. Consider again the birthweight data presented in the previous exercise and in Exercise 6.2.

(a) Instead of using a logistic model as was done previously, fit a classification tree. (Hint: See Example 7.6 and make sure that the `rpart` library has been downloaded, then at your R console, type `libary(rpart)` followed by `?rpart`.)

(b) From the results obtained in part (a), state your conclusions.

(c) Why might a discriminant analysis be a poor choice to classify the observations in this dataset?

CHAPTER 8

Analysis of Repeated Measures Data

8.1 Introduction

Often in clinical, epidemiological, or laboratory studies the response of each individual is recorded repeatedly over time. In such experiments, comparisons among treatments (or other characteristics of primary interest) are being assessed over time. This type of experiment is often called a *repeated measures* experiment. Repeated measures designs can be employed when, for example, each subject receives many different treatments so that different subjects may receive different *sequences* of treatments. Such studies are often referred to as "crossover" designs and will not be our focus in this chapter. Two excellent references for readers who have an interest in crossover designs are Fleiss [49] and Brown and Prescott [14]. In this chapter, we'll consider repeated measures data where each subject, animal, or experimental unit receives the same treatment and is measured repeatedly or is repeatedly observed under the same condition. We're interested in either characterizing a single group of such subjects or possibly comparing two or more groups of subjects where the individuals within a given group each receive the same treatment over time. Treatments or conditions *across groups* of subjects will be different. Because each subject is associated with only one treatment, the subject effect is said to be *nested within* the treatment effect. The general data structure for such studies is depicted below.

Data Structure

| | Treatment 1 | | | ... | | Treatment g | | |
Subject	Time 1	...	Time s	...	Subject	Time 1	...	Time s
1	y_{111}	...	y_{1s1}	...	1	y_{g11}	...	y_{gs1}
2	y_{112}	...	y_{1s2}	...	2	y_{g12}	...	y_{gs2}
$\vdots$	$\vdots$	...	$\vdots$	...	$\vdots$	$\vdots$	...	$\vdots$
n_1	y_{11n_1}	...	y_{1sn_1}	...	n_g	y_{g1n_g}	...	y_{gsn_g}

It should be noted here that "Subject 1" within treatment group 1 is a different individual than "Subject 1" within group g. Also, note that the different treatment groups may have different numbers of subjects.

189

8.2 Plotting Repeated Measures Data

With modern software, there are many ways to plot repeated measures data. Such plots are useful for visualizing different features of the data. For our examples, we'll consider rat weight data from a toxicity study originally presented by Box [11] and later reproduced by Rao (1958) [118].

8.2.1 Arranging the Data

A practical issue that arises when one is writing programs to plot or analyze repeated measures data is how best to arrange the data so that the algorithm of interest can be implemented. One way to arrange repeated measures data is to arrange all of each individual's data on the same data line so that repeated measures and other covariates each are defined as a variable for each record of the data file. Such an arrangement is useful for plotting programs such as matplot in R and S-plus and for analysis algorithms like that used in the REPEATED statement of PROC GLM. An alternative arrangement is employed by arranging one of the repeated observations per line along with a time indicator and the other variables for a given individual. Such arrangements are useful for SAS algorithms like PROC GPLOT and PROC MIXED or algorithms in R or S-plus such as plot and lme.

Example 8.1. We now consider the rat weight data presented visually by Box (1950) [11] and later, in full detail, by Rao (1958) [118]. The data represent the weekly weights of 27 rats who are split into three groups: a control group consisting of 10 rats, a group consisting of 7 rats receiving the compound thyroxin, and a group consisting of 10 rats receiving the compound thiouracil. The data as presented by Rao consisted of intial weights of the rats and their weekly weight gains over the four weeks after the initiation of the administration of the compounds. For the purposes of this example, in Table 8.1, we present the actual weights of the rats over the five week period. As is demonstrated in Example 8.3, the rates of change for each animal could be converted into their actual weights at each time point by using the DATA step of SAS. These data are typical of toxicity studies done on compounds which may affect growth rates of young animals.

Table 8.1. Weekly Weights for Rats Receiving Different Compounds

Control					Thyroxin					Thiouracil				
Wt 0	Wt 1	Wt 2	Wt 3	Wt 4	Wt 0	Wt 1	Wt 2	Wt 3	Wt 4	Wt 0	Wt 1	Wt 2	Wt 3	Wt 4
57	86	114	139	172	59	85	121	156	191	61	86	109	120	129
60	93	123	146	177	54	71	90	110	138	59	80	101	111	122
52	77	111	144	185	56	75	108	151	189	53	79	100	106	133
49	67	100	129	164	59	85	116	148	177	59	88	100	111	122
56	81	104	121	151	57	72	97	120	144	51	75	101	123	140
46	70	102	131	153	52	73	97	116	140	51	75	92	100	119
51	71	94	110	141	52	70	105	138	171	56	78	95	103	108
63	91	112	130	154						58	69	93	114	138
49	67	90	112	140						46	61	78	90	107
57	82	110	139	169						53	72	89	104	122

To initiate the analysis of these data, one might first plot each rat's weight trajectory over time. The plots often enhance the interpretation of the results of the formal statistical analyses of the data. As was indicated above, two arrangements of the data are implemented for graphing and analytic purposes. In the first arrangement, each record contains all of the data for a given rat including the group indicator (gp), the rat id number (rat) and the weekly weight measurements (wt0–wt4 in SAS or ratwts in *S*-plus / R). Hence, the first arrangement is said to be in a "multivariate form" [68]. In the second arrangement, each record contains a single observation (measurement) (ratwt) along with the group indicator (gp) and animal ID number (rat). Because each record represents a single observation, the second arrangement is said to be in a "univariate form" in Hedeker and Gibbons [68]. The two arrangements of these data are depicted below.

Arrangement # 1

```
1  01  57  86  114  139  172
                 ⋮
1  10  57  82  110  139  169
2  01  59  85  121  156  191
                 ⋮
2  07  52  70  105  138  171
3  01  61  86  109  120  129
                 ⋮
3  10  53  72   89  104  122
```

Arrangement # 2

```
1  01   57  0
1  01   86  1
1  01  114  2
1  01  139  3
1  01  172  4
1  02   60  0
1  02   93  1
            ⋮
3  10   53  0
3  10   72  1
3  10   89  2
3  10  104  3
3  10  122  4
```

8.2.2 *Some Plot Routines*

In the examples below, several plots are demonstrated. These particular plots were implemented using R but similar plots could be produced using SAS or a number of other high level computer programs.

Example 8.2. The first program below allows one to display overlayed plots of the rat weights. The format of the input data is according to arrangement # 1 given above. Individual trajecteory plots for the rat weight data by treatment group are given in Figure 8.1. Each rat's data is represented by an interpolated curve.

```
par(mfrow=c(3,1))
y.lim<-c(50,200)
matplot(0:4,t(ratwts[1:10,3:7]),xlab="Week",ylab="Weight (g)",
        main="Control Group",type="b",lwd=rep(2,10),ylim=y.lim)
matplot(0:4,t(ratwts[11:17,3:7]),xlab="Week",ylab="Weight (g)",
        main="Thyroxin Group",type="b",lwd=rep(2,7),ylim=y.lim)
matplot(0:4,t(ratwts[18:27,3:7]),xlab="Week",ylab="Weight (g)",
        main="Thiouracil Group",type="b",lwd=rep(2,10),ylim=y.lim)
```

Figure 8.1 *Plots of Weekly Weight Measurements for Rats Receiving Different Compounds*

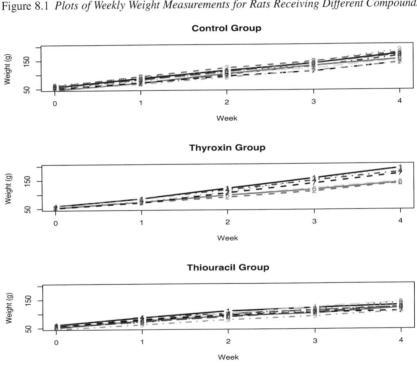

Since the plots are all on the same scale, one can see that the trajectories of weights for the group of rats receiving the Thiouracil compound appear to be lower than those in the other two groups. A second set of plots, called "panel plots," can be created, allowing each rat's data to be displayed separately. Within R, these plots are created using the library called "lattice" (Sarkar [129]). The data for this type of plot should be formatted similar to arrangement #2 above. A panel plot for the rat weight data is given in Figure 8.2.

```
# Panel plots
library(lattice)
# Plots for each rat
colr<-"black"
xyplot(wt~week|rat+gp,data=ratwt, type="b",
       lwd=2,ylab="Weight (g)",xlab="Week",col=colr)
```

Figure 8.2 *Panel Plots of Rat Weights for Groups 1 [Lowest] Through Group 3 [Highest]*

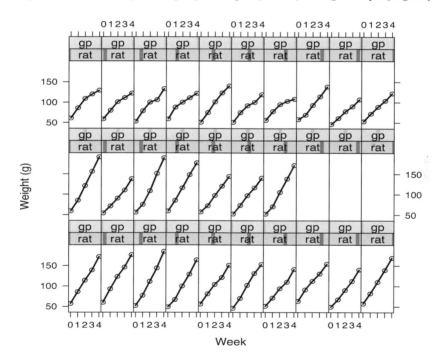

Another plot of interest in cases where the measurements over time are the same for each individual within a study is that of means plus or minus either their standard errors or means and their pointwise 95% confidence intervals (see Figure 8.3).

Figure 8.3 *Mean Weight Plots and Standard Error Bars over Weeks by Group*

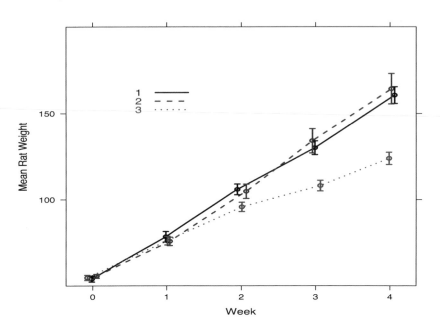

8.3 Univariate Approaches

One method of analyzing repeated measures data that is still used but with limited utility is the "classical method," which is accomplished using analysis of variance (ANOVA) methods or a "univariate approach." Consider the following model:

$$Y_{ijk} = \mu + \alpha_i + \pi_{k(i)} + \tau_j + (\alpha\tau)_{ij} + \epsilon_{ijk} , \quad (8.1)$$

where Y_{ijk} is the response of kth subj. within i^{th} group at time j, μ is the grand mean, α_i is the offset due i^{th} group, $\pi_{k(i)}$ is the offset due k^{th} subj. in group i, τ_j is the offset due to time j, $(\alpha\tau)_{ij}$ is the group × time interaction, and ϵ_{ijk} is the error for k^{th} response of jth subj. in i^{th} group.

Note that we assume that the responses among individuals are independent and that the responses within each treatment have the same correlations across time (an assumption that may not hold in practice). In this hypothesized experiment, there are a total of $n = n_\bullet = \sum_{i=1}^{g} n_i$ subjects (individuals) in g groups or treatments measured at s time points.

ANOVA Table

Source	d.f.	Sums of Squares	MSE	F
Treatment [TRT]	$g-1$	$\sum_{i=1}^{g} n_i(\bar{y}_{i\bullet\bullet} - \bar{y}_{\bullet\bullet\bullet})^2$	$\frac{SS_{TRT}}{g-1}$	$\frac{MS_{TRT}}{MS_{SUBJ}}$
SUBJ within TRT	$n_\bullet - g$	$s\sum_{i=1}^{g}\sum_{k=1}^{n_i}(\bar{y}_{i\bullet k} - \bar{y}_{i\bullet\bullet})^2$	$\frac{SS_{SUBJ}}{n_\bullet - g}$	
Time	$s-1$	$n_\bullet\sum_{j=1}^{t}(\bar{y}_{\bullet j\bullet} - \bar{y}_{\bullet\bullet\bullet})^2$	$\frac{SS_{TIME}}{s-1}$	
TRT $\times$ Time	$(g-1)(s-1)$	$\sum_{i=1}^{g}\sum_{j=1}^{t} n_i(\bar{y}_{ij\bullet} - \bar{y}_{i\bullet\bullet} - \bar{y}_{\bullet j\bullet} + \bar{y}_{\bullet\bullet\bullet})^2$	$\frac{SS_{TRT\times TIME}}{(g-1)(s-1)}$	
Error	$(n_\bullet - g)(s-1)$	$\sum_{i=1}^{g}\sum_{j=1}^{t}\sum_{k=1}^{n_i}(y_{ijk} - \bar{y}_{i\bullet k} - \bar{y}_{\bullet jk} + \bar{y}_{\bullet\bullet k})^2$	$\frac{SS_E}{(n_\bullet - g)(s-1)}$	

The repeated measures design combines several ideas already introduced. First of all, we will assume that the treatment and time effects are fixed, i.e., inference is limited to those levels used in the particular experiment being performed. The "subject or individual effect" is random, i.e., we assume that the subjects (experimental units) in our experiment represent a random sample from a large population. Hence, these types of models are known as *mixed effects* models. Furthermore, the subject effect itself can vary and so is associated with a probability distribution, which is assumed to be normal in our case.

Each observation for each subject form a *block* having the property that variability within the blocks (that is, within subject) is generally smaller than is the variability across subjects. The $\pi_k(i)$ term in equation (8.1) indicates that each subject is nested within a given treatment. Hence, in the repeated measures design described above, each subject is given one and only one treatment over the course of the study.

Using the ANOVA table, a treatment effect is analyzed univariately by dividing the treatment mean square, that is, $\frac{TRT\ SS}{g-1}$ by the subject (within treatment) mean square, that is, $\frac{SUBJ\ SS}{n_\bullet - g}$. If we assume that the errors are independently and identically distributed (i.i.d.) $N(0, \sigma^2)$ and that the subjects $[\pi_{k(i)}]$ are i.i.d. $N(0, \sigma_\pi^2)$ then the ratio, $F = \frac{TRT\ SS/(g-1)}{SUBJ\ SS/(n_\bullet - g)} = \frac{TRT\ Mean\ Square}{SUBJ\ Mean\ Square}$ is distributed as an F-distribution with $g-1$ and $n_\bullet - g$ degrees of freedom.

To test whether or not there is a time effect of whether or not there is a treatment $\times$ time interaction requires more assumptions than the test of the overall treatment effect. To perform valid F-tests for testing time and treatment $\times$ time effects, we must make the following additional assumption: the population of subjects must have uniform or *compound symmetric* correlation structure, that is, two measurements made at adjacent times will have the same correlation as two measurements made at non-adjacent times. That assumption is not unreasonable in studies where there are sufficient lengths between successive measurements. However, in many studies with observations measured at short time intervals, the assumption does not hold because observations made at times closer together tend to be more highly correlated than those which are made at times farther apart. Box (1950) [11] provided a test for the uniformity of the correlation structure in a repeated measures experiment. If the test for uniformity *fails to reject*, that is, the correlation structure is not significantly different from compound symmetry, then one can form the statistics

$$F = \frac{\text{TIME SS}/(s-1)}{\text{Error SS}/(n_\bullet - g)(s-1)} \sim F_{s-1,(n_\bullet-g)(s-1)}, \text{ and}$$

$$F = \frac{\text{Interaction SS}/(g-1)(s-1)}{\text{Error SS}/(n_\bullet - g)(s-1)} \sim F_{(g-1)(s-1),(n_\bullet-g)(s-1)}, \text{ and}$$

to test for time and time $\times$ treatment interaction effects, respectively. In the cases when the correlation structure is significantly different from uniform then correction factors can be added to the degrees of freedom in the F-distributions so that more valid approximations of the distributional form of the statistics are obtained.

Example 8.3. The programs below are used to analyze the rat weight data as in Rao [118]. The program below first reads data that has a format where the initial weight and weekly weight gains are displayed for each rat. The weight gains are then transformed into the actual weights. This would be similar to arrangement # 1 above. The data are then transformed into a format similar to arrangement # 2 above so that analyses can be performed using different SAS procedures.

```
/* ****************************************************************
    Code for analyzing data from Rao, Biometrics, 1958
   ***************************************************************/
OPTIONS LS='77 pageno=1;
title 'Initial weights & weekly gains of rats receiving 3 different treatments';
title2 'Data from Rao (1958), Biometrics, Vol 14(1), p.4';
data rat0;
input rat Y0 Y1 Y2 Y3 Y4 @@;
if _n_ le 10 then gp = 1;
else if _n_ gt 10 and _n_ le 17 then gp = 2;
else gp = 3;
datalines;
  1 57 29 28 25 33     2 60 33 30 23 31     3 52 25 34 33 41
  4 49 18 33 29 35     5 56 25 23 17 30     6 46 24 32 29 22
  7 51 20 23 16 31     8 63 28 21 18 24     9 49 18 23 22 28    10 57 25 28 29 30
  1 59 26 36 35 35     2 54 17 19 20 28     3 56 19 33 43 38
  4 59 26 31 32 29     5 57 15 25 23 24     6 52 21 24 19 24     7 52 18 35 33 33
  1 61 25 23 11 9      2 59 21 21 10 11     3 53 26 21  6 27
  4 59 29 12 11 11     5 51 24 26 22 17     6 51 24 17  8 19
  7 56 22 17 8 5       8 58 11 24 21 24     9 46 15 17 12 17    10 53 19 17 15 18
;    run;

proc format;
value g 1='Control' 2='Thyroxin' 3='Thiouracil';    run;

title 'Weekly weights of rats receiving 3 different treatments';
data rat1; set rat0;
wt0=y0;   wt1=y0+y1;  wt2=wt1+y2;  wt3=wt2+y3;   wt4=wt3+y4;
label wt0='Init Wt' wt1='Week 1 Wt' wt2='Week 2 Wt'
      wt3='Week 3 Wt' wt4='Week 4 Wt';
drop y0--y4; run;

* Put data into Arrangement 2;
```

```
DATA rat2; set rat1;
  array wt 5 wt0-wt4;
   KEEP gp rat week ratwt;
   DO I=1 TO 5;
       ratwt = wt{I};
       week = I-1; output;
   END;
run;

PROC GLM data=rat2;
    CLASS gp week rat;
    MODEL ratwt = gp  rat(gp) week gp*week;
    TEST  H=gp  E=rat(gp);
RUN; quit;
```

```
------------------------- Partial Output ----------------------------
                                Sum of
Source                   DF      Squares    Mean Square  F Value  Pr > F
Model                    38   170008.5507   4473.9092     86.93  <.0001
Error                    96     4940.7086     51.4657
Corrected Total         134   174949.2593

Source                   DF   Type III SS   Mean Square  F Value  Pr > F
gp                        2     6638.6535   3319.3268     64.50  <.0001
rat(gp)                  24    10300.2057    429.1752      8.34  <.0001
week                      4   147539.3064  36884.8266    716.69  <.0001
gp*week                   8     6777.2470    847.1559     16.46  <.0001

 Tests of Hypotheses Using the Type III MS for rat(gp) as an Error Term
Source                   DF   Type III SS   Mean Square  F Value  Pr > F
gp                        2   6638.653545  3319.326772      7.73  0.0026
```

Note that the statement TEST H=gp E=rat(gp); in SAS forces the program to correctly use the "rat within group" mean square error as the error term for testing the overall group effect. Thus, an appropriate F-test is formulated by the test statistic, $F \approx 3319.3268/429.1752 \approx 7.73$, which under $H_0 : \mu_1 = \mu_2 = \mu_3$ is $\sim F_{g-1,n_\bullet - g} = F_{2,24}$. The critical value of $F_{2,24}$ at the $\alpha = 0.01$ is 5.61 so that the group effect is significant at the 1% level ($p = 0.0026$). Thus, it is clear that there is a significant effect due to the group; however, the nature of that effect with the analysis above is not known.

The PROC MIXED program listed below will produce similar output to that given above. This program will also output the estimated pooled subject covariance and correlation matrices across the five time points. One can see the compound symmetric structure in the two matrices. As we will shortly see, PROC MIXED gives us the capability of entertaining a much broader class of models than considered here.

```
proc mixed data=rat2;
class gp week rat;
model ratwt=gp gp*week;
repeated week/subject=rat(gp) type=cs r rcorr;
run;
```

```
------------------------- Partial Output ----------------------------
                        The Mixed Procedure

             Estimated R Matrix for rat(gp) 1 1

      Row      Col1       Col2       Col3       Col4       Col5
       1     127.01     75.5419    75.5419    75.5419    75.5419
       2     75.5419    127.01     75.5419    75.5419    75.5419
       3     75.5419    75.5419    127.01     75.5419    75.5419
       4     75.5419    75.5419    75.5419    127.01     75.5419
```

| 5 | 75.5419 | 75.5419 | 75.5419 | 75.5419 | 127.01 |

Estimated R Correlation Matrix for rat(gp) 1 1

Row	Col1	Col2	Col3	Col4	Col5
1	1.0000	0.5948	0.5948	0.5948	0.5948
2	0.5948	1.0000	0.5948	0.5948	0.5948
3	0.5948	0.5948	1.0000	0.5948	0.5948
4	0.5948	0.5948	0.5948	1.0000	0.5948
5	0.5948	0.5948	0.5948	0.5948	1.0000

Covariance Parameter Estimates

Cov Parm	Subject	Estimate
CS	rat(gp)	75.5419
Residual		51.4657

Fit Statistics

-2 Res Log Likelihood	897.1
AIC (smaller is better)	901.1
AICC (smaller is better)	901.2
BIC (smaller is better)	903.7

Type 3 Tests of Fixed Effects

Effect	Num DF	Den DF	F Value	Pr > F
gp	2	24	7.73	0.0026
gp*week	12	96	247.85	<.0001

In a follow-up analysis, one can use orthogonal polynomial contrasts introduced in Chapter 5 (see Table 5.1) to test whether there are linear, quadratic, cubic, and quartic trends across weeks.

```
PROC GLM data=rat2; *<-- Uses arrangement #2 of data;
    CLASS gp week rat;
    MODEL ratwt = gp  rat(gp)  week  gp*week;
    TEST  H=gp  E=rat(gp);
    CONTRAST 'Linear' week -2  -1   0   1   2;
    CONTRAST 'Quad'   week  2  -1  -2  -1   2;
    CONTRAST 'Cubic'  week -1   2   0  -2   1 ;
    CONTRAST 'Quartic'  week 1   -4  6 -4   1 ;
Title3 'Repeated measures example: Univariate analysis';
RUN; quit;
```

```
-------------------------- Partial Output --------------------------
```

Source	DF	Type III SS	Mean Square	F Value	Pr > F
gp	2	6638.6535	3319.3268	64.50	<.0001
rat(gp)	24	10300.2057	429.1752	8.34	<.0001
week	4	147539.3064	36884.8266	716.69	<.0001
gp*week	8	6777.2470	847.1559	16.46	<.0001

Contrast	DF	Contrast SS	Mean Square	F Value	Pr > F
Linear	1	147449.5501	147449.5501	2865.01	<.0001
Quad	1	20.4188	20.4188	0.40	0.5303
Cubic	1	0.3172	0.3172	0.01	0.9376
Quartic	1	69.0204	69.0204	1.34	0.2497

From the first analysis, it can be seen that a group × week interaction exists. It is clear from the second analysis that only the linear trend is significant. Hence, we can ultimately fit a model that includes only the linear term along with interaction terms, which account for the linear terms being different in the three groups. Consequently, it appears that rat weights in all groups exihibit a linear trend across weeks but the growth rates (slopes) are different across groups.

```
PROC GLM data=rat2;*<--- This uses arrangement #2 of data;
    CLASS gp week rat;
    MODEL ratwt = gp  rat(gp)  week  gp*week;
    TEST  H=gp  E=rat(gp);
    CONTRAST 'Linear' week -2  -1  0  1   2;
    CONTRAST 'grplin' week*gp -2  -1  0  1  2   2  1  0  -1  -2  0 0 0 0 0,
                      week*gp -2  -1  0  1  2   0 0 0 0 0  2  1  0  -1  -2;
RUN; quit;
```

```
------------------------- Partial Output ---------------------------
Contrast              DF     Contrast SS    Mean Square   F Value   Pr > F
Linear                 1     147449.5501   147449.5501   2865.01   <.0001
grplin                 2       6194.4101     3097.2051     60.18   <.0001
```

8.4 Covariance Pattern Models

As was stated in the previous section, the use of standard ANOVA for modeling repeated measures data requires that the covariance structure for observations across
time points be uniform or *compound symmetric*, meaning that the errors between observations adjacent times have the same correlation as those that are further apart.
Such a covariance (correlation) structure is often not apropos for a given situation.
For example, data from repeated measures studies that occur over short time periods
often are associated with error structures that have higher correlations for observations that are closer together in time than for those further apart in time.

One way to modify this is by use of *covariance pattern models* to attempt to correctly specify the within subject covariance structure (Hedeker and Gibbons) [68].
Specifying the correct covariance structure among observations helps to ensure that
the inferences made about repeated measure models are accurate.

In the previous analysis using PROC GLM, the data were assumed to have a compound
symmetric covariance structure, that is,

$$\Sigma = \sigma^2 \begin{pmatrix} 1 & \rho & \rho & \cdots & \rho \\ \rho & 1 & \rho & \cdots & \rho \\ \vdots & & \ddots & & \vdots \\ \rho & \rho & \cdots & \rho & 1 \end{pmatrix}. \tag{8.2}$$

From this covariance structure, one can see that adjacent observations have the same
correlation as do observations made at time points farther apart. Often, however, observations that are closer together in time will have a higher (positive) correlation
than those made at time points that are further apart. A very common covariance
structure having this property is the autoregressive, lag 1 (*AR(1)*) covariance structure. One form of this type of covariance structure is written as

$$\Sigma = \sigma^2 \begin{pmatrix} 1 & \rho & \rho^2 & \cdots & \rho^{s-1} \\ \rho & 1 & \rho & \cdots & \rho^{s-2} \\ \vdots & & \ddots & & \vdots \\ \rho^{s-1} & \rho^{s-2} & \cdots & \rho & 1 \end{pmatrix}. \tag{8.3}$$

This latter structure implies that correlations between observations will attenuate over time because if $0 < \rho < 1$ then $\rho > \rho^2 > \rho^3 > \ldots$. Sometimes observations are correlated but not with any particular pattern so that an *unstructured* covariance is apropos. Thus,

$$
\Sigma = \begin{pmatrix}
\sigma_1^2 & \sigma_{12} & \sigma_{13} & \cdots & \sigma_{1s} \\
\sigma_{21} & \sigma_2^2 & \sigma_{23} & \cdots & \sigma_{2s} \\
\sigma_{31} & \sigma_{32} & \sigma_3^2 & \cdots & \sigma_{3s} \\
\vdots & \cdots & \cdots & \ddots & \vdots \\
\sigma_{s1} & \sigma_{s2} & \sigma_{s3} & \cdots & \sigma_s^2
\end{pmatrix},
\tag{8.4}
$$

where $\sigma_{ij} = \sigma_{ji}$ for all $i \neq j$, $1 \leq i, j \leq s$. Less commonly, the observations within subjects are independent meaning that there is no correlation between observations made at different time points. Many other covariance structures between observations can exist. A few are outlined in the SAS documentation of the REPEATED statement within PROC MIXED (see http://www.sfu.ca/sasdoc/sashtml/stat/chap41 /sect20.htm#mixedrepeat). The four covariance structures described above can be specified in the REPEATED statement in PROC MIXED with the TYPE option and specifying CS, AR(1), UN, or VC, respectively.

So how does one decide on the most appropriate correlation structure? These models may have different numbers of parameters to be estimated and the models don't necessarily have to be "nested." If models are "nested" then one model can be derived by setting some of the parameters equal to 0. For example, if the ρ in equation (8.2) is 0, then the resulting model would have an *independent* error structure. Since the number of parameters to be estimated are not the same for each type of covariance structure, one can decide on the "best model" by evaluating the *Akaike Information Criterion* (AIC) [2] (or corrected versions of it) and choosing the fit with the minimum AIC (or AICC) value. This criterion is similar to a likelihood ratio test but with $2p$ added as a "penalty" for the number of parameters, p, used to fit the model. A similar criteria called the "Schwarz Information Criterion" or "Bayesian Information Criterion" (BIC) [134] can also be used. In the implementation in PROC MIXED, the "best model" using the BIC criterion is that associated with the minimum value. In cases where the models indeed are nested, the "best model" can be obtained by first calculating $-2 * (\ell_1 - \ell_2)$ where $\ell_2 > \ell_1$ and the ℓ_i represent the natural log of the likelihood functions. If $\Delta p = p_2 - p_1 > 0$ then one could compare this statistic to a χ^2 distribution with Δp degrees of freedom at an appropriate α-level. This is known as a *likelihood ratio test*.

Example 8.4. In Example 8.3, suppose that we wish to entertain a model where the covariance structure among observations across the repeated rat weights is AR(1) instead of compound symmetric. Such a model could be implemented with the following PROC MIXED code:

```
proc mixed data=rat2; *<--- This uses arrangement #2 of data;
class gp week rat;
model ratwt=gp gp*week;
repeated week/subject=rat(gp) type=ar(1) r rcorr;
run;
```

```
---------------------------- Partial Output ----------------------------------
                        The Mixed Procedure

                 Estimated R Matrix for rat(gp) 1 1
   Row       Col1          Col2          Col3          Col4          Col5
    1       137.53        121.31        107.01        94.3923       83.2629
    2       121.31        137.53        121.31        107.01        94.3923
    3       107.01        121.31        137.53        121.31        107.01
    4       94.3923       107.01        121.31        137.53        121.31
    5       83.2629       94.3923       107.01        121.31        137.53

              Estimated R Correlation Matrix for rat(gp) 1 1
   Row       Col1          Col2          Col3          Col4          Col5
    1       1.0000        0.8821        0.7781        0.6863        0.6054
    2       0.8821        1.0000        0.8821        0.7781        0.6863
    3       0.7781        0.8821        1.0000        0.8821        0.7781
    4       0.6863        0.7781        0.8821        1.0000        0.8821
    5       0.6054        0.6863        0.7781        0.8821        1.0000

                    Covariance Parameter Estimates
                    Cov Parm      Subject      Estimate
                    AR(1)         rat(gp)       0.8821
                    Residual                   137.53

                           Fit Statistics
                  -2 Res Log Likelihood          819.6
                  AIC (smaller is better)        823.6
                  AICC (smaller is better)       823.7
                  BIC (smaller is better)        826.2

                        The Mixed Procedure
                  Type 3 Tests of Fixed Effects
                           Num      Den
          Effect           DF       DF      F Value      Pr > F
          gp                2       24       5.84        0.0086
          gp*week          12       96      190.61       <.0001
```

Note that, in our case, the error structure is $AR(1)$ (see equation (8.3)) and $\hat{\sigma}^2 = 137.53$ and $\hat{\rho} = 0.8821$. Furthermore, $0.8821^2 \approx 0.7781$, $0.8821^3 \approx 0.6863$ and $0.8821^4 \approx 0.6054$. Also, notice that the AIC, AICC, and BIC are all substantially reduced indicating that the $AR(1)$ covariance structure is better to use for these data than is the compound symmetric covariance structure. One could explore other covariance structures to find a "best structure." Also, the F-values for the group and group $\times$ week interaction terms are altered but the general inference is the same, namely, that there is an overall group effect with respect to rat weight and that the time effect is significantly different across groups.

8.5 Multivariate Approaches

Univariate approaches to the analysis of repeated measures data suffer from several drawbacks. Two such drawbacks are

- If a small number of observations are missing for some individuals, the programs that implement univariate methods usually remove all of those individuals' observations in the analysis. Such a strategy can lead to one making biased and inefficient inferences about group differences over time or the nature of the pooled trajectories of the groups.

- It is difficult to implement univariate models that are useful for predicting group or individual trajectories.

Example 8.5. Another approach to the analysis can be employed in PROC GLM using arrangement # 1:

```
PROC GLM data=rat1;  *<--- This uses arrangement #1 of data;
  CLASS gp; format gp g.;
  MODEL wt0-wt4=gp/NOUNI;
  REPEATED week;
  means gp;
title 'Rat weights for three different treatments';
title3 'Repeated measures example';
run;  quit;
```

The analysis below is excerpted from the output given by PROC GLM. Notice that multivariate tests for both the time (week) and time × group interaction are performed by using multivariate methods such as Wilk's Λ. (Others such as Pillai's trace, Roy's greatest root, etc., are available but not displayed in the partial output given below.) An advantage of this analysis is one can get explicit p-values for testing for a general effect across weeks and whether or not a group × week interaction exists while adjusting for an error structure that is not uniform across time (weeks) [see the G–G (Greenhouse–Geisser) and H–F (Huynh–Feldt) p-values]. A more thorough discussion of these tests is given in Fleiss [49]. However, in many cases, with the use of covariance pattern models, these corrections are not necessary. In any case, one can infer that both the "time trend" and the interaction by group over time are highly significant.

```
------------------------- Partial Output -------------------------------
                  Rat weights for three different treatments
                      From Rao, Biometrics, 1958
                       Repeated measures example


                 Manova Test Criteria and Exact F Statistics
                      for the Hypothesis of no week Effect
                        H = Type III SSCP Matrix for week
                            E = Error SSCP Matrix

                        S=1     M=1     N=9.5
Statistic                     Value   F Value   Num DF   Den DF   Pr > F
Wilks' Lambda             0.01764192   292.34        4       21   <.0001


                 Manova Test Criteria and F Approximations
                    for the Hypothesis of no week*gp Effect
                        H = Type III SSCP Matrix for week*gp
                            E = Error SSCP Matrix

                        S=2    M=0.5    N=9.5

Statistic                     Value   F Value   Num DF   Den DF   Pr > F
Wilks' Lambda             0.26548584     4.94        8       42   0.0002

                 Repeated Measures Analysis of Variance
               Tests of Hypotheses for Between Subjects Effects

Source               DF    Type III SS   Mean Square   F Value   Pr > F
gp                    2     6638.65354    3319.32677      7.73   0.0026
Error                24    10300.20571     429.17524
```

Source	DF	Type III SS	Mean Square	F Value	Pr > F
week	4	147539.3064	36884.8266	716.69	<.0001
week*gp	8	6777.2470	847.1559	16.46	<.0001
Error(week)	96	4940.7086	51.4657		

	Adj Pr > F	
Source	G - G	H - F
week	<.0001	<.0001
week*gp	<.0001	<.0001
Error(week)		

Greenhouse-Geisser Epsilon	0.3316
Huynh-Feldt Epsilon	0.3729

Multivariate techniques are useful for answering more general questions that may arise in repeated measures studies such as: "Is there a difference among treatments in the rates of change over time?" and "Can I use my models to predict the outcomes of groups or of individuals?" Such questions are better answered by regression methods that consider time to be a continuous variable instead of a factor as is assumed in models using ANOVA techniques. Potthoff and Roy [115] considered one such model, which can be written as

$$\underbrace{\mathbf{Y}}_{n_\bullet \times s} = \underbrace{\mathbf{A}}_{n_\bullet \times g} \underbrace{\mathbf{B}}_{g \times q} \underbrace{\mathbf{P}}_{q \times s} + \underbrace{\mathbf{E}}_{n_\bullet \times s}, \tag{8.5}$$

where the $\mathbf{A}$ and $\mathbf{P}$ are fixed matrices that allow one to test hypotheses between and within subjects, respectively and $\mathbf{E}$ is a matrix with rows that are each multinormally distributed and having the property that different rows are independent of each other. Here $n_\bullet = \sum_{i=1}^{g} n_i$ is the total number of subjects, s is the number of time points, g is the number of groups, and q is the number of parameters of the fitted polynomial (e.g., for a line, $q = 2$). As Potthoff and Roy [115] pointed out in their paper, with the structure outlined above, one could test differences across groups with the $\mathbf{A}$ matrix and could fit a polynomial model for the subjects with the $\mathbf{P}$ matrix.

For the rat weight data, the $\mathbf{Y}$, $\mathbf{A}$, $\mathbf{B}$, and $\mathbf{P}$ matrices are

$$\underbrace{\mathbf{Y}}_{27 \times 5} = \begin{pmatrix} Y_{110} & Y_{111} & Y_{112} & Y_{113} & Y_{114} \\ \vdots & \vdots & \vdots & \vdots & \vdots \\ Y_{1,10,0} & Y_{1,10,1} & Y_{1,10,2} & Y_{1,10,3} & Y_{1,10,4} \\ Y_{210} & Y_{211} & Y_{212} & Y_{213} & Y_{214} \\ \vdots & \vdots & \vdots & \vdots & \vdots \\ Y_{270} & Y_{271} & Y_{272} & Y_{273} & Y_{274} \\ Y_{310} & Y_{311} & Y_{312} & Y_{313} & Y_{314} \\ \vdots & \vdots & \vdots & \vdots & \vdots \\ Y_{3,10,0} & Y_{3,10,1} & Y_{3,10,2} & Y_{3,10,3} & Y_{3,10,4} \end{pmatrix} = \begin{pmatrix} 57 & 86 & 114 & 139 & 172 \\ \vdots & \vdots & \vdots & \vdots & \vdots \\ 57 & 82 & 110 & 139 & 169 \\ 59 & 85 & 121 & 156 & 191 \\ \vdots & \vdots & \vdots & \vdots & \vdots \\ 52 & 70 & 105 & 138 & 171 \\ 61 & 86 & 109 & 120 & 129 \\ \vdots & \vdots & \vdots & \vdots & \vdots \\ 53 & 72 & 89 & 104 & 122 \end{pmatrix}$$

$$\underbrace{\mathbf{A}}_{27\times 3} = \begin{pmatrix} 1 & 0 & 0 \\ \vdots & \vdots & \vdots \\ 1 & 0 & 0 \\ 0 & 1 & 0 \\ \vdots & \vdots & \vdots \\ 0 & 1 & 0 \\ 0 & 0 & 1 \\ \vdots & \vdots & \vdots \\ 0 & 0 & 1 \end{pmatrix}, \quad \underbrace{\mathcal{B}}_{3\times 2} = \begin{pmatrix} \mathcal{B}_{10} & \mathcal{B}_{11} \\ \mathcal{B}_{20} & \mathcal{B}_{21} \\ \mathcal{B}_{30} & \mathcal{B}_{31} \end{pmatrix}, \text{ and }$$

$$\underbrace{\mathbf{P}}_{2\times 5} = \begin{pmatrix} 1 & 1 & 1 & 1 & 1 \\ t_0 & t_1 & t_2 & t_3 & t_4 \end{pmatrix} = \begin{pmatrix} 1 & 1 & 1 & 1 & 1 \\ 0 & 1 & 2 & 3 & 4 \end{pmatrix},$$

respectively, and $\mathbf{E}$ is a 27×5 matrix for which each row has a multivariate normal distribution and the different rows are independent of each other.

Example 8.6. Consider the rat data and suppose we wish to fit a line to the weights associated with each of the groups. Four linear models were fit by specifying the residual error covariance structures as compound symmetry, $AR(1)$, independent, and unstructured. The AICs associated with the four models had values 958.1, 889.0, 1014.0, and 840.0, respectively. BIC values had the same pattern. Hence, the "best model" as determined by the AIC and BIC criteria was the linear model associated with an error structure that had a unstructured covariance matrix. That model was specified with the following PROC MIXED code:

```
proc mixed data=rat2; *<--- This uses arrangement #2 of data;
CLASS gp rat;
model ratwt = gp week gp*week /solution;
repeated/ type=un subject=rat(gp);
run;
```

```
------------------------- Partial output ----------------------------
         Data Set                    WORK.RAO2
         Dependent Variable          ratwt
         Covariance Structure        Unstructured
         Subject Effect              rat(gp)
         Estimation Method           REML
         Residual Variance Method    None
         Fixed Effects SE Method     Model-Based
         Degrees of Freedom Method   Between-Within


                       Class Level Information

         Class     Levels    Values
         gp             3    1 2 3
         rat           10    1 2 3 4 5 6 7 8 9 10


                             Dimensions
                 Covariance Parameters            15
                 Columns in X                      8
                 Columns in Z                      0
                 Subjects                         27
                 Max Obs Per Subject               5
                 Observations Used               135
```

```
                    Observations Not Used              0
                    Total Observations              135


                          Fit Statistics
                 -2 Res Log Likelihood            810.0
                 AIC (smaller is better)          840.0

                     Solution for Fixed Effects

                              Standard
  Effect       gp    Estimate    Error     DF    t Value    Pr > |t|
  Intercept          54.0131    1.3065     24     41.34     <.0001
  gp           1      1.7240    1.8476     24      0.93     0.3601
  gp           2      4.4420    2.0360     24      2.18     0.0392
  gp           3           0       .        .        .         .
  week               19.3510    1.0626     24     18.21     <.0001
  week*gp      1      7.3145    1.5027     24      4.87     <.0001
  week*gp      2      5.7303    1.6559     24      3.46     0.0020
  week*gp      3           0       .        .        .         .

                   Type 3 Tests of Fixed Effects
                          Num       Den
                Effect     DF        DF     F Value    Pr > F
                gp          2        24        2.38     0.1138
                week        1        24     1305.75     <.0001
                week*gp     2        24       12.83     0.0002
```

One major difference between this model and those introduced in Examples 8.3 and 8.4 is that the variable week here is not a "class" variable but rather, is considered to be continuous with a coefficient associated with the rate of growth of the weights of the rats over time. Thus, the estimates associated with intercept and week represent the intercept and slope of the line associated with group 3. The other values associated with "gp" and "week*gp" represents the *offsets* from group 3 of the intercepts and slopes associated with groups 1 and 2, respectively. The estimated covariance matrix between observations across the times (not shown in the output) is given by

$$\widehat{\Sigma}_e = R = \begin{pmatrix} 24.5219 & 40.6847 & 36.4190 & 31.6378 & 20.7326 \\ 40.6847 & 89.2421 & 82.3871 & 67.8461 & 43.1063 \\ 36.4190 & 82.3871 & 103.59 & 116.60 & 112.00 \\ 31.6378 & 67.8461 & 116.60 & 194.54 & 219.22 \\ 20.7326 & 43.1063 & 112.00 & 219.22 & 293.49 \end{pmatrix}.$$

From the output, we can derive the model for group 1 as $\widehat{y} = (54.0131 + 1.724) + (19.3510 + 7.3145)(\text{week}) = \widehat{y} = 55.7371 + 26.6655(\text{week})$; the model for group 2 as $\widehat{y} = (54.0131 + 4.442) + (19.3510 + 5.7303)(\text{week}) = \widehat{y} = 58.4551 + 25.0813(\text{week})$; and the model for group 3 as $\widehat{y} = 54.0131 + 19.3510(\text{week})$. The parameters of this model are all considered to be *fixed effects*. In the next section, we will allow these parameters to possibly vary according to individual.

8.6 Modern Approaches for the Analysis of Repeated Measures Data

More recent methodology has focused on multivariate techniques even in the presence of missing and/or unequally spaced observations made on each subject. These approaches explicitly model both the fixed and random effects associated with repeated measures data. The seminal papers by Harville (1976) [65] and Laird and

Ware (1982) [83] were crucial in the development of these models. Consider a model for N subjects given by

$$\underbrace{\mathbf{y}_k}_{s_k \times 1} = \underbrace{\mathbf{X}_k}_{s_k \times q} \underbrace{\boldsymbol{\beta}}_{q \times 1} + \underbrace{\mathbf{Z}_k}_{s_k \times s} \underbrace{\mathbf{b}_k}_{s \times 1} + \underbrace{\mathbf{e}_k}_{s_k \times 1} , \tag{8.6}$$

$i = 1, \ldots, n$ and $s \leq q$; where $\mathbf{b}_k \sim MVN(\mathbf{0}, \mathbf{G}_k)$, $\mathbf{e}_k \sim MVN(\mathbf{0}, \mathbf{R}_k)$ and and $\mathrm{Cov}(\mathbf{e}_k, \mathbf{b}_k) = \mathbf{0}$.

In this formulation, the $\mathbf{y}_k$ represents the observation vector for individual k and s_k is the number of observations (time points) for that individual. Hence, numbers of observations can vary across individuals. This model has a population parameter vector, $\boldsymbol{\beta}$, *which is common to all members of a defined population*, together with parameter vectors, $\mathbf{b}_k$, *which are specific to each individual*. Hence, the model in equation (8.6) can accommodate situations where the population of individuals has a polynomial "growth curve" or "trajectory" over time and where one or more parameters of the polynomial are allowed to vary across individuals. For example, if the trajectory fitted to the complete population is a line then the construct of the model allows either the intercept or the slope or *both* the intercept and slope of that line to vary across individuals. Because $\mathcal{E}(\mathbf{b}_k) = 0$, the $\mathbf{b}_k$'s represent the *offsets* of the population parameters specific to individuals. Hence, each individual has his/her/its own unique trajectory. Models where polynomials are fit at both a group level (fixed effect) and an individual level (random effect) are sometimes referred to as *random coefficients* models, see Brown and Prescott [14] and Hedeker and Gibbons [68]. The covariance matrix of $\mathbf{y}_k$ in equation 8.6 is given by

$$\mathbf{V}_k = \mathbf{\Sigma}_{\mathbf{y}_k} = \mathrm{var}(\mathbf{X}_k \boldsymbol{\beta}) + \mathrm{var}(\mathbf{Z}_k \mathbf{b}_k) + \mathrm{var}(\mathbf{e}_k) = \mathbf{Z}_k \mathbf{G} \mathbf{Z}_k' + \mathbf{R}_k . \tag{8.7}$$

The second equality in equation 8.7 is obtained by observing that for the fixed effect part of the model, $\mathrm{var}(\mathbf{X}_k \boldsymbol{\beta}) = \mathbf{0}$. In most cases, it is assumed that $\mathbf{V}_k$ and $\mathbf{R}_k$ have the same form for all individuals or all individuals within a group but that the individuals may have different numbers of observations, which means that $\underbrace{\mathbf{R}_k}_{s_k \times s_k}$ may have different dimensions for different individuals. This concept is further developed by reexamining the Rao rat weight data.

Example 8.7. We now consider a model that allows for random parameters for the individual rats in the experiment. Since the population models are linear one could potentially (1) allow the intercept to be random (and unique to each rat); (2) allow the slope (associated with the variable week) to be random; or (3) allow both the intercept and slope to be random. In our case, with the unstructured error covariance, the model only converged in case (1), which allows for the intercepts to be random. The partial output from the SAS proc mixed procedure shows how the intercepts for each rat are offset from the mean of 54.0132. The plots are given in Figure 8.4.

```
proc mixed data=rat2;
```

```
CLASS gp rat;
model ratwt = gp week gp*week /solution;
random intercept/subject=rat(gp) solution g v;;*<-Different intercept for each rat;
repeated/ type=un subject=rat(gp);
title3 'Linear model fit with subject specific intercepts';
run;
```

--------------------------- Partial Output ----------------------------
Solution for Fixed Effects

Effect	gp	Estimate	Standard Error	DF	t Value	Pr > |t|
Intercept		54.0132	1.3065	24	41.34	<.0001
gp	1	1.7241	1.8476	24	0.93	0.3600
gp	2	4.4421	2.0360	24	2.18	0.0391
gp	3	0	.	.	.	.
week		19.3510	1.0626	105	18.21	<.0001
week*gp	1	7.3144	1.5028	105	4.87	<.0001
week*gp	2	5.7302	1.6560	105	3.46	0.0008
week*gp	3	0	.	.	.	.

Solution for Random Effects

Effect	gp	rat	Estimate	Std Err Pred	DF	t Value	Pr > |t|
Intercept	1	1	0.9745	1.5060	105	0.65	0.5190
Intercept	1	2	2.2774	1.5060	105	1.51	0.1335
Intercept	1	3	-1.2073	1.5060	105	-0.80	0.4246
Intercept	1	4	-1.1181	1.5060	105	-0.74	0.4595
Intercept	1	5	1.6644	1.5060	105	1.11	0.2716
Intercept	1	6	-7.8305	1.5060	105	-5.20	<.0001
Intercept	1	7	-0.6066	1.5060	105	-0.40	0.6879
Intercept	1	8	6.0489	1.5060	105	4.02	0.0001
Intercept	1	9	-2.3754	1.5060	105	-1.58	0.1177
Intercept	1	10	2.1725	1.5060	105	1.44	0.1521
Intercept	2	1	1.4453	1.6918	105	0.85	0.3949
Intercept	2	2	-0.6498	1.6918	105	-0.38	0.7017
Intercept	2	3	0.8950	1.6918	105	0.53	0.5979
Intercept	2	4	0.5691	1.6918	105	0.34	0.7372
Intercept	2	5	3.2767	1.6918	105	1.94	0.0555
Intercept	2	6	-3.9010	1.6918	105	-2.31	0.0231
Intercept	2	7	-1.6353	1.6918	105	-0.97	0.3360
Intercept	3	1	4.4344	1.5060	105	2.94	0.0040
Intercept	3	2	4.3465	1.5060	105	2.89	0.0047
Intercept	3	3	-1.8956	1.5060	105	-1.26	0.2109
Intercept	3	4	-0.5580	1.5060	105	-0.37	0.7118
Intercept	3	5	-4.4782	1.5060	105	-2.97	0.0037
Intercept	3	6	-4.2349	1.5060	105	-2.81	0.0059
Intercept	3	7	0.2398	1.5060	105	0.16	0.8738
Intercept	3	8	8.3637	1.5060	105	5.55	<.0001
Intercept	3	9	-5.3995	1.5060	105	-3.59	0.0005
Intercept	3	10	-0.8183	1.5060	105	-0.54	0.5880

Visually, the predicted intercepts don't appear to vary that much by individual animal. A plot of the group curves is given in Figure 8.5. One can see that the average trajectory for Group 3 (rats receiving Thiouracil) is quite different from the trajectories of Groups 1 and 2 (control rats and rats receiving Thyroxin, respectively).

Figure 8.4 *Predicted Individual and Population Rat Weights by Group*

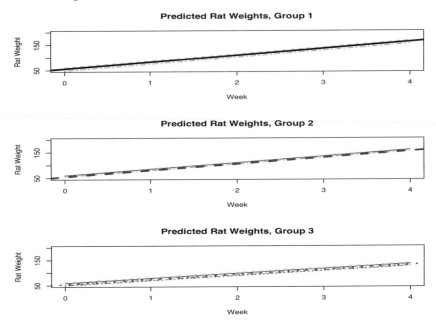

Figure 8.5 *Predicted Population Rat Weights by Group*

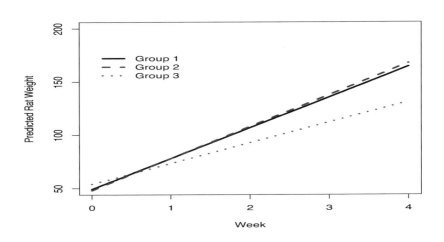

8.7 Analysis of Incomplete Repeated Measures Data

One issue with repeated measures data is that, often, some measurements on individuals are not recorded. Missingness in the data can occur because of errors in the measurement process, incorrect recording of data, individuals who miss appointed times for measurements or because some individuals dropped out of the study.

Cases where there is no relationship between the missingness and the outcomes of interest or any characteristics of the individuals are called *missing completely at random* (MCAR). Cases where the missingness is related to a measurable characteristic are known as *missing at random* (MAR). Examples might include missingness that is related to sex, age, ethnicity or economic status. A third type of missingness is that which is related to outcomes or characteristics which have not (yet) been measured. Such missingness is referred to as *missing not at random* (MNAR) (Molenberghs and Kenward [101]) or earlier as *not missing at random*. The first two types of missing data mechanisms are said to be *ignorable*, that is, the observed data can be modeled (using multiple imputation methods) without having to directly model the missing data mechanism. The third type of missing data, said to be *nonignorable* (Little and Rubin [87]), requires the missing data mechanism to be modeled. That method, which is beyond the scope of this book, often requires the data analyst to use *sensitivity* analyses that compare different models of the missing data mechanism (Molenberghs and Kenward, [101]).

Missing data poses a problem for the classical methods covered earlier in this chapter. For example, in some of the older statistical procedures, if a number of subjects have missing observations, those subjects' data are completely dropped from the analysis. Newer methods implemented in procedures like PROC MIXED use all of the data observed but assume that the underlying missing data mechanism is ignorable. In these procedures, if there are many nonlinear parameters to be estimated (e.g., covariance parameters), the associated computer algorithms sometimes fail to converge.

Example 8.8. Consider the rat weight data from the previous examples in this chapter. Suppose that we arbitrarily remove the second week weight from rat 3 in group 1, the fourth week weights from rats 4 and 7 in group 2 and the third and fourth week weights from rat 8 in group 3. In this case the model estimation algorithm did not converge for an unstructured covariance matrix for the residuals. Hence, the "repeated" statement was dropped, which is equivalent to having uncorrelated errors in the model. As one can see, the AIC and BIC values are higher here than in the model with the full data (see Example 8.6), that used allowed the covariance matrix for the errors to be unstructured across the five time points. The code and partial output are included below.

```
data rat3;
set rat2;
if gp eq 1 and rat eq 3 and week eq 2 then ratwt = .;
if gp eq 2 and rat in (4,7) and week eq 4 then ratwt = .;
if gp eq 3 and rat eq 8 and week in (3,4) then ratwt = .;
run;
```

```
proc mixed data=rat3;
CLASS gp rat;
model ratwt = gp week gp*week /solution;
random intercept/subject=rat(gp) solution g v; *<- Diff. intercept for each rat;
title3 'Linear model fit with subject specific intercepts (5 data pts missing)';
run;
-------------------------- Partial Output -------------------------------
                              Dimensions
                       Covariance Parameters        2
                       Columns in X                 8
                       Columns in Z Per Subject      1
                       Subjects                     27
                       Max Obs Per Subject           5
                       Observations Used           130
                       Observations Not Used         5
                       Total Observations          135

                              Iteration History
        Iteration    Evaluations    -2 Res Log Like        Criterion
            0             1          973.79021858
            1             2          917.68585605        0.00000261
            2             1          917.68494728        0.00000000

                         Convergence criteria met.

                           Estimated G Matrix
                   Row    Effect       gp    rat       Col1
                    1     Intercept     1     1      74.5713

                          The Mixed Procedure
                    Estimated V Matrix for rat(gp) 1 1
      Row      Col1        Col2         Col3         Col4         Col5
       1     127.51      74.5713      74.5713      74.5713      74.5713
       2     74.5713     127.51       74.5713      74.5713      74.5713
       3     74.5713     74.5713      127.51       74.5713      74.5713
       4     74.5713     74.5713      74.5713      127.51       74.5713
       5     74.5713     74.5713      74.5713      74.5713      127.51

                   Covariance Parameter Estimates
                 Cov Parm       Subject      Estimate
                 Intercept      rat(gp)      74.5713
                 Residual                    52.9376

                            Fit Statistics
                  AIC (smaller is better)       921.7
                  BIC (smaller is better)       924.3

                      Solution for Fixed Effects
                                 Standard
   Effect        gp    Estimate      Error     DF    t Value    Pr > |t|
   Intercept            57.9866      3.2653     24     17.76     <.0001
   gp            1      -5.0626      4.6175     24     -1.10     0.2838
   gp            2      -5.5856      5.0966     24     -1.10     0.2840
   gp            3           0          .        .        .         .
   week                16.6932      0.7564    100     22.07     <.0001
   week*gp       1      9.7868      1.0495    100      9.32     <.0001
   week*gp       2     10.2773      1.2047    100      8.53     <.0001
   week*gp       3           0          .        .        .         .

                      Solution for Random Effects
                                 Std Err
   Effect       gp   rat    Estimate      Pred     DF    t Value    Pr > |t|
   Intercept     1    1      6.7567     3.9776    100     1.70      0.0925
   Intercept     1    2     12.1858     3.9776    100     3.06      0.0028
   Intercept     1    3      7.3173     4.1714    100     1.75      0.0825
   Intercept     1    4     -3.5763     3.9776    100    -0.90      0.3708
   Intercept     1    5     -2.8757     3.9776    100    -0.72      0.4714
```

Intercept	1	6	-4.8022	3.9776	100	-1.21	0.2302
Intercept	1	7	-10.9319	3.9776	100	-2.75	0.0071
Intercept	1	8	3.6042	3.9776	100	0.91	0.3670
Intercept	1	9	-12.5081	3.9776	100	-3.14	0.0022
Intercept	1	10	4.8302	3.9776	100	1.21	0.2275
Intercept	2	1	14.0616	4.3237	100	3.25	0.0016
Intercept	2	2	-12.0334	4.3237	100	-2.78	0.0064
Intercept	2	3	8.2822	4.3237	100	1.92	0.0583
Intercept	2	4	7.7652	4.4914	100	1.73	0.0869
Intercept	2	5	-7.3048	4.3237	100	-1.69	0.0942
Intercept	2	6	-9.4064	4.3237	100	-2.18	0.0319
Intercept	2	7	-1.3645	4.4914	100	-0.30	0.7619
Intercept	3	1	8.4301	3.9819	100	2.12	0.0367
Intercept	3	2	2.8258	3.9819	100	0.71	0.4796
Intercept	3	3	2.4756	3.9819	100	0.62	0.5355
Intercept	3	4	4.0518	3.9819	100	1.02	0.3113
Intercept	3	5	5.8031	3.9819	100	1.46	0.1481
Intercept	3	6	-3.4790	3.9819	100	-0.87	0.3844
Intercept	3	7	-2.9536	3.9819	100	-0.74	0.4600
Intercept	3	8	-1.0888	4.4932	100	-0.24	0.8090
Intercept	3	9	-13.1114	3.9819	100	-3.29	0.0014
Intercept	3	10	-2.9536	3.9819	100	-0.74	0.4600

Type 3 Tests of Fixed Effects

Effect	Num DF	Den DF	F Value	Pr > F
gp	2	24	0.83	0.4495
week	1	100	2484.09	<.0001
week*gp	2	100	55.16	<.0001

Notice that the tests in this model yielded roughly the same results, that is, there is no significant group effect at baseline but there is significant growth in all groups over time (in weeks) and a significant group × week effect indicating that the animals in group 3 have lower growth rates than those in groups 1 and 2.

8.8 Exercises

1. Re-run the analysis in Example 8.3 but without a `repeated` statement. (Hint: The SAS code below would work.)

```
proc mixed data=rat3;
CLASS gp rat;
model ratwt = gp week gp*week /solution;
random intercept/subject=rat(gp) solution g v; *<- Diff. intercept for each rat;
title3 'Linear model fit with subject specific intercepts (5 data pts missing)';
run;
```

Compare your result to the results displayed in Examples 8.3 and 8.8.

2. Another way to analyze the Rao rat weight data is to use the baseline value of the rat weights (that is, at week 0; prior to the administration of the three compounds) as a fixed covariate and use the week 1 – week 4 data in the mixed model analysis. Therefore, modify the analysis in Example 8.7 by using the week 0 data as a fixed covariate. Answer the following questions.

(a) Is there a significant time trend overall?

(b) Are the trends across groups significantly different?

(c) Using the AIC as the criterion for the "best model," what is the "best" covariance structure among compound symmetric, AR(1), independent, and unstructured structures?

(d) State your conclusions from the result of this model.

3. As can be seen from Figure 8.4, the variances of the rat weight data appear to increase over the four-week period. To stabilize the variance for such data, a log transformation is often used. Conduct similar analyses to those in Examples 8.5 and 8.6 but on the natural log transformation of the data. Are your conclusions similar to those in the two examples?

4. Consider the data from Potthoff and Roy (1964) [115]. The data are measurements in millimeters (mm) of the distance from the center of the pituitary to the pteryomaxillary fissure in 27 children taken at ages 8, 10, 12, and 14 years by investigators at the University of North Carolina Dental School. Provide a plot of these data similar to either Figures 8.1 or 8.2. (Hint: You can get the data in the library called `nmle`, Pinheiro and Bates [114]. The commands `library(nlme); attach(Orthodont); names(Orthodont)` are needed to get started if you choose to do this in R.

5. For the data referred to in the previous problem, do similar analyses to those in Examples 8.3 – 8.5. Comment on your results and the differences between the two analyses.

6. For the data referred to in problem 8.2, do similar analyses to those in Examples 8.6 and 8.7. Comment on your results and the differences between the two analyses.

CHAPTER 9

Nonparametric Methods

9.1 Introduction

So far, we've investigated statistical inferences assuming that the underlying distribution of the data is <u>known</u>. Furthermore, we've conducted statistical tests by estimating certain parameters of the underlying distribution. For example, if we assume that our sample(s) is (are) being drawn from an underlying normal distribution, then we estimate the *population* mean of group i, μ_i, by taking the *sample* mean, $\bar{x}_i$. Since we have to estimate underlying population *parameters* from our data, we can refer to the overall class of statistical methods as *parametric methods*. When we do <u>not</u> assume any underlying distributional form (and, therefore, do not estimate any population parameters), we refer to *nonparametric* or *distribution-free* statistical methods.

To begin our discussion, it is worthwhile to review some basic data types.

<u>Three basic data types</u> (based on measurement scale):

1. Cardinal data: data measured on a scale where arithmetic is meaningful. Examples: Body weight, blood pressure, cholesterol, age
2. Ordinal data: data that have an natural ordering but on which arithmetic definitions are not meaningful. Examples: Patient health status (Poor, Mediocre, Good); Survey data such as course ratings or polls
3. Nomimal data: data that have no natural ordering (but which may have a bearing on the outcome of another variable). Examples: Demographic data (sex, race, religion, etc.)

Most nonparametric methods can only be appropriately applied to either cardinal or ordinal data.

Six nonparametric methods will be covered in this chapter. The methods and their parametric counterparts (assuming an underlying normal distribution) are listed in Table 9.1 below:

Table 9.1. *Nonparametric Methods and Their Parametric Counterparts*

Method (Test)	Parametric Counterpart
Sign test	Paired t-test
Wilcoxon signed rank	Paired t-test
Wilcoxon rank sum (Mann–Whitney U)	Two-sample t-test
Kolmorgorov–Smirnov	Two-sample t-test
Kruskal–Wallis	One-way ANOVA (Overall F-test)
Spearman's rho (ρ_s)	Correlation test

9.2 Comparing Paired Distributions

9.2.1 Sign Test

The *sign test* is appropriate for testing hypotheses (or making inferences) about *paired* data, (x_i, y_i), where only the following distinctions are made: x_i is better than y_i (+) or x_i is the same as y_i (0) or x_i is worse than y_i (−). However, the cases where the x_i and the y_i are the same are not of interest and are <u>not</u> included in a sign test. We will denote the number of cases where x_i is better than y_i by n_+ and the number of cases where x_i is worse than y_i by n_- and the total number of cases where a "better or worse" distinction is made by $n_\bullet = n_+ + n_-$. We're interested in determining whether or not the x's are "better" (or "worse") than the y's. To do so, we define

$$d_i = \begin{cases} 1, & \text{if } x_i \text{ is better than } y_i; \\ 0, & \text{if } x_i \text{ is worse than } y_i; \end{cases} \quad i = 1, \ldots, n_\bullet \quad (n_\bullet = n_+ + n_-).$$

Define, also, D to be $D = \sum_{i=1}^{n_\bullet} d_i$. Since the d_i, $i = 1, \ldots, n_\bullet$ can only take on values 0 and 1 and the d_i's are assumed to be independent $\Rightarrow D$ is binomially distributed with parameters $n_\bullet$ and p, that is, $D \sim b(n_\bullet, p)$. The formal test of interest here is whether or not the *median*, Δ, of the population *differences* is equal to 0. Thus, for example, we might test $H_0 : \Delta = 0$ versus $H_1 : \Delta \neq 0$. If H_0 is indeed true then we would expect half of the x's to be better than the y's and half to be worse. Thus, under H_0, $D \sim b(n_\bullet, \frac{1}{2})$. Under H_0, the expected value of D is $E(D) = n_\bullet p = n_\bullet(\frac{1}{2}) = \frac{n_\bullet}{2}$ whereas the variance of D, that is, $\text{Var}(D) = n_\bullet pq = n_\bullet(\frac{1}{2})(\frac{1}{2}) = \frac{n_\bullet}{4}$.

For small $n_\bullet$ (say, $n_\bullet < 20$) we can get an exact p-value for a (two-sided) test as

$$p\text{–value} = \begin{cases} 2 \times \sum_{i=0}^{n_+} \binom{n_\bullet}{i}(\frac{1}{2})^{n_\bullet} & \text{if } S = n_+ < \frac{n_\bullet}{2}; \\ 2 \times \sum_{i=n_+}^{n_\bullet} \binom{n_\bullet}{i}(\frac{1}{2})^{n_\bullet} & \text{if } S = n_+ \geq \frac{n_\bullet}{2} \end{cases}.$$

For a one-sided alternative, *drop the factor of 2 in the above equation.*

If $n_\bullet$ is reasonably large, say, $n_\bullet \geq 20$, then we can form a test using a normal approximation to the binomial.

Form

$$Z = \frac{|S - \frac{n_\bullet}{2}| - \frac{1}{2}}{\sqrt{\frac{n_\bullet}{4}}} = \frac{|n_+ - \frac{n_\bullet}{2}| - \frac{1}{2}}{\sqrt{\frac{n_\bullet}{4}}} = \frac{|n_+ - \frac{1}{2}(n_+ + n_-)| - \frac{1}{2}}{\frac{1}{2}\sqrt{n_\bullet}} =$$

$$\frac{|\frac{1}{2}n_+ - \frac{1}{2}n_-| - \frac{1}{2}}{\frac{1}{2}\sqrt{n_\bullet}} = \frac{|n_+ - n_-| - 1}{\sqrt{n_\bullet}}.$$

Under H_0, $Z \dot\sim N(0, 1)$.

Example 9.1. Suppose we have a cohort of 30 mildly depressed adults who are asked to rate their overall quality of life (QOL) before and after problem-solving therapy (PST). Twenty-three individuals indicate that their QOL improves after PST, seven individuals indicate that their QOL declines, and 10 indicate there is no change.

For this analysis, we ignore the 10 cases where the QOL does not change. Hence, $n_\bullet = n_+ + n_- = 23 + 7 = 30 > 20 \Rightarrow$ use the normal approximation method to test $H_0 : \Delta = 0$ versus, say, $H_1 : \Delta \neq 0$. Now, form

$$Z = \frac{|n_+ - n_-| - 1}{\sqrt{n_\bullet}} = \frac{|23 - 7| - 1}{\sqrt{30}} = \frac{15}{\sqrt{30}} \approx \frac{15}{5.4772} \approx 2.7386 \ .$$

$$p-\text{value} \approx 2[1 - \Phi(2.7386)] \approx 2[1 - 0.9969] \approx 0.0062 \ .$$

The exact p-value is $2 \times (0.5)^{30} \sum_{i=23}^{30} \binom{30}{i} \approx 9.313226 \times 10^{-10}[\binom{30}{23} + \binom{30}{24} + \ldots + \binom{30}{30}]$

$\approx 2(0.0026) = 0.0052.$

Thus, we reject H_0 at the $\alpha = .05$ level and conclude that PST significantly improves the QOL of individuals.

<u>R program to do above problem</u>

```
# Sign test on hypothetical QOL data
library(BSDA) #<-- Must FIRST install BSDA
QOL<-c(rep(-1,7), rep(0,10),rep(1,16))
SIGN.test(QOL)
```

Partial Output

```
          One-sample Sign-Test
data:   QOL
s = 23, p-value = 0.005223
```

9.2.2 Wilcoxon Signed Rank Test

Suppose that in the above example, a finer distinction can be made. If the degree of redness is put on a 11-point scale (0 to 10) where 0 is the worst QOL and 10 is the best QOL, then we can use a *Wilcoxon signed rank test* to assess the effectiveness of the intervention. In this case, the d_i's are formed by $d_i = x_i - y_i$ where $x_i =$ QOL score after PST and $y_i =$ QOL score prior to PST. Therefore, if the QOL score improves after PST for person i, then $d_i > 0$ whereas if the QOL score gets worse after PST for person i, then $d_i < 0$. $d_i = 0$ indicates no change in QOL score for person i.

Assumptions for Wilcoxon Signed Rank Test

1. Measurement scale of the d_i's is at least ordinal.
2. The d_i's are independent for the $i = 1, \ldots, n_\bullet$ pairs.
3. *Under H_0, the d_i's have the same median, Δ.*

Algorithm for forming the Wilcoxon Signed Rank Test Statistic

1. Calculate $d_i = x_i - y_i$ for each of the $n_\bullet$ pairs;
2. Arrange the differences, d_i, in order of their absolute values;
3. Count the differences with the same absolute (these are ties);
4. Ignore 0 differences and rank all other absolute values from 1 to $n_\bullet$;
5. For ties, compute the average rank;
6. Calculate the expected value of the sum of the ranks and its associated variance under H_0 for, say, group 1:

$$E(R_1) = \frac{n_\bullet(n_\bullet + 1)}{4}, \quad Var(R_1) = \frac{1}{24}\left\{ n_\bullet(n_\bullet+1)(2n_\bullet+1) - \frac{\sum_{i=1}^{g}(t_i^3 - t_i)}{2}\right\};$$

7. If the number of *nonzero* d_i's < 20 (i.e., $n_\bullet < 20$), then use an exact table; else, form

$$Z = \frac{\left| R_1 - \frac{n_\bullet(n_\bullet+1)}{4}\right| - \frac{1}{2}}{\sqrt{\frac{n_\bullet(n_\bullet+1)(2n_\bullet+1)}{24} - \frac{\sum_{i=1}^{g}(t_i^3 - t_i)}{48}}}.$$

Note: $t_i =$ number of d_i's (in the ith "cluster") with the same absolute value. Under H_0, $Z \dot\sim N(0, 1)$.

Example 9.2.

The perceived quality of life (QOL) scores for our hypothetical example before and after the problem-solving therapy (PST) are given in the table below.

Before	After	Diff	Before	After	Diff	Before	After	Diff	Before	After	Diff
4	4	0	4	3	-1	4	7	3	3	3	0
3	4	1	3	6	3	2	3	1	4	5	1
4	5	1	3	3	0	4	6	2	6	6	0
2	3	1	3	4	1	4	5	1	3	3	0
4	4	0	3	5	2	4	4	0	2	5	3
3	3	0	3	5	2	8	9	1	2	2	0
2	1	-1	3	6	3	6	9	3	5	7	2
4	1	-3	8	7	-1	6	9	3	2	1	-1
3	5	2	5	7	2	3	5	2	4	2	-2
4	2	-2	6	6	0	5	9	4	7	10	3

Non-zero differences in QOL scores before and after PST, $d_i = x_i - y_i$

$|d_i|$	Total number	Range of ranks	Average rank	Number of positive	Number of negative
1	12	1–12	6.5	8	4
2	9	13–21	17	7	2
3	8	22–29	25.5	7	1
4	1	30	30	1	0

This problem is similar to the earlier one except that the QOL score is scored on a scale from 0 to 10 (with 10 indicating the best QOL).

$\Rightarrow V = 8(6.5) + 7(17) + 7(25.5) + 1(30) = 379.5$. Under H_0,

$$E(V) = \frac{n_{\bullet}(n_{\bullet} + 1)}{4} = \frac{30(31)}{4} = 232.5, \quad Var(V) = \frac{n_{\bullet}(n_{\bullet} + 1)(2n_{\bullet} + 1)}{24} - \frac{\sum_{i=1}^{g}(t_i^3 - t_i)}{48}$$

$$\Rightarrow Var(V) = \frac{30(31)(61)}{24} - \frac{[(12^3 - 12) + (9^3 - 9) + (8^3 - 8) + (1^3 - 1)]}{48}$$

$= \frac{56730}{24} - \frac{2940}{48} = 2363.75 - 61.25 = 2302.5$. Now, since $n_{\bullet} > 20$, we can use a normal approximation.

$$\Rightarrow Z = \frac{|379.5 - 232.5| - \frac{1}{2}}{\sqrt{2336.625}} \approx \frac{146.5}{73.822} \approx 3.0531 > 1.96 = z_{.975}$$

$\Rightarrow$ p-value $\approx 2[1 - \Phi(3.0531)] \approx 2(1 - .99887) \approx 0.0023 \Rightarrow$ reject H_0 at the $\alpha = .05$ level $\Rightarrow$ PST has a significant positive effect on QOL score. Notice that in the formula, for the tie adjustment, the "cluster" of ones does not modify the formula at all because $1^3 - 1 = 1 - 1 = 0$. Therefore, the "clusters" of ones can be deleted from the formula. This makes sense, of course, because a "cluster of one" is the same as an untied observation. An R program is given below that performs the analysis. For this program, if only one vector of values is given, or if two vectors are given and the user stipulates "paired=T," then a Wilcoxon signed rank test is performed.

R program to perform a Wilcoxon signed rank analysis

```
# Signed rank test on simulated QOL data
QOL.score<-c(4,rep(3,7),rep(2,7),rep(1,8),rep(0,10),-1,-1,-1,-1,-2,-2,-3)
wilcox.test(QOL.score)
```

Partial Output

```
data:  QOL.score
V = 379.5, p-value = 0.002265
alternative hypothesis: true location is not equal to 0
```

Note that R adjusts for the tie structure in the Wilcoxon signed rank test. By default, a continuity correction is also employed. One can force the program not to use the continuity correction with the command, `wilcox.test(how.red,correct=F)`. For a complete description of the command, type `?wilcox.test` at the R console.

9.3 Comparing Two Independent Distributions

When we have two *independent* groups of observations that are not assumed to be normally distributed, we can again use the *ranks of the observations* instead of the observation values themselves to test whether or not the data from two groups come from the same distribution. The two most common such tests are the Wilcoxon *rank sum* test and the Kolmogorov–Smirnov test. The primary difference between the two methods is that the rank sum test is used to test whether the central tendencies (in this case, the medians) of the two distributions are the same whereas the Kolmogorov–Smirnov test is used to determine whether or not extreme tendencies of the distributions are the same.

9.3.1 Wilcoxon Rank Sum (Mann–Whitney U) Test

The rank sum test was first formulated by Frank Wilcoxon [157]. Later, an equivalent formulation was proposed by Mann and Whitney [91]. Like the signed rank test, the rank sum test is quite simple in concept. It merely compares the sum of the rankings of the two groups. The assumptions of the test are given below.

Assumptions

1. Both samples, $X_1, X_2, \ldots, X_{n_1}$ and $Y_1, Y_2, \ldots, Y_{n_2}$, are independently and identically distributed (but not specified).

2. The X's and Y's are independent from each other.

3. The measurement scale for the original observations is at least ordinal.

For the rank sum statistic, we assume further that the X_i, $i = 1, \ldots, n_1$, have distribution function, $F(x)$, and the Y_i, $i = 1, \ldots, n_2$, have distribution function, $G(y)$. Notice that in assumption 1 from above, the distributions do not have to be specified, that is, F and G are not specified. The test of interest here is whether or not median values are equal, i.e., test $H_0 : \Delta_1 = \Delta_2$ versus, say, $H_1 : \Delta_1 \neq \Delta_2$. Another way of stating the hypothesis associated with a rank sum test is $H_0 : Pr(X < Y) = \frac{1}{2}$ versus $H_1 : Pr(X < Y) \neq \frac{1}{2}$.

Algorithm for ranking the data

1. Combine the data from the two samples and order them from lowest to highest;

2. Assign ranks to the ordered values: ranks go from 1 to $n_\bullet = n_1 + n_2$; and

3. If there are ties, compute the range of ranks for all tied observations, then assign the average rank of the group to each member of the group.

We'll derive a large sample Z statistic for testing H_0. Use this large sample statistic when $n_1 \geq 10$ and $n_2 \geq 10$. Otherwise, use exact tables. Define

$$r_i = \begin{cases} i, & \text{if rank} = i \text{ and the observation is in group 1} \\ 0, & \text{otherwise} \end{cases} , \quad i = 1, \ldots, n_\bullet .$$

Let $R_1 = $ the sum of the r_i's in group 1 and let $n_\bullet = n_1 + n_2$, that is, $n_\bullet$ represents the total number of observations in both groups 1 and 2. Under H_0, $Pr(r_i = i) = \frac{n_1}{(n_1+n_2)} = \frac{n_1}{n_\bullet}$. This is another way of saying that, under H_0, all ranks equally likely to occur so that the probability of a given rank being in group 1 is merely equal to the proportion of the overall observations located in group 1. Thus, under H_0,

$$E(R_1) = \sum_{i=1}^{n_\bullet} i \frac{n_1}{n_\bullet} = \frac{n_1}{n_\bullet} \sum_{i=1}^{n_\bullet} i = \frac{n_1}{n_\bullet} \frac{n_\bullet(n_\bullet + 1)}{2} = \frac{n_1(n_\bullet + 1)}{2}$$

The variance of R_1 is

$$\text{Var}(R_1) = \frac{n_1 n_2}{12}\left[(n_1 + n_2 + 1) - \frac{\sum_{i=1}^{g} t_i(t_i^2 - 1)}{(n_1 + n_2)(n_1 + n_2 - 1)}\right]$$

$$= \frac{n_1 n_2}{12}\left[(n_\bullet + 1) - \frac{\sum_{i=1}^{g} t_i(t_i^2 - 1)}{(n_\bullet)(n_\bullet - 1)}\right]$$

To test H_0 (for large samples), form

$$Z = \frac{|R_1 - \frac{1}{2}n_1(n_\bullet + 1)| - \frac{1}{2}}{\sqrt{\text{Var}(R_1)}}$$

Under H_0, $Z \stackrel{\cdot}{\sim} N(0, 1)$.

Example 9.3. Consider the dopamine data from psychotic and nonpschotic schizophrenic patients presented in Example 1.17. If one does not assume that the data are normally distributed then one could use a rank sum test for comparing the dopamine levels in the two groups of patients. In this case, we are testing whether or not there is a difference in median dopamine levels between the two types of patients.

X	Rank	Y	Rank
0.0104	1	0.0150	7
0.0105	2	0.0204	14
0.0112	3	0.0208	15
0.0116	4	0.0222	17
0.0130	5	0.0226	18
0.0145	6	0.0245	20
0.0154	8	0.0270	22
0.0156	9	0.0275	23
0.0170	10	0.0306	24
0.0180	11	0.0320	25
0.0200	12.5		
0.0200	12.5		
0.0210	16.0		
0.0230	19.0		
0.0252	21.0		
	$R_1 = 140$		$R_2 = 185$

Under H_0, $E(R_1) = \frac{n_1(n_\bullet + 1)}{2} = \frac{15(25+1)}{2} = \frac{15(26)}{2} = 195$.

$$Var(R_1) = \frac{n_1 n_2}{12}\left[(n_\bullet + 1) - \frac{\sum_{i=1}^{g} t_i(t_i^2 - 1)}{(n_\bullet)(n_\bullet - 1)}\right]$$

$$= \frac{15(10)}{12}\left[26 - \frac{2(3)}{25(24)}\right] = \frac{150}{12}\left[26 - 0.01\right] = \frac{150}{12}(25.99) = 324.875$$

$\Rightarrow \sqrt{Var(R_1)} \approx 18.024$. Thus, $|Z| \approx \frac{|140-195|-\frac{1}{2}}{18.024} \approx \frac{54.5}{18.024} \approx 3.024 \Rightarrow$ p-value $\approx$ $2(.00125) = 0.0025 \Rightarrow$ <u>reject H_0</u> $\Rightarrow$ there is a significant difference in median responses between the two groups.

R session for rank sum example

```
> x<-c(.0104,.0105,.0112,.0116,.0130,.0145,.0154,.0156,.0170,.0180,.0200,.0200,
+ .0210,.0230,.0252)
> y<-c(.0150,.0204,.0208,.0222,.0226,.0245,.0270,.0275,.0306,.0320)
> wilcox.test(x,y,correct=T,exact=F)

        Wilcoxon rank sum test with continuity correction
```

```
data:   x and y
W = 20, p-value = 0.002497
alternative hypothesis: true location shift is not equal to 0
```

NOTE: The Mann–Whitney U test, although conducted in a different manner from the Wilcoxon rank sum test, gives the same p-value. Thus, the Wilcoxon rank sum and the Mann–Whitney U tests are considered equivalent tests.

9.3.2 Kolmogorov–Smirnov Test

Another nonparametric test for comparing two independent samples is the Kolmogorov–Smirnov test. This test examines whether or not *extreme* differences between two cumulative distribution functions warrant the conclusion that two samples come from different distributions.

Assumptions

Two independent (unpaired) samples whose underlying distributions are not necessarily normal but whose measurement scale is ordinal or cardinal. We assume that the two populations have the (cumulative) distribution functions, F_1 and F_2. Thus, we wish to test $H_0 : F_1(x) = F_2(x)$ for all x versus, say, $H_1 : F_1(x) \neq F_2(x)$ for some x. If n_1 and n_2 are sample sizes for, say, Groups 1 and 2, then we can outline an algorithm for performing a Kolmogorov–Smirnov test as follows:

Algorithm for the Kolmogorov–Smirnov test

1. Combine the data from the two samples and order them from lowest to highest;
2. Calculate $\hat{F}_1(x)$ and $\hat{F}_2(x)$ by attaching a probability of $\frac{1}{n_1}$ or $\frac{1}{n_2}$ to each observation depending upon whether the observation is from group 1 or group 2;
3. Calculate $D = \max |\hat{F}_1(x) - \hat{F}_2(x)|$; and
4. If one is interested in performing hypothesis testing, and large sample approximations hold, then one would compare D to $\kappa_\alpha \sqrt{\frac{n_1+n_2}{n_1 n_2}}$ where κ_α is the critical value of an appropriate probability distribution. Large sample κ_α values are given in Massey [97] for different α-level. The two sided κ_α values for $\alpha = 0.10$, $\alpha = 0.05$, and $\alpha = 0.01$ are 1.22, 1.36, and 1.36, respectively. These are appropriate for $n_1, n_2 \geq 10$. Exact values for D for $n_1, n_2 \leq 10$ can be again found in Massey [97]. Alternatively, most modern statistical packages produce p-values associated with the test.

Example 9.4. Consider again the schizophrenia data in Example 1.17. Recall that there are 15 observations in group 1 and 10 observations in group 2. The data from the pooled groups can be ordered and arranged so that differences between their empirical distribution functions can be calculated. For the schizophrenia data, the table below summarizes the two empirical distribution functions and their differences. For a Kolmogorov–Smirnov test, we're interested in finding the maximum difference.

Sample cumulative distributions for the two schizophrenia groups

x	$\hat{F}_1$	$\hat{F}_2$	Abs. Diff.	x	$\hat{F}_1$	$\hat{F}_2$	Abs. Diff.
0.0104	0.0667	0.0000	0.0667	0.0204	0.8000	0.2000	0.2000
0.0105	0.1333	0.0000	0.1333	0.0208	0.8000	0.3000	0.3000
0.0112	0.2000	0.0000	0.2000	0.0210	0.8667	0.3000	0.5667
0.0116	0.2667	0.0000	0.2667	0.0222	0.8667	0.4000	0.4667
0.0130	0.3333	0.0000	0.3333	0.0226	0.8667	0.5000	0.3667
0.0145	0.4000	0.0000	0.4000	0.0230	0.9333	0.5000	0.4333
0.0150	0.4000	0.1000	0.3000	0.0245	0.9333	0.6000	0.3333
0.0154	0.4667	0.1000	0.3667	0.0252	1.0000	0.6000	0.4000
0.0156	0.5333	0.1000	0.4333	0.0270	1.0000	0.7000	0.3000
0.0170	0.6000	0.1000	0.5000	0.0275	1.0000	0.8000	0.2000
0.0180	0.6667	0.1000	0.5667	0.0306	1.0000	0.9000	0.1000
0.0200	**0.8000**	**0.1000**	**0.7000**	0.0320	1.0000	1.0000	0.0000

In our example, each $\hat{F}_i(x)$ is calculated by *dividing the rank of the observation within group* by either 15 (if the observation is in group 1) or 10 (if the observation is in group 2). For example, $\hat{F}_1(0.0104) = \frac{1}{15} \approx .0667$ whereas $\hat{F}_2(0.0104) = \frac{0}{10} = 0$. The *maximum absolute difference* occurs at $x = 0.0200$, which is ordered observation 12 in group 1 and lies between ordered observations 1 and 2 in group 2. Therefore,

$$D = \left| \frac{12}{15} - \frac{1}{10} \right| = 0.7 > 1.36\sqrt{\frac{15 + 10}{15(10)}} \approx 0.555$$

$\Rightarrow$ reject H_0 at the $\alpha = .05$ level. The cutoff for a two-sided, $\alpha = .05$ test associated with the exact procedure is 0.5. Using either an exact or asymptotic test yields a significant difference between the two groups. Notice, however, that the significance level is slightly less extreme than that given by the rank sum test in Example 9.3. This is not always the case but is quite common as Kolmogorov–Smirnov tests tend to be more conservative than rank sum tests.

The program below implements a Kolmogorov–Smirnov test in R. The program outputs $D = \max |\hat{F}_1(x) - \hat{F}_2(x)|$. However, it does not output the critical value of the test statistic but does give the associated p-value.

R Program to implement Kolmogorov–Smirnov test

```
> x<-c(.0104,.0105,.0112,.0116,.0130,.0145,.0154,.0156,.0170,.0180,.0200,
.0200,.0210,.0230,.0252)
> y<-c(.0150,.0204,.0208,.0222,.0226,.0245,.0270,.0275,.0306,.0320)
> ks.test(x,y,correct=T)

        Two-sample Kolmogorov-Smirnov test

data:  x and y
D = 0.7, p-value = 0.00559
alternative hypothesis: two-sided

Warning message:
In ks.test(x, y, correct = T) : cannot compute correct p-values with ties
```

Figure 9.1 *Visualization of the Kolmogorov-Smirnov Test*

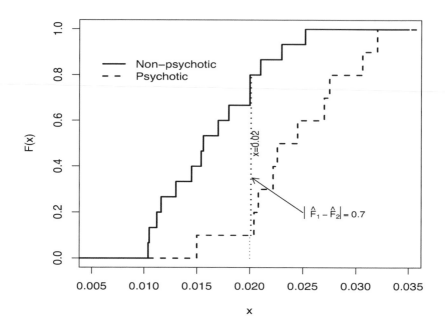

9.4 Kruskal–Wallis Test

The rank sum test can be generalized to testing $k \geq 2$ groups similar to the way an unpaired t-test can be generalized. Accordingly, the analog to the parametric one-way ("Parallel groups") ANOVA F-test is called the *Kruskal–Wallis* test. The difference is that the Kruskal–Wallis procedure is used for testing the heterogeneity of *medians* of different distributions as compared to the parametric F-test, which is used for testing the heterogeneity of *means* of different *normal* distributions. The assumptions associated with the Kruskal–Wallis test and the implementation of the test are outlined below.

Assumptions

1. Basic model is $y_{ij} = \mu + \alpha_i + e_{ij}$, $i = 1, \ldots, k$ subpopulations, $j = 1, \ldots, n_i$ subjects within subpopulation i;

2. Within subpopulation i, the observations are independently and identically distributed;

3. Independence of observations within each subpopulation and across subpopulations;

4. The measurement scale for the original observations is at least ordinal

Notice that no assumption of normality is made above. We want to test the hypothesis, H_0 : All k subpopulations have the same cumulative distribution function, i.e., $H_0 : F_1(x) = F_2(x) = \ldots = F_k(x) = F(x)$ versus H_1 : at least two of the distributions are different, i.e., $F_i(x) \neq F_j(x)$ for some $i \neq j$.

<u>Algorithm for forming the test statistic</u>

1. Combine all $n_\bullet = \sum_{i=1}^{k} n_i = n_1 + n_2 + \ldots + n_k$ observations from the k samples and order them from lowest to highest;

2. Assign ranks 1 to $n_\bullet$ to *all* of the ordered observations in the k samples ($r_{ij} \longleftarrow y_{ij}$). If ties exist, compute the range of ranks and take the average of the ranks in each "tied cluster."

3. Calculate the test statistic: $H = \frac{H^*}{f}$ where

$$H^* = \frac{12}{n_\bullet(n_\bullet + 1)} \left[\sum_{i=1}^{k} \frac{R_i^2}{n_i} \right] - 3(n_\bullet + 1), \ R_i = \sum_{j=1}^{n_i} r_{ij}, \text{ and } f = 1 - \frac{\sum_{j=1}^{g}(t_j^3 - t_j)}{n_\bullet^3 - n_\bullet},$$

and where t_j is the number of observations in the jth cluster of tied observations. If there are no tied observations, then $f = 1 \implies H = H^*$

4. Under H_0, $H \stackrel{\cdot}{\sim} \chi_{k-1}^2 \Rightarrow$ if $H > \chi_{k-1,1-\alpha}^2$, reject H_0; else, do not reject H_0.

5. Procedure should be used when $\min(n_1, n_2, \ldots, n_k) \geq 5$.

Note: another form of H^* is $H^* = \frac{12}{n_\bullet(n_\bullet + 1)} \sum_{i=1}^{k} n_i(\overline{R}_i - \overline{R}_\bullet)^2$, where $\overline{R}_i$ is the mean rank of group i and $\overline{R}_\bullet$ is the mean rank over all groups.

<u>Pairwise comparisons</u>: Typically, a pairwise comparison involves testing $H_0 : F_i(x) = F_j(x)$ versus $H_1 : F_i(x) \neq F_j(x)$ for two groups, i and j, $i \neq j$.

To do pairwise comparisons between subpopulations, we proceed as in steps 1 and 2 for the overall Kruskal–Wallis test, then

3. Calculate

$$\lambda_{KW} = \frac{\overline{R}_i - \overline{R}_j}{\sqrt{\frac{n_\bullet(n_\bullet + 1)}{12}\left(\frac{1}{n_i} + \frac{1}{n_j}\right)}}$$

4. Under H_0, $\lambda_{KW} \stackrel{\cdot}{\sim} N(0, 1)$

Example 9.5. (from Rosner, 5th. ed., pp. 550 [123]) Ophthalmology: Eye lid closure in albino rats — 0 = eyelid completely open; 1, 2 =intermediate closures; 3 = eyelid completely closed.

ΔScores: Baseline score − 15-minute score

	Indomethin		Aspirin		Piroxicam		BW755C	
Rabbit	Δ Score	Rank	Δ Score	Rank	Δ Score	Rank	Δ Score	Rank
1	2	13.5	1	9	3	20	1	9
2	3	20	3	20	1	9	0	4
3	3	20	1	9	2	13.5	0	4
4	3	20	2	13.5	1	9	0	4
5	3	20	2	13.5	3	20	0	4
6	0	4	3	20	3	20	-1	1
		$R_1 = 97.5$		$R_2 = 85$		$R_3 = 91.5$		$R_4 = 26$

$$H^* = \frac{12}{24(25)}\left[\frac{(97.5)^2}{6} + \frac{(85)^2}{6} + \frac{(91.5)^2}{6} + \frac{(26)^2}{6}\right] - 3(25) = 10.932$$

$$f = 1 - \frac{(1^3-1)+(5^3-5)+(5^3-5)+(4^3-4)+(9^3-9)}{24^3-24} = 1 - \frac{1020}{13800} \approx 0.926$$

$$\Rightarrow H \approx \frac{10.932}{0.926} \approx 11.806 > \chi^2_{3,.95} \text{ (because } \chi^2_{3,.95} \approx 7.81)$$

$\Rightarrow$ reject H_0 at the $\alpha = .05$ level $\Rightarrow$ at least two of the distributions are significantly different. Since $\chi^2_{3,.99} = 11.34 < 11.806 < 12.34 = \chi^2_{3,.995}$, we can also conclude that $.005 < p < .01$.

In the above example, there were $\frac{4(3)}{2} = 6$ total pairwise comparisons. Methods like those proposed by Bonferroni (if all pairwise comparisons are considered), Tukey or Scheffé will be conservative in that they won't detect as many differences as other multiple comparisons procedures (see Chapter 5 for more details). Of course, one could use a less conservative procedure. In the above problem, if we were interested in whether or not the Indomethin group is "significantly" different than the $BW755C$ group in the given experiment, then we would form

$$\lambda_{KW} = \frac{\frac{97.5}{6} - \frac{26}{6}}{\sqrt{\frac{24(25)}{12}\left(\frac{1}{6} + \frac{1}{6}\right)}} = \frac{97.5 - 26}{5\sqrt{24}} \approx 2.92.$$

Now, $z_{1-.05/12} \approx z_{.9958} \approx 2.635 \Rightarrow \lambda_{KW} > z_{1-\alpha^*} \Rightarrow$ reject H_0 and conclude that the Δ score cumulative distributions are significantly different in the two groups.

9.5 Spearman's Rho

Spearman's rho (Spearman (1904) [142]), ρ_s, is a measure that allows one to establish whether or not a linear relationship exists between the *ranks* of two variables. It is useful when data are skewed or when outliers exist that unduly influence results or when the data are otherwise not normally distributed.

Assumptions

1. $(X_1, Y_1), \ldots, (X_n, Y_n)$ are n independent bivariate observations.
2. The measurement scale for the original observations is at least ordinal.

Suppose now that we're interested in testing whether or not there is a significant linear association between the ranks of the X's and Y's. We first form an estimate of the *rank correlation*, $r_s = \hat{\rho}_s$, by ranking the n observations *within each group* from smallest to largest. Tied ranks within group are computed by calculating the range of ranks over each cluster and assigning the average rank to each member of the cluster. Then, within each *pair* of observations, (X_i, Y_i), form the difference of their ranks, $d_i = r_{X_i} - r_{Y_i}$ for $i = 1, \ldots, n$ pairs. Next, form

$$r_s = \frac{\frac{n^3-n}{6} - T_X - T_Y - \sum_{i=1}^n d_i^2}{2\sqrt{[\frac{n^3-n}{12} - T_X][\frac{n^3-n}{12} - T_Y]}},$$

where $T_X = \sum_{i=1}^{g_X} \frac{t_i^3-t_i}{12}$, $T_Y = \sum_{i=1}^{g_Y} \frac{t_i^3-t_i}{12}$, t_i = the # of observations in the ith tied group, g_X = # of tied groups in X, g_Y = # of tied groups in Y.

If the number of pairs, $n \geq 10$, we can then test, for example, $H_0 : \rho_s = 0$ versus $H_1 : \rho_s \neq 0$ by forming

$$t = \frac{r_s\sqrt{n-2}}{\sqrt{1-r_s^2}},$$

and noting that under H_0, $t \overset{\cdot}{\sim} t_{n-2}$. So, for hypothesis testing, if $|t_s| > t_{n-2,1-\frac{\alpha}{2}}$, then we reject H_0; otherwise we do not reject H_0. If $n < 10$, then we can use an exact table provided by many authors, e.g., Hollander and Wolfe [69], Rosner [124], and Fisher and van Belle [46], or programs in R or SAS to identify the critical values of a hypothesis test.

Example 9.6. ATP levels in Youngest and Oldest Sons for 17 Families (From Fisher and van Belle, pp. 386–387 [46])

Fam #	ATP–Youngest	Rank	ATP–Oldest	Rank	d_i	d_i^2
1	4.18	4	4.81	11	−7	49.00
2	5.16	12	4.98	14	−2	4.00
3	4.85	9	4.48	6	3	9.00
4	3.43	1	4.19	3	−2	4.00
5	4.53	5	4.27	4	1	1.00
6	5.13	11	4.87	12.5	−1.5	2.25
7	4.10	2	4.74	10	−8	64.0
8	4.77	7	4.53	7	0	0.00
9	4.12	3	3.72	1	2	4.00
10	4.65	6	4.62	8	−2	4.00
11	6.03	17	5.83	17	0	0.00
12	5.94	15	4.40	5	10	100.0
13	5.99	16	4.87	12.5	3.5	12.25
14	5.43	14	5.44	16	−2	4.00
15	5.00	10	4.70	9	1	1.00
16	4.82	8	4.14	2	6	36.0
17	5.25	13	5.30	15	−2	4.00
					$\sum$ =	298.5

Since $n = 17 > 10$, we can use a large sample approximation. To calculate the t-statistic, we proceed as follows:

$$r_s = \frac{\frac{4896}{6} - \frac{8-2}{12} - 298.5}{2\sqrt{[\frac{4896}{12}][\frac{4896}{12} - \frac{8-2}{12}]}} \approx \frac{816-298.5}{2\sqrt{408(407.5)}} \approx \frac{517}{815.5} \approx 0.6340$$

$$t_s \approx \frac{0.6340\sqrt{15}}{\sqrt{1 - .6340^2}} \approx \frac{2.4553}{\sqrt{0.5981}} \approx \frac{2.4553}{0.7734} \approx 3.1749 > t_{15,.975} = 2.131.$$

The p-value associated with this approproximate test is $p \approx .0063$. The problem could have also been solved by using Table A.16 on page 950 in Fisher and van Belle [46]. For $n = 17$, the critical Spearman rho values are $r_s = .490$ for $\alpha = .05$ and $r_s = .645$ for $\alpha = .01 \Rightarrow$ $.01 < p < .05$. Thus, using either the approximate or exact test, we reject $H_0 : \rho_s = 0 \Rightarrow$ Spearman's rho is significantly different from 0.

9.6 The Bootstrap

One of the most popular computer intensive, nonparametric methods of recent years is known as the *bootstrap*. The bootstrap, originally developed by Bradley Efron [35], works on a simple but very clever principle. Suppose that we have a sample, $\mathbf{x} = (x_1, x_2, \ldots, x_n)$, which is independently and identically distributed from an unknown distribution, $F(x)$. We are interested in estimating and drawing an inference about some *feature* of $F(x)$, call it $\theta = t(F)$. We usually think of the feature, $t(F)$, as being a parameter associated with the distribution but it doesn't have to be. Now, suppose that $T(\mathbf{x})$ is an estimator of $t(F)$, which may be quite complicated. The idea is to make an inference on the space of independently and identically distributed samples $\mathbf{x}^* = (x_1^*, x_2^*, \ldots, x_n^*)$ drawn from an empirical approximation, $F^*(x)$, of $F(x)$. Hence, one would sample from the empirical distribution, $F^*(x)$, by weighting the n points. In the case of a nonparametric bootstrap, the weights might typically be $\frac{1}{n}$ for each of the n points. For a parametric estimator, a parameter, θ, may be estimated as $\hat{\theta}$ and consequently, one would sample from $F_n^*(x) = F_{\hat{\theta}}(x)$. The bootstrap *correspondence principle* asserts that not only do $T(\mathbf{x})$ and $T(\mathbf{x}^*)$ have similar distributions but also, $T(\mathbf{x}) - t(F)$ and $T(\mathbf{x}^*) - t(F_n^*)$ have similar distributions. In some cases, $t(F_n^*) = T(\mathbf{x})$.

Given a probability distribution function, $F(x)$, three important features $t(F)$, are

1. $\mu_k(F) = \int x^k dF(x)$ (k^{th} moment);

2. $\omega_k(F) = \int [(x^k - \mu_1(F)]^k) dF(x)$ (k^{th} *central* moment); and

3. $\xi_p = \min\{x : F(x) \geq p\}$ (p^{th} quantile).

Their natural estimators across parametric forms are:

1. $\widehat{\mu}_k(F) = \dfrac{1}{n} \displaystyle\sum_{i=1}^{n} x^k;$

2. $\widehat{\omega}_k(F) = \dfrac{1}{n} \displaystyle\sum_{i=1}^{n} [x_i - \mu_1(F_n^*)]^k;$ and

3. $\widehat{\xi}_p = \min\{x : F_n^*(x) \geq p\}$, respectively.

These were coined as "plug-in estimators" by Efron and Tibshirani [36] and obey the rule $T(\mathbf{x}) = t(F_n^*)$. However,

$$s_n = \frac{1}{n-1} \sum_{i=1}^{n} (x_i - \overline{x})^2 = \frac{n}{n-1} \widehat{\omega}_2(F_n^*)$$

is *not* a plug-in estimator.

The plug-in estimators are useful for constructing point estimates for quantities of complex form. However, what the bootstrap allows us to do is to further estimate variances of these estimators or more generally, their associated approximate distribution functions, which in turn can be used to construct confidence intervals.

Example 9.7. Bootstrapping in a Regression Setting Suppose, for a linear regression model, $\mathbf{y} = \mathbf{X}\boldsymbol{\beta} + \boldsymbol{\epsilon}$, we are interested in obtaining a certain feature (which may not be linear) and its distribution function. For example, suppose that $\boldsymbol{\beta} = \begin{pmatrix} \beta_0 \\ \beta_1 \end{pmatrix}$ and we wish to solve an *inverse prediction* or *calibration* problem, that is, find an x_0 and its distribution which is associated with a prespecified, y_0, that is, $y_0 = \beta_0 + \beta_1 x_0 \Rightarrow x_0 = \frac{y_0 - \beta_0}{\beta_1}$, which is not linear in the parameters of the model, β_0 and β_1. Actually, this problem can be solved analytically given $\boldsymbol{\epsilon} \sim N(\mathbf{0}, \mathbf{I}\sigma^2)$ using what is referred to as "Fieller's Theorem", see Fieller [41] or Finney [43].

To start the bootstrap approximation, one must first fit the model, that is, $\widehat{\mathbf{y}} = \mathbf{X}\widehat{\boldsymbol{\beta}}$ where $\widehat{\boldsymbol{\beta}} = (\mathbf{X}'\mathbf{X})^{-1}\mathbf{X}'\mathbf{y}$ or equivalently, $\mathbf{y} = \mathbf{X}\widehat{\boldsymbol{\beta}} + \widehat{\boldsymbol{\epsilon}} \Rightarrow \mathbf{y} = \widehat{\mathbf{y}} + \widehat{\boldsymbol{\epsilon}}$. So, a question arises: should one create bootstrap samples on the $\widehat{\mathbf{y}}^*$s, or alternatively, on the $\widehat{\boldsymbol{\epsilon}}^*$s? The answer is that the elements of $\widehat{\mathbf{y}}$ are highly correlated so would not be good to use the $\widehat{\mathbf{y}}^*$s for resampling because they would not be independent. On the other hand, if $\boldsymbol{\Sigma}_{\boldsymbol{\epsilon}} = \mathbf{I}\sigma^2$ then the underlying errors are assumed to be *uncorrelated* and therefore, the elements of the *estimated* residual (error) vector, $\widehat{\boldsymbol{\epsilon}}$ are *nearly* uncorrelated. Hence, it would be wise to bootstrap by sampling the residuals. The idea is to fit the model and obtain the estimated residuals. Then, resample the residuals *with replacement* and add them back to the predicted values from the original fit and refit the model obtaining new parameter estimates and a new estimate of x_0, denoted $\widehat{x}_0^*$ indicating it is an *estimated value from a bootstrap sample*. This process should be repeated B times where B is a large number. Efron and others recommend that B should be ≥ 1000. From this, a distribution of the $\widehat{x}_0^*$ can be constructed and confidence intervals can be constructed by either the *percentile method*, that is, estimating the $(\alpha/2)^{th}$ and the $(1 - \alpha/2)^{th}$ percentiles of the distribution $\underline{or}$ by invoking the central limit theorem and taking the mean of the B $\widehat{x}_0^*$'s $\mp$ $z_{1-\alpha/2}$ where "z" denotes the standard normal distribution.

The code below fits simulated data generated as $y = x + 1 + \epsilon$ where x takes on the integer

values from 0 to 10 and $\epsilon \sim N(0, .7^2)$. The program below calculates a bootstrap sample
for $\widehat{x}_0 = \frac{y_0 - \widehat{\beta}_0}{\widehat{\beta}_1}$ where y_0 is chosen to be equal to 5. The fitted regression is $\widehat{y}_i = 1.2487 +$
$0.9637\, x_i$, $i = 0, \ldots, 10$ so that $\widehat{x}_0 \approx 3.004$. The issue, however is to find the 95% "fiducial"
(confidence) limits associated with x_0 (see Fisher [47] and comments by Neyman [109]). The
results of the analysis are summarized in Figure 9.2.

```
> ###########################################################################
> #   This program implements a bootstrap estimate along with confidence
> #   intervals for solving an inverse prediction (calibration) problem
> #   Specifically, given E(y|x)=b0 + b1*x, find x0 for a given y0.
> ###########################################################################
> x<-0:10
> #Y<-x+1
> #y<-Y+rnorm(11,0,.7)
> y<-c(1.117,2.3116,3.085,4.02,5.78,6.074,6.559,8.394,8.166,10.086,11.144)
> #cbind(x,y)
> fit0<-lm(y~x)
> fit0
Call:
lm(formula = y ~ x)

Coefficients:
(Intercept)              x
     1.2487         0.9637
> e<-fit0$resid
>
> B<-5000 #<-- Set the number of bootstrap samples
> pred.y<-fit0$fitted.values
> coef.B<- matrix(rep(NA,length(fit0$coef)*B),ncol=2)
> x0.B<-rep(NA,B)
> y0<-5
>
> for (i in 1:B) {
+     e.boot<-sample(e,replace=T)
+     new.fit<-lm(pred.y+e.boot~x)
+     coef.B[i,]<-new.fit$coef
+     x0.B[i]<-(y0-coef.B[i,1])/coef.B[i,2]
+     }
>
> point.est.x0<-mean(x0.B)
> boot.sd<-sd(x0.B)
>
> alpha<-0.05
> l.ci<-quantile(x0.B,alpha/2)
> u.ci<-quantile(x0.B,1-alpha/2)
> norm.l.ci<-point.est.x0-qnorm(1-alpha/2)*boot.sd
> norm.u.ci<-point.est.x0+qnorm(1-alpha/2)*boot.sd
>
> hist(x0.B,main="Bootstrapped distribution of x0")
> segments(l.ci,0,l.ci,B/10,lty=2,lwd=2)
> segments(u.ci,0,u.ci,B/10,lty=2,lwd=2)
> text(l.ci-.02,B/20,paste(100*alpha/2,"percentile"),srt=90,cex=.75)
> text(u.ci+.02,B/20,paste(100*(1-alpha/2),"percentile"),srt=90,cex=.75)
> results<-matrix(cbind(point.est.x0,l.ci,u.ci,norm.l.ci,norm.u.ci),
+ nrow=1,dimnames=list(c(""),
+ c("point.est.x0","95% LCI","95% UCI","asymptotic 95% LCI",
+ "asymptotic 95% UCI")))
> results
 point.est.x0  95% LCI   95% UCI asymptotic 95% LCI asymptotic 95% UCI
     3.886901 3.646983 4.140691           3.640935           4.132867
```

Figure 9.2 *Bootstrap Sample for Fieller's Theorem Problem*

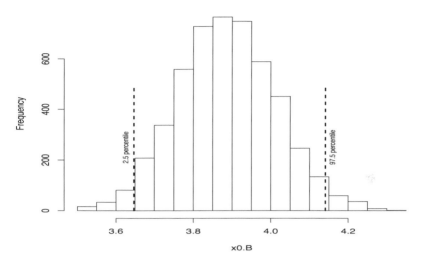

Bootstrapped distribution of x0

9.7 Exercises

1. Demonstrate that, for the data in Example 9.2, if the "Negatives" had been chosen as group 1, then you would have gotten *exactly the same* "Z" statistic and therefore, both the p-value and resulting conclusion would have been the same.

2. Consider a dataset from *Small Data Sets*, dataset # 295 [61]. That data, originally published by Rothschild et al. [126] consists of cortisol levels, in mg/dl in a control group versus four other groups. The data can be obtained from http://www.stat.ncsu.edu/sas/sicl/data/.

(a) Provide a boxplot of each group.

(b) Perform a Kruskal–Wallis analysis on the data.

(c) Compare each of the four psychotic groups to the control group with respect to the cortisol level.

    (I) Do the first set of comparisons without adjustment for multiple comparisons.

    (II) Suggest a multiple comparisons procedure to do this and comment on its validity given that the sample sizes are highly variable.

(d) Comment on the results of your analyses in parts **(a)** – **(c)**.

3. For the data in Example 9.3, demonstrate that if the groups had been labeled in reverse order, then you would have gotten *exactly the same* "Z" statistic and, therefore, the p-value and resulting conclusion would have been the same.

4. Consider the following data

x	0.00	1.00	2.00	3.00	4.00	5.00	6.00	7.0	8.00	9.00	10.00	11.00
y	3.26	0.05	-0.24	1.32	2.46	8.89	16.53	24.3	35.76	50.12	63.02	80.49

(a) Assume that y's are a quadratic function of the x's (which as assumed to be fixed), that is, assume the underlying model is

$$y_i = \beta_0 + \beta_1 x_i + \beta_2 x_i^2 + \epsilon_i, \quad \epsilon_i \text{ i.i.d } N(0, \sigma^2)$$

fit the appropriate model.

(b) Find the value of x, call it $x_{\min}$, which is associated with the minimum y-value for the underlying quadratic model.

(c) Using 5000 bootstrap samples, find the point estimate of $x_{\min}$ and find the point estimate of $x_{\min}$ and its "percentile" and "large sample" 95% bootstrapped confidence intervals.

CHAPTER 10

Analysis of Time to Event Data

10.1 Incidence Density (ID)

The odds ratio and relative risk, which were introduced earlier, give us good measures for risk *assuming that the follow-up times averaged over individuals within two or more cohorts are equal.* In reality, some individuals are followed for shorter periods of time due to the death of some subjects, withdrawals from the study, and general loss of follow-up of some individuals. The crudest measure that can give us information about risk while adjusting for follow-up is known as *incidence density*. The statistic associated with this measure is simply defined as the number of events in a cohort divided by the total person–time in that cohort. The total person–time in a given group is calculated by summing over all persons' follow-up (or exposure) times. If X_j is a random variable with a value of 0 if individual j is event-free and 1 if individual j has had an event, and T_j is the follow-up time for that individual then ID is formally defined as

$$ID \equiv \frac{\sum_{j=1}^{n} X_j}{\sum_{j=1}^{n} T_j}, \qquad (10.1)$$

where n is the number of individuals. To calculate this quantity we need only the number of events of interest and the total person–time for the n individuals.

Example 10.1. Relationship between death from AIDS and gender. The following data is from an AIDS database from Australia and is available in `library(MASS)` in R [see dataset `Aids2`]. The data were originally produced by Dr P. J. Solomon and the Australian National Centre in HIV Epidemiology and Clinical Research and presented by Venables and Ripley [153]. The data consists of $n_1 = 2754$ males and $n_2 = 89$ females who were diagnosed with AIDS. One question of interest is whether or not there is a difference in the mortality rate between the two genders after having been diagnosed with AIDS.

Sex Procedure	# of deaths	# of person–months of follow-up	Incidence Density
Male	$d_1 = 1708$	$t_1 = 36640.43$	$\widehat{ID_1} = \frac{d_1}{t_1} \approx .0466$
Female	$d_2 = 53$	$t_2 = 1275.01$	$\widehat{ID_2} = \frac{d_2}{t_2} \approx .0416$
Total	$d_1 + d_2 = 1761$	$t_1 + t_2 = 37915.43$	$\widehat{ID}_{\text{tot}} = \frac{d_1+d_2}{t_1+t_2} \approx .0464$

231

Notice that the incidence density is in units of events per period of time. To compare therapies, we might test $H_0 : ID_1 = ID_2$ versus $H_1 : ID_1 \neq ID_2$, where the subscript "1" refers to males and the subscript "2" refers to females and "ID" means incidence density.

The theory for comparing IDs between two or more cohorts is very similar to that of contingency table analysis. For example, when comparing incidence densities between two cohorts, it is necessary to determine whether or not to use an approximate or large sample (normal) test versus an exact test. Typically, we determine whether or not the variance, V, is large enough to use the large sample test. Letting $d_\bullet = d_1 + d_2$, and $t_\bullet = t_1 + t_2$ the expression for V is

$$V = d_\bullet \left(\frac{t_1}{t_\bullet} \right) \left(\frac{t_2}{t_\bullet} \right) = \frac{d_\bullet t_1 t_2}{(t_\bullet)^2}. \tag{10.2}$$

This expression is much like the expression for the variance of a binomial distribution, that is, if $Y \sim b(n, p)$ then $\text{Var}(Y) = np(1 - p)$. For an incidence density calculation, the "p" in a given group is the proportion of the overall follow-up time in that group. Similar to what was introduced in Chapter 1, if $V \geq 5$, then we tend to use a normal theory test; otherwise, we use an exact test.

<u>Normal theory test for testing $H_0 : ID_1 = ID_2$ versus $H_1 : ID_1 \neq ID_2$</u>

1. Calculate the expected number of deaths for a cohort, say, $E_1 = d_\bullet \times (t_1/t_\bullet)$.
2. Calculate the variance, V.
3. Form $Z = \dfrac{|d_1 - E_1| - \frac{1}{2}}{\sqrt{V}} \mathbin{\dot\sim} N(0, 1)$ under H_0.
4. p-value = $\begin{cases} 2 \times \left[1 - \Phi(z) \right] & \text{if } Z \geq 0; \\ 2 \times \left[\Phi(z) \right] & \text{if } Z < 0. \end{cases}$

If the alternative hypothesis, H_1 had been one-sided, then the 2 would be dropped from the formula above.

<u>Exact test for testing $H_0 : ID_1 = ID_2$ versus $H_1 : ID_1 \neq ID_2$</u>

1. Calculate the expected proportion of events in, say, cohort 1, under H_0, i.e., $p_0 = \frac{t_1}{(t_\bullet)}$. Also, calculate $q_0 = 1 - p_0$.
2. p-value = $\begin{cases} 2 \times \sum_{k=0}^{d_1} \binom{d_\bullet}{k} p_0^k q_0^{d_\bullet - k} & \text{if } d_1 < E_1 = (d_\bullet)p_0; \\ 2 \times \sum_{k=d_1}^{d_\bullet} \binom{d_\bullet}{k} p_0^k q_0^{d_\bullet - k} & \text{if } d_1 \geq E_1 = (d_\bullet)p_0. \end{cases}$

As before, if the alternative hypothesis had been one-sided, then the 2 would be dropped from the formula.

Example 10.2. Continuing Example 10.1, we can calculate the expected numbers of deaths by sex under H_0 as

Male: $E_1 = d_\bullet(t_1/t_\bullet) = 1708(36640.43)/37915.43 \approx 1701.78$.

Female: $E_2 = d_\bullet(t_2/t_\bullet) = 53(1275.01)/37915.43 \approx 59.218$.

$V = \frac{(d_\bullet)t_1t_2}{(t_\bullet)^2} \approx 57.23 \gg 5 \Rightarrow$ *it is OK to use an approximate test.* Using the males to calculate our Z-statistic, we have

$$Z = \frac{\left| 1708 - 1701.78 \right| - 1/2}{\sqrt{57.23}} \approx 0.756$$

$$\Rightarrow p - \text{value} \approx 2 \times 0.225 = 0.45$$

$\Rightarrow$ do <u>not</u> reject $H_0 \Rightarrow$ there is not enough evidence to show a significant difference (at the $\alpha = .05$ level) in the number of deaths after the diagnosis of AIDS between males and females.

10.2 Introduction to Survival Analysis

It is of interest in many clinical trials (such as those in cancer and heart disease) to estimate the survival time of a cohort of patients. Thus, we are interested in not only *what proportion* of patients survive (or die) but also *how long* they survive (or until they die). If we are interested in only the proportion of patients surviving, we can draw an inference by assuming that the proportion surviving has some underlying binomial distribution. Furthermore, we could compare the survival proportions of two different populations at some given time by using a simple two-sample test comparing binomial proportions. However, overall proportions at a fixed point in time do not give us any information about whether the *patterns of survival over time* between two populations differ. To do this, we will use methods in <u>survival analysis</u>.

Two methods of estimating survival are life table (Actuarial) and Kaplan–Meier estimation procedures. Both methods are simple to carry out and are simple to interpret. Also, both adjust for *censoring* which, roughly speaking, refers to observations made on patients with only partial information. For example, some patients will withdraw from a study before the designated completion of the study. One way to handle "incomplete" observations is to drop them out of the analysis of the data. However, this is an inefficient way of doing things because we can use the information for each patient or subject *up to the point of his/her withdrawal or other type of censorship.*

To begin the formal discussion of these methods, we must first define some terms. By an "event" or a "failure," we usually mean the death of a patient or the occurrence (or recurrence) of a disease or condition of interest in a person. (However, a "failure" to an engineer might be the failure of a certain component of a system.) The *time to failure* is denoted as T. The probability density function for the time to failure is denoted as $f(t) = Pr(T = t)$. The (cumulative) distribution function for time to failure is denoted as $F(t) = Pr(T \leq t)$. The *survival function* is denoted as $S(t) = Pr(T > t)$. The survival function is related to the distribution function for the time to failure as follows:

$$S(t) = Pr(T > t) = 1 - Pr(T \leq t) = 1 - F(t). \tag{10.3}$$

The unadjusted instantaneous failure rate, $f(t)$, can be derived from the cumulative failure function by examining the steepness of the slope of the that curve at a given time. The *hazard function* or *hazard rate* is defined as

$$\lambda(t) = \lim_{\Delta t \to 0} \frac{Pr\{\text{failure in the interval } (t, t + \Delta t) \mid \text{survived past time } t\}}{\Delta t} \quad (10.4)$$

We can think of the hazard function as being related to the probability of a person "failing," for example, getting a specified disease (or dying) in a very short time period *given that the person has not yet contacted the disease (or died).* (We'll see later that the hazard is not a probability because the cumulative hazard distribution can be, e.g., greater than 1.) The hazard function is related to the failure and cumulative survival functions as follows:

$$\lambda(t) = \frac{f(t)}{1 - F(t)} = \frac{f(t)}{S(t)}. \quad (10.5)$$

The hazard function is not a true probability density function in the technical sense because $\int_0^\infty \lambda(t) dt \neq 1$. However, it is a very useful quantity since it gives us an indication of the failure rate of a population over time *adjusted for only those people at risk* at any given time. For this reason, it is also referred to as the *conditional mortality (morbidity) rate* or the *age-specific failure rate.*

10.3 Estimation of the Survival Curve

There are two general methods of estimating the survival curve. The first, called the *actuarial* or *life table* method, was first proposed by Cutler and Ederer (1958) [25] and is most appropriate when data are grouped according to some interval. For example, in some cases, only yearly survival information is available. The second was proposed by Kaplan and Meier [75] and uses all of the event (or death) times to estimate the survival curve.

10.3.1 Life Table Estimation

In some cases, for example, in a journal article or a source of actuarial information provided by the government, the summary of the proportion of a particular cohort that is event- or mortality-free is grouped over quarterly or yearly time intervals. Also, there are cases where the event status of individuals are not known exactly but are known in an interval of time. This latter situation is now referred to as *interval censoring.* An early method of estimating the event-free status for grouped data was proposed by Cutler and Ederer [25] and is known as the life table or actuarial method of estimating the "survival." An example of how this method works, similar to that in Miller [99] is given below.

Example 10.3. Consider the results of a hypothetical clinical trial summarized in Table 10.1. This type of table is known as an *Actuarial* Life Table.

Table 10.1. *Life Table Estimation of Survival for a Hypothetical Study*

(1)	(2) At risk at the beginning of interval	(3) # dying	(4) # withdrew or loss to follow-up	(5) Effective # at risk	(6) Prop. dying	(7) Prop. surviving	(8) Cum. Prop. surviving
Year							
$[0, 1)$	60	19	4	58.0	0.3276	0.6724	0.6724
$[1, 2)$	37	8	4	35.0	0.2286	0.7714	0.5187
$[2, 3)$	25	9	2	24.0	0.3750	0.6250	0.3242
$[3, 5)$	14	5	2	13.0	0.3846	0.6154	0.1995
$[4, 5)$	7	3	0	7.0	0.4286	0.5714	0.1140
≥ 5	4	0	4	2.0	0.0000	1.0000	0.1140

The Life Table (Actuarial) method gives estimates for survival for *fixed time intervals*. The values in columns (1)–(4) are used to calculate the values in columns (5)–(8). In this particular example, the survival rates are given at one-year intervals. The number of deaths in each year represent the total from the beginning of each interval until up to (but not including) the end of the interval. For example, a death at month 12 would be classified in the "Year 1–2" interval.

In the first year of this hypothetical study, the actuarial estimate of the proportion surviving is $1 - 0.3276 = 0.6724$. This is obtained by dividing the number who died in the time period, 19, by the *effective number exposed to the risk of dying*, 58.0, and subtracting the resulting value from 1. In the above calculations, we don't know exactly *when* the people were lost or withdrew. Thus, we make an assumption that, in each interval, those who were lost or withdrew were at risk for *half* of the interval. This premise is a consequence of assuming that the time to loss to follow-up or withdrawal is *uniformly* distributed across each subinterval. Thus, in our case, over the course of a year, the expected sum of the number of people lost to follow-up or who withdrew alive is one half of the sum of the observed values over a year. The effective number at risk is then calculated by subtracting $1/2 \times$ (number lost to follow-up or who withdrew alive) from the total number of patients entering the time interval.

The values in column (6) above are crude estimates of the hazard function in each of the yearly intervals. These values represent one of many possible ways of estimating the hazard function. The values in column (7) are the intervalwise survival rates and are obtained by subtracting the values in column (6) from 1.0. Finally, the values in column (8) are the *cumulative* proportion surviving and are obtained by multiplying all of the values in column (7) up to and including the time interval of interest. For example, the three-year cumulative survival rate is calculated as

$$0.3242 = 0.6724 \times 0.7714 \times 0.6250.$$

10.3.2 The Kaplan–Meier Method

The Kaplan–Meier method of estimation of survival uses *actual event times* for the intervals in which the cumulative survival functions are calculated. Thus, the survival rate estimates are made at variable times according to the failure and censoring times. The Kaplan–Meier (K–M) method uses what is called a *natural filtration* to estimate the survival function because it uses the actual times to death (event) to calculate the cumulative survival curve. Consequently, an estimated cumulative survival function via the K–M method does not change when a censor occurs. It changes only when failures (deaths) occur. In contrast, the actuarial method uses predefined intervals to calculate cumulative survival. Those predefined intervals should be defined so that at least five deaths (events) occur in almost all of the intervals. A demonstration of how the K–M method is employed and how it compares to the life-table method is given in Example 10.4.

Example 10.4. Suppose the following are the actual times (in months after entry into the study) to death in the previous study:

$1, 1, 1, 2, 2, 2, 3, 4+, 5, 5, 5, 6, 6, 7, 7, 8, 8+, 8, 9+, 10, 11, 11, 11+, 12, 12+, 13+, 15, 15,$
$15+, 16, 17, 17, 18, 21, 21+, 24, 24+, 25, 25, 25, 25, 26, 27, 33, 33+, 34, 36, 37+, 38,$
$38+, 40, 41, 43, 51, 52, 59, 61+, 64+, 65+, 73+$

The "+" refers to a time when a patient withdrew, was lost to follow-up, or was alive after the amount of time observed. Thus, for example, 13+ means that the patient was observed as being alive for 13 months then was "censored." The Kaplan–Meier estimates are summarized in Table 10.2. In most computer programs, these data are entered using two variables: (1) a time variable; and (2) a "censoring variable," which typically is coded with a value of 1 indicating a death and a value of 0 indicating that the person was censored (but alive or event-free) at the time point.

The d_i in Table 10.2 refer to the number of failures or events that occurred and the n_i refer to the number at risk. A death in this hypothetical trial is considered a failure. For example, at time 2 months, two patients died. The estimated hazard function in the time interval $(t_{i-1}, t_i]$ is d_i/n_i. t_i refers to the ith distinct time at which at least one event (or failure) occurs. Thus, the estimated hazard rate in our example in the interval $(1, 2]$ is $2/61 = 0.0328$. The estimated pointwise survival function in the above table is $1 - (d_i/n_i)$. The estimate of the *cumulative* survival function at each time point, $\hat{S}(t_i)$, is obtained by multiplying the estimated pointwise survival function at the given time point by the previous value of the estimated cumulative survival function. Because of this multiplicative property, $\hat{S}(t)$ is sometimes called the *Kaplan–Meier product-limit* estimator of the cumulative survival function.

Table 10.2. *Example of the Kaplan–Meier Estimation of Cumulative Survival*

Time	At Risk	d_i/n_i	$1 - d_i/n_i$	$\hat{S}(t_i)$	
0	60	0/60	60/60	60/60	= 1.0000
1	60	3/60	57/60	57/60 × 1.0000	= 0.9500
2	57	3/57	54/57	54/57 × 0.9500	= 0.9000
3	54	1/54	53/54	53/54 × 0.9000	= 0.8833
5	52	3/52	49/52	49/52 × 0.8833	= 0.8324
6	49	2/49	47/49	47/49 × 0.8324	= 0.7984
7	47	2/47	45/47	43/47 × 0.7984	= 0.7644
8	45	2/45	43/45	43/45 × 0.7644	= 0.7304
10	41	1/41	40/41	40/41 × 0.7304	= 0.7126
11	40	2/40	38/40	38/40 × 0.7126	= 0.6770
12	37	1/37	36/37	36/37 × 0.6770	= 0.6587
15	34	2/34	32/34	32/34 × 0.6587	= 0.6200
16	31	1/31	30/31	30/31 × 0.6200	= 0.6000
17	30	2/30	28/30	28/30 × 0.6000	= 0.5600
18	28	1/28	27/28	27/28 × 0.5600	= 0.5400
21	27	1/27	26/27	26/27 × 0.5400	= 0.5200
24	25	1/25	24/25	18/19 × 0.5200	= 0.4992
25	23	4/23	19/23	19/23 × 0.4992	= 0.4124
26	19	1/19	18/19	18/19 × 0.4124	= 0.3907
27	18	1/18	17/18	17/18 × 0.3907	= 0.3689
33	17	1/17	16/17	16/17 × 0.3689	= 0.3472
34	15	1/15	14/15	14/15 × 0.3472	= 0.3241
36	14	1/14	13/14	13/14 × 0.3241	= 0.3009
38	12	1/12	11/12	11/12 × 0.3009	= 0.2759
40	10	1/10	9/10	9/10 × 0.2759	= 0.2483
41	9	1/9	8/9	8/9 × 0.2483	= 0.2207
43	8	1/8	7/8	7/8 × 0.2207	= 0.1931
51	7	1/7	6/7	6/7 × 0.1931	= 0.1655
52	6	1/6	5/6	5/6 × 0.1655	= 0.1379
59	5	1/5	4/5	4/5 × 0.1379	= 0.1103

SAS Program to Produce Kaplan–Meier Estimates

```
data survdata;
input day dth@@;
datalines;
1  1   1 1   1 1   2 1   2 1   2 1   3 1   4 0   5 1   5 1   5 1   6 1   6 1   7 1   7 1
8 1   8 0   8 1   9 0   10 1   11 1   11 1   11 0   12 1   12 0   13 0   15 1   15 1
15 0   16 1   17 1   17 1   18 1   21 1   21 0   24 1   24 0   25 1   25 1   25 1
25 1   26 1   27 1   33 1   33 0   34 1   36 1   37 0   38 1   38 0   40 1   41 1
43 1   51 1   52 1   59 1   61 0   64 0   65 0   73 0
;  run;

proc lifetest method=km;
time day*dth(0);    *<--- Identify censored observations as 0's;
run;
```

R Program to Produce Kaplan–Meier Estimates

```
library(survival)
daydth<-matrix(c(1,1,1,1,1,1,2,1,2,1,2,1,3,1,4,0,5,1,5,1,5,1,6,1,6,1,7,1,7,1,
8,1,8,0,8,1,9,0,10,1,11,1,11,1,11,0,12,1,12,0,13,0,15,1,15,1,
15,0,16,1,17,1,17,1,18,1,21,1,21,0,24,1,24,0,25,1,25,1,25,1,
25,1,26,1,27,1,33,1,33,0,34,1,36,1,37,0,38,1,38,0,40,1,41,1,
43,1,51,1,52,1,59,1,61,0,64,0,65,0,73,0),byrow=T,ncol=2)
day<-daydth[,1]; dth<-daydth[,2]
summary(survfit(Surv(day,dth)~1))
```

As was mentioned above, the Kaplan–Meier estimator provides a finer estimate of the survival curve than the actuarial estimator does. This can be easily seen by examining Figure 10.1 where plots for both the Actuarial and Kaplan–Meier cumulative survival curves for our hypothetical clinical trial are plotted versus time. The actuarial estimates are plotted by interpolating between points whereas the Kaplan–Meier estimates are plotted as a step function. The latter plot is the typical way of graphically representing Kaplan–Meier survival estimates over time. In cases where the predefined intervals are small [i.e., small enough that the hazards are relatively constant in each interval] and the total number of deaths (events) is large, then the two estimators will give very similar results.

Figure 10.1 *Plots of Survival Curve (Actuarial and Kaplan–Meier) for Hypothetical Data*

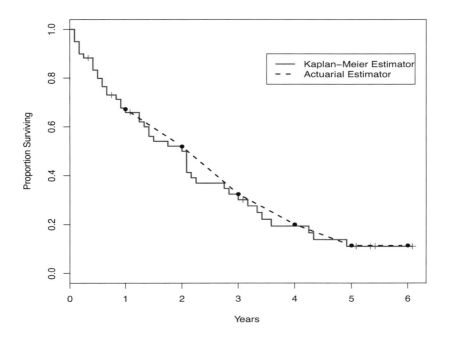

10.4 Estimating the Hazard Function

The estimation of the instantaneous hazard function is usually difficult to do when individual event times are available due to the sparseness of the number of events at any small interval of time. An alternative method for the estimation of a hazard is via the use of a cumulative sum of hazard values over each time. This cumulative quantity is known as the *Nelson–Aalen (N–A) estimator* (Nelson [105, 106] and Aalen [1]) and is written as

$$\widehat{\Lambda}(t) = \sum_{i:\, t_i \leq t}^{r} \frac{d_i}{n_i} \tag{10.6}$$

where d_i and n_i are the numbers of events and individuals at risk, respectively, at the ith unique event time. Hence, like the K–M estimator, the N–A estimator only increments at each failure time. Unlike estimates from the K–M method, however, N–A estimates increment *upwards* not downwards. By examining its slope, one can get a feel for how the instantaneous hazard rate is changing over time.

Example 10.5. The following is a table of the Nelson–Aalen estimates for the hypothetical survival data displayed in Example 10.4.

Time (mos)	0	1	2	3	5	6	7	8	10
$\widehat{\Lambda}(t)$	0.00	0.0500	0.1026	0.1212	0.1788	0.2197	0.2622	0.3067	0.3310

Time (mos.)	11	12	15	16	17	18	21	24	25
$\widehat{\Lambda}(t)$	0.3810	0.4081	0.4669	0.4992	0.5658	0.6015	0.6386	0.6786	0.8525

Time (mos.)	26	27	33	34	36	38	40	41	43
$\widehat{\Lambda}(t)$	0.9051	0.9607	1.0195	1.0862	1.1576	1.2409	1.3409	1.4520	1.5770

Time (mos.)	51	52	59
$\widehat{\Lambda}(t)$	1.7199	1.8866	2.0866

For these calculations, note that we are merely summing up the proportions of patients failing over time points. For example,

$$\widehat{\Lambda}(5) = \frac{3}{60} + \frac{3}{57} + \frac{1}{54} + \frac{3}{52} \approx 0.1788 \ .$$

Also, note that the hazards do *not* sum to 1, which is consistent with the fact that hazards are not probabilities. The cumulative hazard curve for the data above is given in Figure 10.2.

Figure 10.2 *Nelson–Aalen Cumulative Hazard Curve for Hypothetical Data*

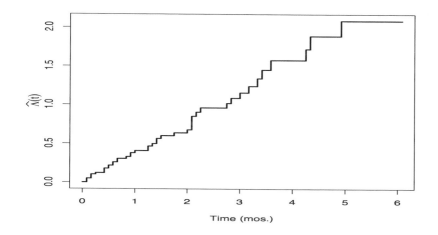

10.5 Comparing Survival in Two Groups

10.5.1 The Log-Rank Test

The most common way that a simple comparison of two groups is performed involves the use of the *log-rank* test (Peto and Peto [112]). This test is equivalent to performing a Mantel–Haenszel procedure on two-by-two tables formed at each unique event time (see Figure 10.3).

Figure 10.3 *Arrangement of 2 × 2 Tables by Event Times for the Log-Rank Test*

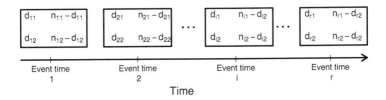

If there are no ties in the event times, then the number of tables would equal r, the total number of events in both groups. Also, the d_{ij} in the tables would always be 0 or 1.

Example 10.6. The following example was taken from a dataset available in the survival library in R. The data are survival (or censoring) times in months for 23 patients who were diagnosed with Acute Myeloid Leukemia (AML) and who were treated either by a "maintenance" therapy or not. These data were earlier presented in Miller (1981) [99]. The variables are time (in months) until death or censor, status (0=alive, 1=dead), and x (Maintained or Nonmaintained). The associated Kaplan–Meier curves are displayed in Figure 10.4.

```
library(survival)
attach(aml)
leg.x<-
  c(paste("Maintained: n =",length(x[x=="Maintained"]),"patients,",
    sum(status[x=="Maintained"]),"events"),
    paste("Nonmaintained: n =",length(x[x=="Nonmaintained"]),"patients,",
    sum(status[x=="Nonmaintained"]),"events"))
xx<-survfit(Surv(time,status)~x)
summary(xx)

plot(xx,lty=1:2,col=1:2,lwd=c(2,2),mark="")
legend(60,0.9,leg.x,lty=1:2,col=1:2,lwd=c(2,2),cex=.8)
```

10.5.2 Extensions and Variations of the Log-Rank Test

The log-rank test is one that gives equal weight to the tables at each event time regardless of the number of individuals at risk at a given time. An alternative to that

Figure 10.4 *Simple Example of Comparison of Survival for Two Groups*

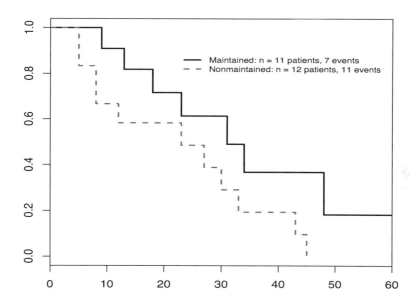

scheme was proposed by Gehan (1965) [53]. For Gehan's test, a rank sum test is performed to compare the event times in each of the two groups. It was later shown by Tarone and Ware (1977) [147] that this scheme is equivalent to weighting each table by the sample size just before the event at the given time point occurs. Thus, earlier survival differences between the two groups are given more weight than later ones. Tarone and Ware suggested that weighting each table by the square root of the sample size might have better properties. However, as pointed out by Harrington and Fleming [63], even more general weighting schemes such as ones that weight later observations more than earlier ones can be employed to accommodate any type of application.

Example 10.7. The SAS program and partial output below provide an analysis of the AML comparing the maintained versus nonmaintained groups. Note that none of the tests yield a statistically significant value and the Wilcoxon (Gehan) test yields a smaller test statistic in this case.

```
proc lifetest;
time time*status(0);
strata x;
run;
```

```
--------------------------PARTIAL OUTPUT--------------------------------
              Test       Chi-Square     DF    Chi-Square
              Log-Rank      2.0342        1      0.1538
              Wilcoxon      1.2080        1      0.2717
              -2Log(LR)     0.3962        1      0.5291
```

10.6 Cox Proportional Hazards Model

As was emphasized in earlier chapters, when making an inference about the comparison of two treatments or interventions, it is important to try to adjust for factors *other than* the treatment or intervention, which may have an influence on the outcome of interest. In survival analysis, such multivariate models are most easily implemented using the hazard function. In most medical and biological settings, the most widely used such model is known as the *Cox Proportional Hazards (Cox PH)* model (Cox [22]).

Letting $\lambda(t)$ denote the hazard rate of a specific event, the Cox proportional hazard model can be written as

$$\lambda(t) = \lambda_0(t) \exp\left(x_1\beta_1 + \ldots + x_r\beta_r\right) = \lambda_0(t) \exp\left(\mathbf{x}'\boldsymbol{\beta}\right) \qquad (10.7)$$

where $\lambda_0(t)$ is the *baseline* hazard [applicable when all covariates are set to 0], $\mathbf{x}' = (x_1 \quad x_2 \quad \ldots \quad x_r)$ is a vector of covariates, and $\boldsymbol{\beta} = (\beta_1 \quad \ldots \quad \beta_r)'$ is a vector of parameters that must be estimated from the data. The salient characteristics of the Cox proportional hazards models can be summarized as follows:

- It is a multiplicative model in that a hazard rate adjusted for the effect of each factor used in a model is calculated by multiplying the baseline hazard by the effect of that factor.

- It is *proportional* because the effect of each factor is assumed to stay constant over time, that is, $\dfrac{\lambda(t)}{\lambda_0(t)} = \exp\left(\mathbf{x}'\boldsymbol{\beta}\right)$ does not vary over time.

- It is called a *semiparametric* model because the specification of the model involves the estimation of the parameters, $\boldsymbol{\beta} = (\beta_1 \quad \beta_2 \quad \ldots \quad \beta_r)'$ but does not require one to parameterize the baseline hazard function, $\lambda_0(t)$.

The Cox proportional hazard model is a very convenient way of quantifying effects of any factor (treatment or otherwise) on a time-to-event outcome. For example, a common way to quantify the average effect of a covariate on an outcome is via the *hazard ratio*. If, for example, one wishes to compare the relative effect of having a value of $X_1 = y$ versus $X_1 = z$ while holding all other covariates fixed, then

$$\text{HR} = \frac{\lambda_y(t)}{\lambda_z(t)} = \frac{\lambda_0(t) \exp\left(y\beta_1 + x_2\beta_2 + \ldots + x_r\beta_r\right)}{\lambda_0(t) \exp\left(z\beta_1 + x_2\beta_2 + \ldots + x_r\beta_r\right)} = \exp\left((y - z)\beta_1\right).$$

Example 10.8. Consider again the Australian AIDS data. A Cox PH model of days until death was fit using the state of residence (three dummy variables with "NSW" as the baseline group); sex (F,M); reported transmission category; and age and age squared as continuous variables as potential predictors of the outcome. In the data step, the age variable was "centered" at 30 years (useful for model interpretation) and an age-squared variable (also "centered" at 30) was introduced to explore a quadratic relationship between age at diagnosis and the time to mortality. The centering allows one to compare the hazard to a 30-year old. Transmission type was constructed so that all forms of AIDS transmission other than direct homosexual contact were compared to direct homosexual contact. The SAS code and results of the analysis are given below.

```
options nonumber ls=75 nodate;
data aids;
infile 'C:\Aids2.txt' firstobs=2; *<-- pasted into "Aids2.txt" from R;
input    patnum  state $ sex $ diag death status $ Tcateg $ age;
run;

data aids1; set aids;
agem30=age-30;
agem30sq=(age-30)**2;
days=death-diag;
dth=0; if status eq 'D' then dth=1;
gender=0; if sex eq 'M' then gender=1;
state12=0; if state eq 'Other' then state12 = 1;
state13=0; if state eq 'QLD' then state13 = 1;
state14=0; if state eq 'VIC' then state14 = 1;
trantyp=(Tcateg ne 'hs');
run;

proc phreg;
model days*dth(0)=agem30 agem30sq gender trantyp state12 state13 state14/rl ;
tstate: test state12,state13,state14;
run;
```
```
--------------------------- Partial output ---------------------------
            Summary of the Number of Event and Censored Values
                                                     Percent
                   Total      Event    Censored     Censored
                   2843       1761       1082         38.06

                          Convergence Status
            Convergence criterion (GCONV=1E-8) satisfied.

                         Model Fit Statistics
                                   Without          With
                   Criterion      Covariates      Covariates
                   -2 LOG L       24954.236       24901.357
                   AIC            24954.236       24915.357
                   SBC            24954.236       24953.672

               Testing Global Null Hypothesis: BETA=0
            Test                Chi-Square      DF     Pr > ChiSq
            Likelihood Ratio      52.8792        7       <.0001
            Score                 60.5861        7       <.0001
            Wald                  59.8900        7       <.0001

             Analysis of Maximum Likelihood Estimates
                         Parameter      Standard
Variable      DF         Estimate        Error      Chi-Square    Pr > ChiSq
agem30        1          0.00476        0.00388       1.5089        0.2193
agem30sq      1          0.0004752      0.0001448    10.7627        0.0010
gender        1          0.15724        0.15698       1.0033        0.3165
```

trantyp	1	-0.04769	0.08582	0.3088	0.5784
state12	1	-0.10827	0.08945	1.4649	0.2261
state13	1	0.13890	0.08763	2.5124	0.1130

Analysis of Maximum Likelihood Estimates

Variable	DF	Parameter Estimate	Standard Error	Chi-Square	Pr > ChiSq
state14	1	-0.03202	0.06112	0.2744	0.6004

The results of the above analysis indicate that only age squared was a significant predictor of time to death in the AIDS patients and hence, if one were to construct a model, would warrant dropping all variables except age squared and age and then re-fitting the model. The linear age term would be retained so that the nature of the risk by age was properly calculated. It could also be eliminated by centering the data at the mean value of age.

From the above results, we can also compare the hazard of mortality of say, a 50-year old AIDS patient compared to a 30-year old AIDS patient as $\exp\left[0.00538(20) + 0.0004456(400)\right] \approx 1.331$. Hence, a 50-year old with AIDS in Australia was about 33% more likely to die (presumably from the disease) than a 30-year old with AIDS.

The effect of proportionality assumption of the Cox proportional hazards (PH) model on strength of relationship of covariates has been a topic of much research (Grambsch and Therneau [57]). In such cases, the Cox PH models characterize effects of covariates on the survival outcome averaged over time, which may or may not be of interest in a particular application. In the survival library in R, there is a function called cox.zph that allows one to test the proportionality assumption (Therneau and Grambsch[150]). Accelerated risk, parametric models using, e.g., the Weibull distribution, and other types of models can be developed if the proportionality assumption fails. Such models are beyond the scope of this book but an excellent overview can be found in Therneau and Grambsch[150].

10.7 Cumulative Incidence for Competing Risks Data

Another approach to modeling failure is via the use of *cumulative incidence* curves. The most common use of cumulative incidence is when one is interested in modeling *cause–specific* time to event endpoints (Gray [58]). Suppose, for example, we are interested in characterizing the cancer-related deaths in a cohort of women with breast cancer while simultaneously characterizing their noncancer-related deaths (Gaynor et al. [52]). These two causes of death are said to be "competing risks" of death. If we index the two causes of death as $j = 1, 2$ for the cancer-related and noncancer-related deaths, respectively, then the cumulative incidence associated with cause j is estimated as

$$\hat{F}_j(t) = \sum_{i:\, t_i \leq t} \frac{d_i}{n_i} \hat{S}(t-)$$ (10.8)

where $\hat{S}(t-)$ is typically the Kaplan–Meier estimate of the cumulative survival at an instant just before time t. The events used to calculate $\hat{S}(t-)$ include all deaths regardless of the cause.

Cumulative incidence curves have a direct relationship to probability distributions. Technically, the cumulative incidence of each cause-specific failure process has a *subprobability* distribution (Gray [58]). What this means is that due to other causes of failure, the overall probability for one particular cause-specific failure over time is ≤ 1 whereas for a process that is described by a "true" probability distribution, the probability of having a failure over all time would be *equal to* 1.

Example 10.9. Consider the simulated data given below where the given times represent either deaths from breast cancer (cause = 1), deaths from another cause (cause = 2), or time last seen alive. The cause—specific for two different treatment groups are given in the table below.

Time	Cause	$\hat{S}$	$\hat{F}_{11}(t)$	$\hat{F}_{12}(t)$	Time	Cause	$\hat{S}$	$\hat{F}_{21}(t)$	$\hat{F}_{22}(t)$
0.00	–	1.000	0.000	0.000	0.00	–	1.000	0.000	0.000
0.14	1	0.967	0.033	0.000	0.08	2	0.960	0.000	0.040
0.37	1	0.933	0.067	0.000	0.17	1	0.920	0.040	0.040
0.81	1	0.900	0.100	0.000	0.89	0	0.920	0.040	0.040
1.28	2	0.867	0.100	0.033	3.10	1	0.878	0.082	0.040
1.79	2	0.833	0.100	0.067	3.23	1	0.836	0.124	0.040
2.18	2	0.800	0.100	0.100	3.43	2	0.795	0.124	0.082
2.34	1	0.767	0.133	0.100	3.51	2	0.753	0.124	0.124
2.61	2	0.733	0.133	0.133	5.18	2	0.711	0.124	0.165
3.34	1	0.700	0.167	0.133	5.25	2	0.669	0.124	0.207
3.48	1	0.667	0.200	0.133	5.55	1	0.627	0.165	0.207
3.59	2	0.633	0.200	0.167	5.82	2	0.585	0.165	0.249
4.01	1	0.600	0.233	0.167	6.62	2	0.544	0.165	0.291
4.17	1	0.567	0.267	0.167	8.41	2	0.502	0.165	0.291
4.34	1	0.533	0.300	0.167	8.50	1	0.460	0.207	0.333
4.68	2	0.500	0.300	0.200	11.29	2	0.418	0.207	0.375
4.96	1	0.467	0.333	0.200	12.63	2	0.376	0.207	0.416
4.99	1	0.433	0.367	0.200	12.87	0	0.376	0.207	0.416
7.63	1	0.400	0.400	0.200	13.57	2	0.329	0.207	0.463
7.73	1	0.367	0.433	0.200	14.36	2	0.282	0.207	0.510
7.80	1	0.333	0.467	0.200	15.52	1	0.235	0.254	0.510
7.94	2	0.300	0.467	0.233	17.79	2	0.188	0.254	0.557
11.18	1	0.267	0.500	0.233	18.88	1	0.141	0.301	0.557
12.18	2	0.233	0.500	0.267	23.64	1	0.094	0.348	0.557
18.61	1	0.200	0.533	0.267	27.29	2	0.047	0.348	0.605
21.85	1	0.167	0.567	0.267	49.56	2	0.000	0.348	0.652
27.98	0	0.167	0.567	0.267					
29.23	1	0.125	0.608	0.267					
29.90	1	0.083	0.650	0.267					
50.02	1	0.042	0.692	0.267					
54.62	1	0.000	0.733	0.267					

For a given time, say, $t = 3.10$ years, the cumulative incidence of breast cancer related deaths (cause = 1) in group 2 is $\hat{F}_{21}(3.10) = 0.082$ whereas the cumulative incidence for other deaths is $\hat{F}_{22}(3.10) = 0.04$. The corresponding cumulative incidences in group 1 for deaths due to breast cancer or due other causes at that time point are $\hat{F}_{11}(3.10) = 0.133$ and $\hat{F}_{12}(3.10) = 0.133$. Note that there is no death at time $t = 3.10$ years in group 1 so the values for the last time in that group where a death occurred ($t = 2.61$ years) are carried forward. As can be seen from the graphs, patients in group 1, on average, appear to do better with respect to cancer-related deaths but do not do as well with respect to noncancer-related deaths.

```
library(survival)
library(cmprsk)

sim.data<-read.table(file="C:\\data\\cuminc-example-1.txt",header=T)
```

```
attach(sim.data)
names(sim.data)
cum.inc.obs<-cuminc(time,cause,trt)
names(cum.inc.obs)
timepoints(cum.inc.obs,1:10)

leg.trt.1<-c(paste("Trt A: n =",length(trt[trt==1]),"patients",
                     length(cause[cause==1&trt==1]),"events"),
             paste("Trt B: n =",length(trt[trt==2]),"patients",
                     length(cause[cause==1&trt==2]),"events"))
leg.trt.2<-c(paste("Trt A: n =",length(trt[trt==1]),"patients",
                     length(cause[cause==2&trt==1]),"events"),
     paste("Trt B: n =",length(trt[trt==2]),"patients",
                     length(cause[cause==2&trt==2]),"events"))

par(mfrow=c(1,2))
plot(cum.inc.obs$"1 1"$time,cum.inc.obs$"1 1"$est,xlim=c(0,10),type="l",
   lty=1,col=1,ylim=c(0,.6),xlab="Time (Years)",ylab="Probability of event",lwd=2)
lines(cum.inc.obs$"1 2"$time,cum.inc.obs$"1 2"$est,lty=2,col=2,lwd=2)
title("Cumulative Incidence of Cause 1")
legend(0,.6,leg.trt.1,col=1:2,lty=1:2,lwd=c(2,2),bty="n",cex=.7)

plot(cum.inc.obs$"2 1"$time,cum.inc.obs$"2 1"$est,xlim=c(0,10),type="l",
   lty=1,col=1,ylim=c(0,.6),xlab="Time (Years)",ylab="Probability of event",lwd=2)
lines(cum.inc.obs$"2 2"$time,cum.inc.obs$"2 2"$est,lty=2,col=2,lwd=2)
title("Cumulative Incidence of Cause 2")
```

Figure 10.5 *Cumulative Incidence Plots of Two Competing Causes of Death*

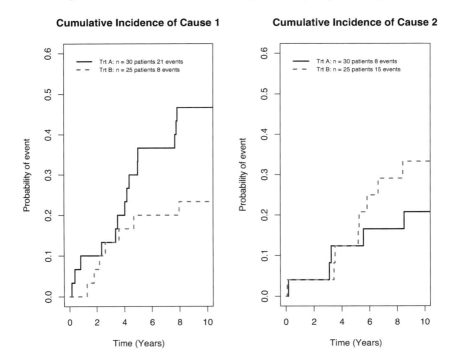

After entering the program above, one can type cum.inc.obs (which is the data frame with the cumulative incidence information in it) to obtain the following (partial) output:

```
Tests:
        stat          pv df
1 4.381057 0.03634046   1
2 5.030727 0.02490144   1
```

This indicates that the differences in each cause are significantly different by treatment arm (although the p-values are only approximations as the number of cause–specific events are small). However, the benefit for Treatment B with respect to breast cancer-related deaths is almost completely offset by the benefit for Treatment A with respect to nonbreast cancer-related deaths. By using the code given below, one can see that the overall treatment difference between all causes is neglible, that is, not even close to being significant.

```
> survdiff(Surv(time,(cause>0))~trt)
Call:
survdiff(formula = Surv(time, (cause > 0)) ~ trt)
         N Observed Expected (O-E)^2/E (O-E)^2/V
trt=1 30       29     28.1    0.0277    0.0642
trt=2 25       23     23.9    0.0326    0.0642

 Chisq= 0.1  on 1 degrees of freedom, p= 0.8
```

Work by Fine and Gray [42] allows one to model competing risks while accommodating covariate information. To see how this is implemented in R, first, install and load the library, cmprsk, then type ?crr at the R console.

10.8 Exercises

1. Consider the dataset called "Melanoma" (found in the R library called "MASS") characterizing 205 patients in Denmark with malignant melanoma. It was originally presented by Andersen et al. [5]. The variables are the survival time in days (time); status (1 = died from melanoma, 2 = alive, 3 = dead from other causes); sex (0 = female, 1 = male); age (in years), year of operation; tumor thickness in mm (thickness), and ulcer (0 = absence, 1 = presence). Create Kaplan–Meier curves for all causes of death (status = 1 or 3) for the patients with ulcers versus those without ulcers. Do a log-rank test to test if there was a significant difference in time to mortality between the two groups. What are your conclusions?

2. Using the Melanoma dataset, create two cumulative incidence curves similar to those in Figure 10.5: one for those dying of melanoma (status = 1) and one for those dying of other causes (status = 3). In each graph, overlay a curve for the patients with ulcers by one for the nonulcer patients. Does the relation between ulcer status and time to mortality appear to be different for those dying of melanoma versus those dying of other causes? Elaborate on your results.

3. Using the Melanoma dataset, fit a Cox proportional hazards model for time to all cause mortality in SAS (using PROC PHREG) or R/S-plus (using coxph). Use all of the available covariates in the initial fit of the (time, status) outcome. Then, discard the variables not significant at the $\alpha = .05$ level and re-fit the model. Continue to do this until all fitted variables are associated with p-values < 0.05. Comment on your results.

4. In the Melanoma dataset, fit separate cause–specific proportional hazard models for those dying of melanoma and those dying of other causes. For the two causes, use a strategy similar to that outlined in the previous problem. Do you get similar results for the two types of cause of death? Comment.

## CHAPTER 11

Sample Size and Power Calculations

A fundamental part of the design of any study is the consideration of the number of individuals, "experimental units," or observations that are required to appropriately conduct the research. Unfortunately, even if we think some treatment or intervention is successful as compared to another with respect to an outcome, due to sampling and variability within a population of interest, we may not actually observe that success statistically in a given study. However, for example, one may specify that they want to have an 80% chance of detecting some effect if that effect truly exists. Such a study would be said to have 80% *power* to detect the specified effect. The investigators would then calculate a sample size based on this power requirement (along with other statistical and scientific considerations). In contrast, due to cost constraints, ethical or other considerations, one may be only able to obtain a *fixed sample size* for a study. In this latter case, one may wish to calculate the power of a test, given the prespecified sample size.

An important distinction that must be made in calculating power and sample size is whether or not the investigators wish to frame their results in terms of confidence intervals or in terms of hypothesis testing. Another consideration is whether or not the investigators hope to establish that one treatment is more efficacious than other treatments or, alternatively, whether or not two or more treatments are equivalent. In the latter case, one would hope that one of the treatments or interventions would have fewer side effects than the other.

In this chapter, we will focus on issues related to sample size and power calculations when hypothesis testing is employed and when the question of interest is whether or not efficacy is different between two treatments or interventions. If one wishes to test some hypothesis in a "comparative treatment efficacy" (CTE) trial (Piantadosi[113]), then a projected sample size can only be calculated after the following information is specified:

- a specific "primary" outcome has been defined; and
- a specific null hypothesis has been specified; and
- an alternative hypothesis [one-sided or two-sided] has been specified; and
- a scientifically or clinically meaningful difference (ratio) has been specified; and

- the power to detect the pre-specified "meaningful difference" is established; and
- an α-level has been specified.

Notational Note: Recall that Z_α denotes the value, Z. of a normal distribution having the property that $\Pr(Z < z) = \alpha$. For example, $\Pr(Z < Z_{1-\frac{\alpha}{2}}) = 1 - \frac{\alpha}{2}$. In the case that $\alpha = .05$, $Z_{1-\frac{\alpha}{2}} = Z_{.975} = 1.96$. Also, Φ represents the cumulative standard normal distribution, that is, $\Phi(z) = Pr(Z \le z)$ where $Z \sim N(0,1)$.

11.1 Sample Sizes and Power for Tests of Normally Distributed Data

11.1.1 Paired t-Test

For a paired t-test, one analyzes the differences among pairs and uses a one-sample test to determine if the magnitude of the average difference is significantly different from some value (which is 0 in most cases). We define a scientifically significant difference as $\Delta = \mu_1 - \mu_0$, where μ_0 and μ_1 are the population mean differences under the null and alternative hypotheses, respectively.

To derive a formula for power and sample for a one-sample t-test (or z-test), consider Figure 11.1. By declaring the α-level, setting say, a one-sided alternative hypothesis, and estimating the standard deviation, σ, of the average population difference, one creates a cutoff or decision point where one declares whether or not the null hypothesis is rejected. For the normally distributed data depicted in Figure 11.1 this cutoff value can be written as *either* $\mu_0 + Z_{1-\alpha}\frac{\sigma}{\sqrt{n}}$ or $\mu_1 - Z_{1-\beta}\frac{\sigma}{\sqrt{n}} = \mu_1 - \delta$, where $\delta \equiv Z_{1-\beta}\frac{\sigma}{\sqrt{n}}$. Thus, we can set $\mu_0 + \frac{\sigma}{\sqrt{n}} = \mu_1 - \delta$. From this relationship, we have

$$\mu_0 + Z_{1-\alpha}\frac{\sigma}{\sqrt{n}} = \mu_1 - Z_{1-\beta}\frac{\sigma}{\sqrt{n}} \Rightarrow \mu_1 - \mu_0 = \Delta = (Z_{1-\alpha} + Z_{1-\beta})\frac{\sigma}{\sqrt{n}} \quad (11.1)$$

$$\Rightarrow Z_{1-\alpha} + Z_{1-\beta} = \frac{\sqrt{n}}{\sigma}\Delta \Rightarrow Z_{1-\beta} = \frac{\sqrt{n}}{\sigma}\Delta - Z_{1-\alpha}$$

$$\Rightarrow 1 - \beta = \Phi\left\{\frac{\sqrt{n}}{\sigma}\Delta - Z_{1-\alpha}\right\},$$

where $1 - \beta$ is the power of the test. If the test were two-sided then $Z_{1-\alpha}$ would simply be replaced by $Z_{1-\frac{\alpha}{2}}$.

Alternatively, if one were to solve for n (the required sample size) in equation (11.1), the derivation would change to

$$\sqrt{n} = \frac{(Z_{1-\alpha} + Z_{1-\beta})\sigma}{\Delta} \Rightarrow n = \frac{\sigma^2(Z_{1-\alpha} + Z_{1-\beta})^2}{\Delta^2}.$$

Again, if the test were two-sided then $Z_{1-\alpha}$ would be replaced by $Z_{1-\frac{\alpha}{2}}$. In all cases covered in this chapter, power and sample size can be solved by similar arguments (although the algebra is messier in some cases).

Figure 11.1 *Setup for Power and Sample Size Calculations*

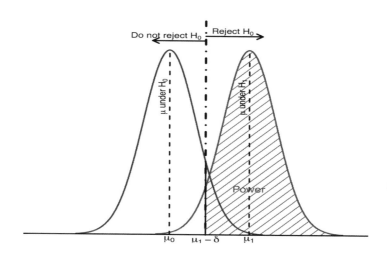

For the case of a paired t-test we'll replace σ by σ_d to denote that we are taking the standard deviation of the *differences* of the pairs. So, by declaring the α-level, setting, say, a two-sided alternative hypothesis and estimating the standard deviation, σ_d, of the average population difference, a power formula can be written as

$$\text{Power} = 1 - \beta = \Phi\left\{\Delta\frac{\sqrt{n}}{\sigma_d} - Z_{1-\frac{\alpha}{2}}\right\}, \tag{11.2}$$

where n is the number of *pairs* of observations.

A sample size formula is as follows:

$$n = \frac{\sigma_d^2(Z_{1-\frac{\alpha}{2}} + Z_{1-\beta})^2}{\Delta^2}, \tag{11.3}$$

where, once again, n is the number of *pairs* of observations one would have to accrue to achieve a certain power.

11.1.2 Unpaired t-test

In the two-sample "parallel groups" design, patients, subjects or experimental units are assumed to be randomly assigned to the two arms of the study. Thus, sample 1 may contain n_1 observations and sample 2 may contain n_2 observations and there is no requirement that n_1 equals n_2. In cases where the outcome of interest is continuous and distributed approximately normally (or where the central limit theorem holds), then an "unpaired" t-test is used to determine whether the mean values, μ_1 and μ_2, of the two arms are "significantly" different.

To test $H_0 : \mu_1 = \mu_2$ versus $H_1 : \mu_1 \neq \mu_2$ for a given alternative hypothesis, $\mu_1 = \mu_2 - \Delta$ (or $\mu_1 = \mu_2 + \Delta$) at significance level, α, the large sample expression for power is

$$\text{Power} = 1 - \beta = \Phi\left\{ \frac{\sqrt{n_1}\Delta}{\sqrt{\sigma_1^2 + \sigma_2^2/k}} - Z_{1-\frac{\alpha}{2}} \right\}, \tag{11.4}$$

where $k = \frac{n_2}{n_1}$ (Rosner [122]). For a one-sided alternative, substitute $Z_{1-\alpha}$ for $Z_{1-\frac{\alpha}{2}}$ in equation 11.4.

Example 11.1. Suppose that, in a problem of determining the relationship of blood pressure to race, we want to detect a difference of 7 mm Hg between Caucasian and African-American populations assuming that the Caucasian population has 64 observations and the African-American population has 32 observations ($\Rightarrow k = 1/2$). Also, assume that $\sigma_1^2 = 10^2 = 100$ and $\sigma_2^2 = 11^2 = 121$ and that $\alpha = .05$.

$$\Rightarrow 1 - \beta = \Phi\left\{ \frac{\sqrt{64} \cdot 7}{\sqrt{100 + \frac{144}{(1/2)}}} - 1.96 \right\} = \Phi\left\{ 1.2413 \right\} \approx 0.8573 .$$

Hence, with the given sample sizes, one would have about 85.7% power to detect the difference of interest.

For testing $H_0 : \mu_1 = \mu_2$ versus $H_1 : \mu_1 \neq \mu_2$ at a given α-level and power and for a given difference, Δ:

$$n_1 = \frac{(\sigma_1^2 + \sigma_2^2/k)(Z_{1-\frac{\alpha}{2}} + Z_{1-\beta})^2}{\Delta^2}$$

$$n_2 = \frac{(k\sigma_1^2 + \sigma_2^2)(Z_{1-\frac{\alpha}{2}} + Z_{1-\beta})^2}{\Delta^2}, \tag{11.5}$$

where $k = \frac{n_2}{n_1}$. For testing a one-sided alternative hypothesis, replace $Z_{1-\frac{\alpha}{2}}$ by $Z_{1-\alpha}$ in equation 11.5.

It should be noted that if $\sigma_1^2 = \sigma_2^2$, then the smallest $N = n_1 + n_2$ is achieved if $n_1 = n_2$, that is, by setting $k = 1$. Also, observe that the sample size of each group is *inversely proportional to the square of the difference that is designated to be scientifically or clinically relevant*. Thus, for example, if one investigator specifies a particular "scientifically relevant" difference and a second investigator specifies a difference that is half of the difference specified by the first investigator, then the sample size required by the second investigator will be *four times* that of the first investigator. This relationship is very important to consider when designing a study and eliciting scientific or clinical input about specifying "meaningful" differences.

11.2 Power and Sample Size for Testing Proportions

It is often of interest in clinical trials to determine whether or not the proportion of some event is different among two arms of a study. These events could be death, relapse, occurrence of side effects, occurrence of myocardial infarction, etc. If the patients or subjects within two groups have relatively equal follow-up and the expected time to an event of interest is relatively short, then one may want to test whether the proportions of events are different between two arms of a study. Given the sample sizes, n_1 and n_2, and pre-specifying the α-level, one can calculate the power of a test comparing two binomial proportions as

$$1 - \beta = \Phi\left\{ \frac{\Delta - Z_{1-\frac{\alpha}{2}}\sqrt{\overline{p}\,\overline{q}\left(\frac{1}{n_1} + \frac{1}{n_2}\right)}}{\sqrt{\frac{p_1 q_1}{n_1} + \frac{p_2 q_2}{n_2}}} \right\}, \tag{11.6}$$

where p_1 and p_2 are the projected probabilities for the two groups, $q_1 = 1 - p_1$, $q_2 = 1 - p_2$, $k = \frac{n_2}{n_1}$, $\Delta = p_1 - p_2$, $\overline{p} = \frac{p_1 + k p_2}{1+k}$ and $\overline{q} = 1 - \overline{p}$.

From equation 11.6, one can derive the sample sizes, n_1 and n_2 as

$$n_1 = \frac{\left[\sqrt{\overline{p}\,\overline{q}\left(1 + \frac{1}{k}\right)}\, Z_{1-\frac{\alpha}{2}} + \sqrt{\frac{p_1 q_1 + p_2 q_2}{k}}\, Z_{1-\beta}\right]^2}{\Delta^2}, \quad n_2 = k n_1, \tag{11.7}$$

where the definition of the variables are the same as in equation (11.6).

Example 11.2. A study comparing diet + a compound, which lowers blood pressure to diet + a placebo, uses the proportion of hypertensive individuals converting to normotensive in a 6-week period of time as its primary outcome. In the diet + placebo group, 15% of the individuals are expected to become normotensive whereas it is projected that 25% of the individuals in the diet + compound group will convert. The design of the study requires 80% for a one-sided test with $\alpha = 0.05$. Equal numbers of individuals will be randomized into the two groups. Find the sample size required for this study to meet its design requirements. In this case, $p_1 = 0.15$, $q_1 = 0.85$, $p_2 = 0.25$, $q_2 = 0.75$, $k = 1$ so that $\overline{p} = \frac{0.15 + 0.25}{2} = 0.2$, $\overline{q} = 1 - 0.2 = 0.8$ and $\Delta = 0.1$. Also, since the proposed test is one-sided, we substitute $Z_{1-\alpha}$ in for $Z_{1-\alpha/2}$ in equation (11.6) yielding $Z_{0.95} \approx 1.6449$ and $Z_{0.8} \approx 0.8416$. Hence,

$$n_1 = n_2 = \frac{\left[\sqrt{0.2\,(0.8)\left(1 + 1\right)}\,1.6449 + \sqrt{\frac{(0.15)(0.85) + (0.25)(0.75)}{1}}\,0.8416\right]^2}{0.1^2}$$

$$\approx \frac{0.9305 + 0.4726}{0.01} \approx 196.8 \longrightarrow 197 \,.$$

Consequently, the study will require about 197 patients per group for a total of 394 patients required for the study. This calculation assumes no patients drop out during the six week period. Formulas for dropout in studies can be found in, e.g., Rosner [124].

11.3 Sample Size and Power for Repeated Measures Data

Sample size and power calculations for longitudinal data can be quite complex depending on the nature of the question being asked, that is, the nature of the null and alternative hypotheses. However, if the question being asked is "Is there a group difference averaged across time?" and if the number of observations per subject are the same, say t, are equally spaced and the covariance structure is compound symmetric, then, as is outlined in Brown and Prescott [14] one can easily develop a sample size formula. First, for the sum of observations on individual i, we note that

$$\mathrm{var}\left(\sum_{j=1}^{t} y_{ij} \right) = t\mathrm{var}(y_{ij}) + t(t-1)\mathrm{cov}(y_{ij}, y_{ik}) \ (k \neq j)$$

$$= t\sigma^2 + t(t-1)\rho\sigma^2 = t\sigma^2[1 + (t-1)\rho] \ .$$

so that for each patient mean,

$$\mathrm{var}(\overline{y}_{i\bullet}) = \frac{\sigma^2[1 + (t-1)\rho]}{t},$$

where

$t =$ the number of repeated measurements per individual,

$\sigma^2 =$ the between subject variation (assuming a compound symmetric covariance pattern), and

$\rho =$ correlation of observations within a subject (again, assuming a compound symmetric covariance pattern).

Letting $\Delta = (z_{1-\alpha/2} + z_\beta) \times \mathrm{SE}(\alpha_i - \alpha_k)$, then the number, n, of individuals required per group (for a two-sided test) is given by

$$n = \frac{2(z_{1-\alpha/2} + z_{1-\beta})^2 \sigma^2[1 + (t-1)\rho]}{t\Delta^2}, \tag{11.8}$$

where

$\alpha =$ the significance level of the test

$1 - \beta =$ the desired power,

$\Delta =$ the difference to be detected, and

$\alpha_i =$ the ith group effect.

For small trials where n is expected to be ≤ 10, the appropriate value is obtained from the t-distribution.

Example 11.3. Consider the rat data presented in Example 8.3. For that data, $\sigma^2 = 51.5$ and, for a compound symmetric covariance pattern, $\rho = 0.595$. If we wanted to design a new study with 4 time points and 90% power to detect a difference of 5 mm Hg at the 5% significance level then the number of patients per group would be

$$n = \frac{2(1.96 + 1.281552)^2 \times 51.5 \times [1 + 3(0.595)]}{4(5)^2} \approx \frac{3014.11}{100} \approx 30.14 \longrightarrow 31$$

An R program can be easily written to determine sample sizes (see below).

```
> alpha <- 0.05 #<-- Type I error probability
> beta <- 0.1   #<-- Type II error probability (1-beta = Power)
> Power<- 1 - beta
> sgmsq <- 51.5 #<-- Between individual variance
> tt<-4 #<-- Number of time points
> rho <- 0.595  #<-- Correlation among observations
> Delta <- 5   #<--- Difference of interest
> N <- 2*(qnorm(1-alpha/2)+ qnorm(1-beta))^2*sgmsq*(1+(tt-1)*rho)/(tt*Delta^2)
> n <- floor(N)+1
> cbind(alpha,sgmsq,m,rho,Delta,Power,n)
       alpha sgmsq m   rho Delta Power  n
[1,]   0.05  51.5  4 0.595     5   0.9 31
```

11.4 Sample Size and Power for Survival Analysis

In many clinical studies, the primary interest is to determine whether one treatment as compared to another increases the time to some event (e.g., death, relapse, adverse reaction). Such studies are analyzed using survival analysis as outlined in Chapter 10. In most such studies, at the time of the definitive analysis, the outcome (time to event) is not observed for all of the individuals. Those observations for which no event is observed are considered to be "censored." In analyzing such data, each individual is still "at risk" up until the time they were observed to have an event or censoring has occurred.

In survival analysis, the power and sample size are determined by the number of events observed in the study. From the timing of these events, one can estimate *hazard rates*. The hazard rate is related to the instantaneous failure rate in a time interval $(t, t + \Delta t)$ given that an individual is still at risk at the beginning of the interval (i.e., at time t). In studies of chronic diseases, it is often the case that the hazard rate, denoted λ, is constant over time. This property is equivalent to saying that the failure distribution is exponential. In such studies, the hazard rate for a cohort can be estimated simply by taking the number of events and dividing through by the total number of person–years in the cohort, i.e., $\hat{\lambda} = n_e / \sum_i t_i$.

For a study comparing the event-free rates in two groups, where the hazard rates, λ_1 and λ_2 are assumed to be constant, George and Desu [54] showed that the total number of events, n_e, in the study is given by

$$n_e = \frac{4 \left(Z_{1-\frac{\alpha}{2}} + Z_{1-\beta} \right)^2}{\left[\ln(\Delta) \right]^2}, \tag{11.9}$$

where "ln" denotes the natural logarithm (that is, log base e), $\Delta = \frac{\lambda_1}{\lambda_2}$ and the α-level and power, $1 - \beta$, are prespecified. From equation 11.9, we can derive a formula for power as

$$1 - \beta = \Phi\left\{\frac{\sqrt{n_e}\,|\ln(\Delta)|}{2} - Z_{1-\frac{\alpha}{2}}\right\} \tag{11.10}$$

Example 11.4. In a breast cancer trial, two therapeutic regimens are being tested. The first is a combination of two active drugs with a placebo and the second is the same two drugs but with an active agent added. The trial is testing whether or not the third active agent is associated with a reduction of mortality. A survey of the literature indicates that the 5-year survival of patients taking the two-drug commbination is about 88%. It is speculated that the addition of the third active agent might improve this 5-year survival to 92%. One way to get the average hazard ratio over the 5-year period is to calculate the average hazards. First, on average, the failure rates are $0.92 = \exp\{-\lambda_1 t\}$ and $0.88 = \exp\{-\lambda_2 t\}$ where $t = 5$ years and solve for λ_1 and λ_2, that is $\lambda_1 = \frac{-\ln(0.88)}{5}$ and $\lambda_2 = \frac{-\ln(0.92)}{5}$. Then

$$\frac{\lambda_2}{\lambda_1} = \frac{\frac{-\ln(0.92)}{5}}{\frac{-\ln(0.88)}{5}} = \frac{\ln(0.92)}{\ln(0.88)}$$

corresponding to a reduction of the 5-year the hazard rate of mortality of about 25.8%.

Thus, $\Delta = \frac{\lambda_2}{\lambda_1} \approx 0.742$. Based on recent studies, a very rough approximation of the patients in such a population fail according to an exponential distribution. In order to have 90% power to detect the above-mentioned 25.8% reduction in hazard rates for a two-sided test, one must observe

$$n_e = \frac{4(1.96 + 1.28155)^2}{\left(\ln(.742)\right)^2} \approx \frac{26.278}{.08899} \approx 295.3 \longrightarrow 296 \text{ deaths.}$$

Notice here that the number 295.3 was "rounded up" instead of rounded to the nearest number. This builds in a slightly conservative sample size estimate to guarantee that we have the stated power to detect a difference of interest. Consequently, in order to ensure that we have 90% or greater power in our case to detect a reduction in hazard of 25.8% (given an exponential failure rate), one would need to observe 296 deaths in the study. This, of course, does not give the sample size needed but, rather, the number of deaths that would have to be observed in the study to achieve the stated power.

R Program to calculate number of events to obtain 90% power to detect a reduction in survival from 92% to 88% (exponential failure)

```
surv.5yr.1<-.92   #<-- 5-year survival in control group
surv.5yr.2<-.94   #<-- 5-year survival in experimental group

alpha <- .20 # alpha-level
beta <- .1   # Beta error rate
r <- 1    # Number of comparisons
sided <- 2   # 1 or 2
Z.a <- qnorm(1-alpha/(r*sided))
Z.b <- qnorm(1-beta)
num <- 4*(Z.a+Z.b)^2
HR <- log(surv.5yr.2)/log(surv.5yr.1) #<-- logs here are base e

num.events.exp <- floor(num/(log(HR)^2)) + 1
print(paste("Number of comparisons being made=",sided,"- sided test"))
print(paste(sided,"-sided test"))
```

```
print(alpha/r)

# Give numbers of deaths to be observed in the proposed study
cbind(surv.5yr.1,surv.5yr.2,HR,num.events.exp)
```

It is sometimes the case that we wish to project the number of events in each arm of the study. One can derive the number of events in each arm using the expression

$$\frac{1}{n_{e_1}} + \frac{1}{n_{e_2}} = \frac{\left[\ln(\Delta)\right]^2}{\left(Z_{1-\frac{\alpha}{2}} + Z_{1-\beta}\right)^2}, \tag{11.11}$$

where n_{e_1} and n_{e_2} are the numbers of events (deaths) in arms 1 and 2, respectively.

In cases where there is no assumption about the event-time distributions but where the hazards are assumed to be proprotional, i.e., $\frac{\lambda_1(t)}{\lambda_2(t)} = $ constant for all t, Freedman [50] derived an expression for the total number of deaths as

$$n_e = \frac{\left(Z_{1-\frac{\alpha}{2}} + Z_{1-\beta}\right)^2(\Delta+1)^2}{(\Delta-1)^2}. \tag{11.12}$$

This latter expression gives a slightly more conservative estimate than the expression in equation 11.9 for determining the number of events needed to achieve a specified power.

Example 11.5. In the cancer study described above, one would need to observe

$$n_e = \frac{(1.96+1.2855)^2(1.7421)^2}{(0.0.2579)^2} \approx 299.7$$

$\Rightarrow$ 300 deaths would have to be observed to achieve 90% power to detect a 25.8% mortality reduction.

From equation 11.12, an expression for power can be derived as

$$1 - \beta = \Phi\left\{\sqrt{n_e}\left|\frac{\Delta-1}{\Delta+1}\right| - Z_{1-\frac{\alpha}{2}}\right\} \tag{11.13}$$

Equations 11.9–11.13 are written with the assumption that the alternative hypotheses are two-sided. For a one-sided alternative hypothesis, $Z_{1-\frac{\alpha}{2}}$ is replaced by $Z_{1-\alpha}$.

11.4.1 Incorporating Accrual

A practicality in designing a clinical trial is that it may take months or years to accrue patients to the trial. Suppose, for example that a trial accrues patients over the interval $(0, T)$ where T is the time interval from the beginning of the accrual period to the end of the accrual period. Obviously, the follow-up time, τ, to achieve a specified power would be longer in this case than a (hypothetical) trial, which accrued its complete population instantaneously. In addition, one would expect that, as in most clinical studies, there would be a proportion, μ, of patients who are lost to follow-up. Using these parameters and the assumption of Poisson accrual rates, Rubinstein, Gail, and Santner [127] showed that the study parameters must satisfy

$$\sum_{i=1}^{2} \frac{2(\lambda_i^*)^2}{n\lambda_i} \frac{1}{\lambda_i^* T - e^{-\lambda_i^* \tau}(1 - e^{-\lambda_i^* T})} = \frac{\left(Z_{1-\frac{\alpha}{2}} + Z_{1-\beta}\right)^2}{\left[\ln(\Delta)\right]^2} , \qquad (11.14)$$

where n = the total accrual into the study, $\lambda_i^* = \lambda_i + \mu$ and the other parameters are defined as above. Numerical techniques are often needed to solve for parameters of interest in this equation. Programs such as PASS and Survpower use such techniques in solving for these parameters.

11.5 Constructing Power Curves

A useful way to help determine how large a study should be is to construct plots of "power curves." The idea behind the construction of such plots is to graph the power of the study to detect differences of interest as a function of a range of parameters. The parameter values are chosen as possible values that would be reasonable scientifically. For example, one might be interested in relating power to the magnitude of the differences or ratios being projected (that is, the *effect size*), the number of events in each arm of the study, the proportions occurring in each arm of the study, the sample size of the overall population, or by how long one wishes to follow a particular cohort.

Example 11.6. Suppose we are designing a cancer study with time to event as the primary outcome and we wish to construct a power curve as a function of the number of relapses that we observe over a five-year period of time. Using equation 11.13, we will write a program to calculate power curves for two cases: (1) where α is set at 0.05; and (2) where α is set at 0.01. The tests will be two-sided.

```
alpha <- .05 # alpha
alpha.2<- .01
sided <- 2  # 1 or 2
Z.a <- qnorm(1-alpha/sided)
Z.a.2 <- qnorm(1-alpha.2/sided)
HR <- .75
num.events<-100:600
Power.curve<-pnorm(sqrt(num.events)*abs((HR-1)/(HR+1))-Z.a)
Power.curve.2<-pnorm(sqrt(num.events)*abs((HR-1)/(HR+1))-Z.a.2)

leg<-c(expression(paste(alpha,"=.05")),expression(paste(alpha,"=.01")))
plot(num.events,Power.curve,type="l",xlab="Number of events",lwd=2,
     xlim=c(100,600),ylim=c(0,1),ylab="Power")
lines(num.events,Power.curve.2,lty=2,col="blue",lwd=2)
#title(paste("Power to detect a hazard reduction of",round((1-HR)*100,3),"%"))
abline(h=0:5/5,lty=5,col="lightgray")
legend(400,.35,leg,col=c("black","blue"),lty=1:2,bty="n",lwd=2:2)
```

Figure 11.2 *Power curves to Detect a 25% Hazard Ratio Reduction by Number of Events for Two-Sided Tests with* $\alpha = .05$ *and* $\alpha = .01$

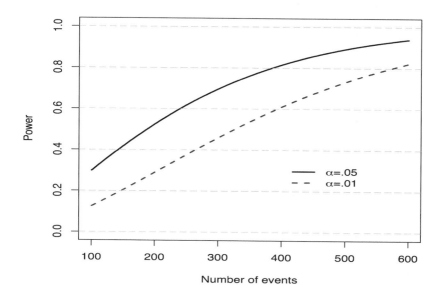

Notice that these calculations involve the number of events that one would be required to observe in the study rather than the total sample size necessary. More sophisticated software is required to calculate total sample sizes in this setting, especially if considerations like accrual rates and differential failure rates are taken into account.

The software for creating power or sample-size curves has vastly improved over the last 10 years. Such software can be used to calculate power and sample sizes in very complex cases and for nearly every type of study endpoint and design. The software also is useful for instances when equivalence of outcomes across groups is of interest and in other cases where one is more interested in sample size/power for studies where confidence intervals of point estimates are of interest to the investigator. Although there are now many such packages/algorithms, two that I find to be quite useful is the statistical package called PASS and the SAS algorithm (PROC) called PROC POWER.

11.6 Exercises

1. Consider again the study in Example 11.1. Suppose now, however, that $k = 1$, that is, $n_1 = n_2$, and the difference of interest is 5 mm Hg. As before, $\sigma_1^2 = 100$, $\sigma_i^2 = 121$ and a two-sided $\alpha = 0.05$. As part of the study design team, you've been asked to construct a *power curve* as a function of total sample size ($N = n_1 + n_2$).

(a) Write a program in your favorite package to calculate the power as a function of total sample size ($N = n_1 + n_2$) for $N = 30$ to $N = 150$ by increments of 10.

(b) Repeat part **(a)** but change α to 0.01.

(c) Plot your results from part **(a)**, that is, plot the required sample size (y-axis) versus k (x-axis). Overlay the results from part **(b)** on the same plot.

2. Consider again the study in example **11.1**. Suppose again that detecting a difference of 7 mm Hg between the two groups is of interest, and that $\sigma_1^2 = 100, \sigma_2^2 = 121$ and $\alpha = .05$. Suppose now, however, that you've been asked to calculate the *sample size required to achieve a power of 80%* to detect a difference. Furthermore, you have been asked to do a sensitivity analysis of how the proportion, $k = \frac{n_2}{n_1}$ affects the required sample size.

(a) Write a program in your favorite package to calculate the sample size as a function of k for $k = 1/5, 1/4, 1/3, 1/2, 2/3, 3/4, 4/5$, and 1.

(b) Plot your results from part **(a)**, that is, plot the required sample size (y-axis) versus k (x-axis).

3. Consider Example 11.2. In that example, we found that 197 people were needed per group to detect the proportion of people converting to normotensive status being reduced from 25% to 15% in the placebo and active groups, respectively. This works out roughly to 50 people being converted to normotensive in the diet + blood pressure lowering compound group and 30 people being converted to normotensive in the diet + placebo for a total of 80 "events" in the six-week period.

(a) Suppose that you are able to measure people daily on whether or not they've converted to being normotensive over the six-week period of time. You decide that a more efficient way to analyze the data is via survival analysis by using a log-rank test. You reason that the hazard ratio of interest is $\Delta = \frac{.25}{.15} \approx 1.667$ and keep the α-level and power the same as in Example 11.2. Using these assumptions and the formula in equation (11.12), calculate the number of "events" required using the survival analysis approach.

(b) Comment on the validity of your method in part **(a)** and the differences (or similarity) in the number of "events" required.

4. You've been asked by investigators in the department of internal medicine to design a repeated-measures study that compares two different medications for adults with hypertension. The plan is to have 30 subjects per group. The investigators wish to follow the subjects over five time points to measure Systolic Blood Pressure (SBP) and want to be able to detect a mean difference, Δ, of 7 mm Hg between the two groups. The between-patient variation is assumed to be $\sigma^2 = 81$, and the correlation between all time points is $\rho = 0.4$.

(a) Assuming that the investigators want to perform a two-sided test with $\alpha = 0.05$, calculate the power to detect the proposed difference for this design.

(b) The principle investigator (PI) of the study has decided that the cost for bringing a group of subjects into the clinic five times is too high. She indicates that only three time points are necessary to complete the goals of the study. Recalculate the power if all parameters given in part (a) are the same except that the number of time points is now three.

5. Suppose we have two different studies where, for power calculations, we assume that the 5-year differences in event-free survival in both studies will be 5%. The failure distributions in both studies are roughly exponential. Assume also that we wish to conduct two-sided tests with $\alpha = .05$ and we wish to obtain 80% power to detect such differences. Given this information, calculate the number of events that would be required if the percent event-free at five years are

(a) 75% versus 70%; and

(b) 50% versus 45%.

Compare the answers you got in (a) and (b) and comment on the differences in the number of events required. (Hint: Use Example 11.4 to help guide your calculations.)

6. Your supervisor has asked you to do a set of simulations regarding the robustness of a particular test statistic. One issue is how close the α-level is to the true α-level of 0.05. To do this, you'll simulate N datasets under the null hypothesis, record a 1 for every time the test rejects and a 0 otherwise. Your supervisor wants to know how big N should be to adequately perform the task.

(a) How large should N be in order to be 95% confident that you'll be within ∓ 0.01 of $\alpha = .05$? (Hint: Construct a 95% confidence interval for p, i.e., $\hat{p} \mp Z_{1-\alpha/2} \sqrt{\frac{\hat{p}\hat{q}}{N}}$ where your "$\hat{p}$" is 0.05 and $\hat{q} = 1 - \hat{p}$ [see equation (1.10)]. Your "$Z_{1-\alpha/2}$" will constructed according to the stipulation of being "95% confident". Plug these numbers into the confidence interval equation and solve for N.)

(b) Answer the question in part (a) if you wish to be 99% confident that you'll be within ∓ 0.01 of $\alpha = .05$?

CHAPTER 12

Appendix A: Using SAS

12.1 Introduction

Statistical packages typically consist of three basic components: (1) procedures to input and output data and results; (2) procedures that perform both summary and inferential statistical analyses; and (3) procedures that help the user to create high quality graphics. Most modern statistical packages also include an internal programming language that allows users to customize analyses and graphics, perform simulations and create reports to suit their particular needs.

The SAS System provides the user with tools to retrieve and store information, perform statistical analyses, produce an array of graphical displays of data summaries and results, and write reports. Because SAS has a complete set of programming statements, data may be read in almost any form and organized into a SAS dataset. In the same way, the data may be written in almost any form, providing the user with flexibility for writing reports. In between reading in data and writing out reports, SAS may be used to statistically analyze the data using procedures that range from simply descriptive to highly complex modeling.

In this chapter, I will only give examples of the SAS syntax-based commands. SAS also provides menus for many operations. However, the presentation of the syntax-based language allows the user to reproduce the results of the chapter. The examples given here provide a very brief introduction to how commands are built in the SAS language. For example, one requirement in the SAS language is that each command is followed by a semicolon (;) and that certain "key words" are necessary for the program to be successfully implemented. However, unlike the languages of many other statistical packages, the commands are *case insensitive*, that is, SAS will process commands given in lower case, upper case, or in a mixture of the two. Hence, in my SAS examples, I will mix upper case and lower case commands. The SAS language also requires that certain elements (such as a DATA step) be present in most programs.

Example 12.1. To start learning about SAS, it is useful to consider an example of a simple SAS program. Consider the following SAS session, which is used for constructing a *t*-test table look-up.

263

```
options pageno=1 ls=72; *<-- Sets first page at 1, linesize is 72 characters;
data a;
input t df;
probt=(1-probt(abs(t),df))*2;
cards;
  2.5 31
  3.1 44
  1.98 68
; run;

data b; set a;
label t='Value of t statistic'
      df='Degrees of freedom' probt='2-sided p-value';
run;
title 'SAS Example #1';
title2 'Printing out two-sided p-values for different t-statistics with d.f.';
proc print label; run;
```

An extract from the output file obtained by executing the commands listed above is given below.

```
----------------------- PARTIAL OUTPUT ---------------------------------
                       Example of using SAS                          1
    Printing out two-sided p-values for different t-statistics with d.f.
                                   14:03 Saturday, December 29, 2007

                            Degrees
                Value of t     of        2-sided
         Obs    statistic    freedom     p-value
          1       2.50         31        0.017920
          2       3.10         44        0.003370
          3       1.98         68        0.051752
```

Another example of a simple program is given in Example 12.2 below. This example introduces the user to the notion of the structure of a SAS programming session complete with "SAS," "LOG," and "LST" files.

Example 12.2. Consider the SAS program called RAT.SAS listed below. The purpose of the program is to compare two groups of rats with respect to heart rate using a *t*-test.

```
options linesize=72 nonumber nodate;*<- pages not numbered or dated;
data a;
 input group $ rat hr;
 datalines;
 control 1 268
 control 2 282
 control 3 276
 control 4 290
 treated 1 142
 treated 2 153
 treated 3 151
 treated 4 160
 ;
proc ttest;
 class group;
 var hr;
title 'Example of the use of PROC TTEST';
run;
```

One can run this program by hitting the "F3" key or by clicking on the icon in SAS, which looks like an individual who is running. When any SAS program is run, a LOG file is created that can be viewed in the LOG window within the SAS session. This file gives information about possible errors made within the program or processing information about the run if the SAS programming statements are indeed, correct. Processing time is typically summarized in the LOG using the terms "real time" and "cpu time." The term "real time" refers to the actual amount of time it takes to perform a given action. The processing time by your computer's CPU is given by "cpu time." Sometimes, on machines with multiple processors, the cpu time is greater than the real time. This can happen when a machine has multiple processors and the SAS procedure being run accesses parallel processes so multiple CPUs may run at the same time. A very nice discussion by Russell of how SAS calculates CPU time can be found online at http://www.nesug.org/proceedings/nesug07/cc/cc31.pdf.

The LOG file can be stored externally. By default, the external file will be an ASCII-type file which, by default, can be saved as a file with the same name as the SAS program file but with the extension, ".LOG." The LOG file is readable by most editors (e.g., Notepad) as well as in SAS.

We first examine the contents of the resulting LOG file (called RAT.LOG).

```
NOTE: Copyright (c) 2002-2008 by SAS Institute Inc., Cary, NC, USA.
NOTE: SAS (r) Proprietary Software 9.2 (TS2M2)
      Licensed to UNIVERSITY OF PITTSBURGH-T&R, Site 70082163.
NOTE: This session is executing on the X64_VSPRO  platform.


NOTE: SAS initialization used:
      real time            0.88 seconds
      cpu time             0.76 seconds

1     options nonumber nodate;*<- pages not numbered or dated;
2     data c;
3       input group $ rat hr;
4       datalines;

NOTE: The data set WORK.C has 8 observations and 3 variables.
NOTE: DATA statement used (Total process time):
      real time            0.03 seconds
      cpu time             0.04 seconds

13    ;
14    proc ttest data=c;
15      class group;
16      var hr;
17      title 'Example of the use of PROC TTEST'; run;

NOTE: PROCEDURE TTEST used (Total process time):
      real time            0.02 seconds
      cpu time             0.03 seconds
```

The above information indicates that the program ran properly and that 8 observations and 3 variables were read into the dataset WORK.C. "WORK" is a temporary storage area on your computer that SAS designates to cache data while processing the program. If no action is taken, this will go away after one exits SAS. An extensive description of log files can be found online at http://support.sas.com/documentation/. If, indeed, errors were made in the program then one can examine the LOG file to debug the program. We will return to the issue of programming errors in SAS later in this chapter.

If there are no errors in the program, then a listing of the results is created. Results from each procedure can be accessed in the SAS session itself (in the OUTPUT window) or can be stored externally. By default, this file will be an ASCII-type file saved as a file with the same name as the SAS program file but with the extension, ".LST." Like the LOG file, the LST file is readable by most editors (e.g., Notepad) as well as in SAS. The output can also be saved in a rich text format (with a ".RTF" extension), which is readable by programs like Word or WordPerfect.

Listed below is the output of the program called RAT.SAS, which is stored in a file called RAT.LST.

```
                    Example of the use of PROC TTEST

                         The TTEST Procedure

                         Variable:  hr

   group        N       Mean     Std Dev     Std Err    Minimum    Maximum

   control      4      279.0      9.3095      4.6547     268.0      290.0
   treated      4      151.5      7.4162      3.7081     142.0      160.0
   Diff (1-2)          127.5      8.4163      5.9512

   group       Method                 Mean        95% CL Mean      Std Dev

   control                            279.0     264.2    293.8      9.3095
   treated                            151.5     139.7    163.3      7.4162
   Diff (1-2)   Pooled                127.5     112.9    142.1      8.4163
   Diff (1-2)   Satterthwaite         127.5     112.8    142.2

            group        Method              95% CL Std Dev

            control                        5.2737   34.7109
            treated                        4.2012   27.6516
            Diff (1-2)   Pooled            5.4234   18.5331
            Diff (1-2)   Satterthwaite

     Method              Variances       DF     t Value     Pr > |t|

     Pooled              Equal            6      21.42       <.0001
     Satterthwaite       Unequal       5.7145    21.42       <.0001

                         Equality of Variances

     Method          Num DF     Den DF     F Value    Pr > F

     Folded F           3          3        1.58      0.7178
```

12.2 Data Input, Manipulation, and Display

Three key pieces of information needed when reading data into SAS (or any other program) are:

1. Where can SAS find my data, for example, is the data entered *internally* in my SAS program or somewhere *external* to the SAS program?;
2. What is the structure of my data?; and
3. How do I want my data processed, for example, do I want all of my data or only part of it processed by SAS?

12.2.1 The DATA Step

The DATA step is used to retrieve, edit, and mathematically manipulate data. In each DATA step, an *internal* set of dataset is created, which SAS can use. For data with a simple structure, it is easiest to view the DATA step as creating a table where each row is an observation and each column a variable. The DATA step is a very powerful tool for reading data in almost any form, editing the data into almost any form, and writing the data in almost any form. Additionally, the data may then be passed onto SAS procedures that provide a large variety of statistical analyses.

The DATA statement tells SAS that you're reading, writing, or manipulating data. In earlier versions of SAS (Version 6 or before), each SAS dataset (variable) had to have a name that was no longer than 8 characters. Later versions have no such requirement. Examples of dataset (or variable) names are MARY, Titanic, A_OK, NUMBER_1, and antidisestablishmentarianism.

The CARDS, DATALINES, FILE and INFILE Statements

There are two ways to read our data initially into SAS. The first way is to enter data *internally* into SAS via either the cards; or datalines; statement as in Examples 12.1 and 12.2. The second is to read the data into SAS from an *external* file via either an FILE or an INFILE statement as will be demonstrated in Examples 12.3 and 12.4. The first way is the simpler of the two because the data that the program "sees" is in the program itself. The second way requires that SAS be given information about the external environment. This information tells SAS *where* the data is located. This is equivalent to asking "what folder the does the data file reside in?"

The INPUT Statement

After SAS knows where the data is, it must be told what the structure of the data is. This is accomplished via the INPUT statement. The INPUT statement allows us to give meaningful names to the data we are reading with SAS. It also tells SAS whether or not each data item is a <u>numeric</u> value (such as $1.7, -0.067$, or 1000) or a *nonnumeric* value (such as M, Y, rat, Mouse1, or MISFIT). A *numeric* value is one that can be manipulated arithmetically whereas a *nonnumeric* or *character* value is not manipulated arithmetically in the usual sense.

NOTE: A very helpful programming device is to add comments to remind yourself of what you're doing in the program. The comments are not processed by SAS but provide invaluable information to someone who is trying to understand your program. SAS provides two ways for you to put comments in your program:

1. Enclose the comment using / * and * /. The "block" commenting in SAS is *one of the few instances in SAS where a statement is <u>not</u> followed by a semicolon.*

2. Put a * in front of the commented text followed by a semicolon (;) that closes the statement.

Example 12.3. The following program implements reading an internal file into SAS.

```
/* This program demonstrates a simple example of HOW to read data into SAS.
   NOTE THAT ALL SAS STATEMENTS ARE FOLLOWED BY A SEMICOLON (;) except for
   this type of a comment statement.  */
DATA BEAST1;
 INPUT ANID SEX $ GROUP CREAT TOT_BILI;
 CARDS;
   0001 M 01 0.65 164
   0003 M 01 0.54 191
   0002 M 01 0.50 153
   0004 M 01 0.61 170
   0011 M 02 0.66 147
   0012 M 02 0.59 170
   0013 M 02 0.51 151
   0014 M 02 0.58 164
   0021 M 03 0.63 162
   0022 M 03 0.64 138
   0023 M 03 0.56 159
   0024 M 03 0.52 166
   0031 M 04 0.55 184
   0032 M 04 0.56 149
   0033 M 04 0.52 153
   0091 F 04 0.63 175
   0034 M 04 0.49 168
   0061 F 01 0.61 140
   0062 F 01 0.58 173
   0063 F 01 0.60 182
   0064 F 01 0.51 186
   0072 F 02  .   155
   0073 F 02 0.53 148
   0074 F 02 0.59 171
   0075 F 02 0.55 177
   0081 F 03 0.58 189
   0082 F 03 0.57 164
   0083 F 03 0.52 169
   0084 F 03 0.58 149
   0092 F 04 0.60 176
   0093 F 04 0.51 158
   0094 F 04 0.56 160
 ;
TITLE 'TYPICAL BIOCHEMISTRY VALUES IN A TOXICOLOGY STUDY OF RATS';
* COMMENT 1: NOTE when inputting data using the CARDS statement, a semicolon is placed
  after ALL of the data is entered (semicolon not needed if "datalines;" is used);
* COMMENT 2: Notice that the data are sorted by SEX and by GROUP within SEX except
  for animal id numbers 0003 and  0091.  Also notice the sequence of the DATA,
  INPUT and CARDS statements;
* COMMENT 3: Note also that this type of comment statement MUST begin with an '*'
  and end with a semicolon;
```

We now look at an example where the data of interest is located in a file external to the SAS program. The data may be read in either fixed or free format. Fixed format specifies the columns for each value. Free format simply reads the data in order expecting that the values are delimited by at least one blank (or, say, tabs or commas). In the previous example, a free format was used.

Example 12.4. Suppose a file called RAT2.txt is located in a folder called c:\example. The data in the file are listed below.

```
0001M0165 24 164
0003M0154 17 191
0002M0150 19 153
0004M0161 22 170
0011M0266 24 147
0012M0259 21 170
0013M0251 19 151
0014M0258 24 164
0021M0363 22 162
0022M0364 22 138
0023M0356 20 159
0024M0352 19 166
0031M0455 24 184
0032M0456 21 149
0033M0452 24 153
0091F0463 24 175
0034M0449 25 168
0061F0161 27 140
0062F0158 24 173
0063F0160 25 182
0064F0151 22 186
0072F02    25 155
0073F0253 20 148
0074F0259 21 171
0075F0255 17 177
0081F0358 26 189
0082F0357 22 164
0083F0352 19 169
0084F0358 22 149
0092F0460 18 176
0093F0451 21 158
0094F0456 20 160
```

We can read in the data above with the following code:

```
* The code below demonstrates HOW to read an external data;
* file into SAS using a FIXED format in the INPUT statement. ;
FILENAME IN 'c:\example\RAT2.txt';
DATA BEAST2;
    INFILE IN; *<-- Could also use 'c:\example\RAT2.txt' instead of IN;
    INPUT ANID 1-4 SEX $5 GROUP 7-8 CREAT 8-9 .2 BUN 10-12 TOT_BILI 13-16; run;
* COMMENT: The last statement is an example of a FIXED format.  Once again, notice
    the sequence of the DATA, INFILE and INPUT statements;
```

One can then create a permanent file external to SAS, which can be read into SAS without the use of an INPUT statement. This is known as a SAS *system* file and will be directly readable only by SAS. The files typically have names designated as the dataset name with an extension of ".sas7bdat" if the SAS version is 7 or above or ".sd2" for version 6.

```
* This program demonstrates  HOW to create a permanent SAS system file which is
* stored in an external file designated by the libname and DATA statements;
libname out 'c:\example';
DATA out.BEAST2; *<-create a SAS system file: 'c:\example\BEAST2.sas7bdat';
   set beast2;
run;
```

Manipulating the Data

In most cases, it is important to preprocess or transform the data in some way before we actually display it or do a descriptive or statistical analysis. What do we mean by "preprocessing?" This means manipulation of the data in such a way that the program can use it for the task at hand. An example of preprocessing is to perform some operation on the raw data. For instance, if we were given hydrogen ion concentration, $[H^+]$, but were really interested in printing out the pH, we would then use a $-log$ transformation of the original data. Another example of preprocessing is subsetting the data. For instance, in a human diet study, we may only be interested in doing descriptive statistics for females weighing over 300 lbs. Sometimes preprocessing means sorting, merging, or concatenating datasets. When new study data is to be added to old study data, the new data may be merged or added to existing data.

Use of the SET Statement

In order to perform operations such as arithmetic operations, subsetting datasets, concatenating and/or any other operations, you must first use the SET statement. *None of the above operations can be performed <u>without</u> the SET statement.* The SET statement moves a copy of a designated "old" dataset into a "new" dataset. The "new" dataset will contain the any manipulations of the "old" dataset that the user designates.

Example 12.5. The following is an example of the use of the SET statement to create a subset of a bigger SAS dataset. In this case, we may wish to create a dataset of only the female rats.

```
* This set of code demonstrates a simple example of HOW to use the SET
* statement to make a subset of female rats only. ;
DATA FEMBEAST;
   SET BEAST2;
   IF SEX EQ 'F'; run;
```

Assignment Statements

The DATA step provides for creation of new variables. This is accomplished with assignment statements. In assignment statements, a value is assigned to a variable with an equal sign. The assigned value may be a number, a character or string of characters, another variable, or a function of another variable. Several assignment statements are listed below:

```
x=3;
group='Treated';
y=x;
ph=-log(hion);
y=a+b*x;
```

Subsetting Data

Subsetting a dataset is very simple. One way to do it is to use an IF to condition inclusion into the set on the value of a given variable or variables. For instance, the statement "IF GROUP=1" includes only those observations for which the value of the variable GROUP is 1. A compound IF statement may be constructed. The statement "IF GROUP=1 AND CREAT GT .5" includes only those observations for which the variable GROUP is 1 *and* the variable CREAT is greater than .5. This method was demonstrated in Example 12.5 above.

Another way to subset when one is doing an analysis is to use a WHERE statement. This is very useful because one does not have to save a dataset but rather can perform a SAS procedure on only the part of the data designated in the WHERE statement.

Example 12.6. The following is an example of the use of the where statement. This allows one to do, say, an F-test on the female rats "on the fly" instead of using an extra data step.

```
proc glm data=BEAST2;
where sex eq 'F';
class group;
model creat=group;
title 'Analysis of variance of creatinine values done only on female rats'; run;
```

In SAS, for the convenience of the users, there are several different ways to designate equalities or inequalities in order to create or refer to a subset of a larger dataset. For example, "GT" is the same as >. The list below shows some symbol equivalencies:

```
EQ =

NE ¬= or in

GT >

GE >=

LT <

LE <=
```

Sorting Data

Sorting datasets is necessary whenever another procedure needs the data in a specific order. For example, sorting is done when analysis is to be done BY some grouping variable. Also, it is often desirable to print data in a sorted format to make verification of the data values easier. A procedure in SAS called PROC SORT (separate from the DATA step) is used to sort datasets. An example of its use is shown in Example 12.7.

Example 12.7. Data Manipulation using PROC SORT

```
* This is a continuation of the program in Example 2.1 and demonstrates the ;
* use of PROC SORT command;
PROC SORT DATA=BEAST1;
   BY SEX GROUP; run; *<-- This sorts the data by SEX and GROUP within SEX;
```

Besides sorting, one can also concatenate one DATA set to the end of another using the SET command. Another function is that two DATA sets can be merged using the MERGE statement. Neither of these procedures will be covered here but is explained at http://support.sas.com/documentation/cdl/en/basess/58133/HTML/default/viewer. htm#a001318477.htm.

Data Types for Statistical Analyses

For the purpose of performing statistical analyses, it is important to distinguish class variables from ordinary measurement variables. "Class" variables in SAS are those which are used to classify or categorize the rest of the data into a discrete number of groups. For example, the value of F in the variable SEX would classify the values of the data as being measured on females. The value of 01 in the variable GROUP may classify data as being from the Control Group in a clinical trial. Hence, the variables SEX and GROUP are called *class* variables and they have to be designated as such using a CLASS statement. This distinction between class and nonclass variables was very important in Chapters 4, 5, 7, and 8 for implementing analysis of variance, regression, and mixed models.

12.2.2 Displaying Data

Displaying data can mean many things. The simplest display of data is a listing of the data values themselves. Unless the dataset of interest is very large, this practice is almost always a good idea. Listing the data provides a check for data entry, and when displayed in a clean format, is an excellent hard copy archive.

Another way of displaying data is graphically. A graphical display of data can provide answers at a glance. It is also a good way to validate statistical testing. When a statistical test cannot be interpreted with respect to graphical display, something has gone wrong.

Listing Data: PROC PRINT and the PUT Statement

There are two primary methods of listing data in SAS. The simplest method of displaying data is to use PROC PRINT. This method is used if there is no reason to print the data output in a very specific manner. This procedure has a variety of options that allow considerable flexibility. Note that PROC PRINT is *not* part of the DATA step but a separate procedure.

Example 12.8. The following program demonstrates a simple use of proc print.

```
PROC PRINT DATA=BEAST1 noobs; *<-- "noobs" suppresses line numbering;
BY SEX GROUP;
TITLE2 'PROC PRINT Default Output Format';
```

If the user wants to list data in a specific format, then the PUT statement is used within the DATA step. The PUT statement allows the user to define which file the output is sent to and in what format it is output to that file. Example 12.9 shows use of the PUT statement in the DATA step. The output file has been defined by the FILENAME and the FILE statements. The FILENAME statement assigns an internal SAS name to an external file. The FILE statement opens the file and the PUT statement then writes to it.

Example 12.9. The following example demonstrates the use of the PUT statement. A "null" dataset is created and then the information is written to an external file called "beast.out."

```
FILENAME BEASTOUT 'BEAST.OUT';
DATA _NULL_;
    SET BEAST1;
    FILE BEASTOUT;
    PUT ANID SEX GROUP CREAT TOT_BILI; run;
* NOTE: The _NULL_ name is given to create a data set only as a temporary file.
* After the data step finishes, it is deleted. This type of statement is often
* used when one is ONLY interested in creating an external file;
```

12.3 Some Graphical Procdures: PROC PLOT and PROC CHART

Graphical display of data can be done using several procedures in SAS. We will review two of these, PROC PLOT and PROC CHART. The PLOT procedure is used to produce scatterplots of the data, when one variable is plotted against another. It may also be used to produce contour plots for three-dimensional data.

Example 12.10. The following code demonstrates how the GPLOT procedure may be used to create a simple scatterplot. The output is displayed in Figure 12.1.

```
PROC GPLOT DATA=BEAST1;
TITLE2 'A Scatterplot of Creatinine versus Total Bilirubin';
    PLOT CREAT*TOT_BILI;
PROC PLOT DATA=BEAST1;
    PLOT CREAT*TOT_BILI;
    BY SEX;
TITLE2 'A Scatterplot of Creatinine versus Total Bilirubin by SEX';
* NOTE: The SORT procedure must be used before plots may be done BY SEX;
run;
```

The above code using the GPLOT procedure produces a high quality graph that can be rendered in many different format types (e.g., BMP, EMF, WFF, PS, PDF, or TIF, among others). One can also produce a lower quality graph that can be printed on an older dot matrix-type printer. This is done by shortening the "GPLOT" to "PLOT." An example of how this is done and the subsequent output is given below Figure 12.1. In that example, the plots are created for each sex.

Figure 12.1 *Example of a Simple Scatterplot using PROC GPLOT*

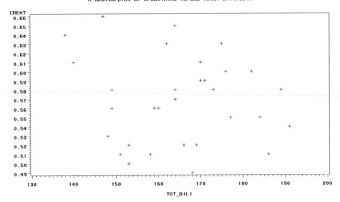

TYPICAL BIOCHEMISTRY VALUES IN A TOXICOLOGY STUDY OF RATS
A Scatterplot of Creatinine versus Total Bilirubin

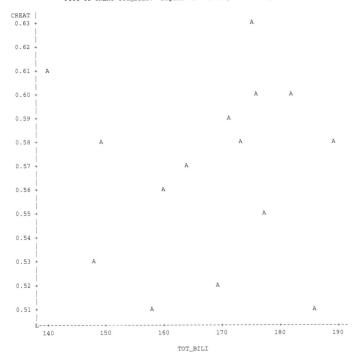

TYPICAL BIOCHEMISTRY VALUES IN A TOXICOLOGY STUDY OF RATS
A Scatterplot of Creatinine versus Total Bilirubin by SEX
----------------------------------- SEX=F --
Plot of CREAT*TOT_BILI.   Legend: A = 1 obs, B = 2 obs, etc.

NOTE: 1 obs had missing values.

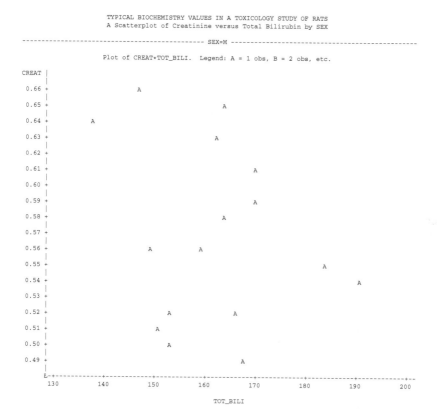

TYPICAL BIOCHEMISTRY VALUES IN A TOXICOLOGY STUDY OF RATS
A Scatterplot of Creatinine versus Total Bilirubin by SEX

The CHART (or GCHART) procedure is used to produce vertical or horizontal bar charts. This type of data display is useful in a variety of situations.

Example 12.11. In addition to the scatterplots given of the variables, CREAT versus TOT_BILI in Example 12.10, we can use PROC GCHART to provide a visual description (bar chart) of the distribution of TOT_BILI. A graphical display of the barchart produced by GCHART is given in Figure 12.2.

```
PROC GCHART DATA=BEAST1;
    VBAR TOT_BILI;
TITLE2 'Total Bilirubin Distribution'; run;
PROC CHART DATA=BEAST1;
    VBAR TOT_BILI;
    BY SEX;
TITLE2 'Total Bilirubin Distribution by Sex'; run;
```

Figure 12.2 *Example of a Bar Plot in SAS*

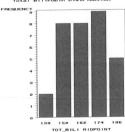

TYPICAL BIOCHEMISTRY VALUES IN A TOXICOLOGY STUDY OF RATS

12.4 Some Simple Data Analysis Procedures

12.4.1 Descriptive Statistics: PROC MEANS and PROC UNIVARIATE

Descriptive statistics allow us to summarize the characteristics of a <u>group</u> of animals measured under similar conditions. If we can assume that the animals in a study are a fair representation of any animal of the species of interest, then the descriptive statistics will give us a rough idea of how a particular compound affects certain biological characteristics.

The most commonly used descriptive statistics are (1) the number of observations, (2) the mean, (3) the median, (4) the minimum, (5) the maximum, (6) the standard deviation, and (7) the standard error of the mean (sometimes abbreviated to the standard error).

One way to obtain descriptive statistics using SAS is to use a procedure called PROC MEANS. Another useful procedure for obtaining descriptive statistics is PROC UNIVARIATE. This procedure provides a lot of information that helps to characterize the distribution of the data.

Example 12.12. Examples of the use of PROC MEANS and PROC UNIVARIATE are given in the code below. (The output is not shown.) The use of upper- and lower-case values in the SAS commands below is arbitrary. The SAS language is not case sensitive *unless* one is inputting or outputting a "literal" string of characters such as in a title statement. Thus, as was mentioned earlier, the commands can be either lower case, upper case, or a mixture of both. (This is *not true* in *S*-plus or in *R* where most commands are lower case.)

```
* This program demonstrates simple examples of PROC MEANS & PROC UNIVARIATE;
PROC MEANS data=beast2;
   CLASS SEX GROUP;
   VAR CREAT TOT_BILI;
TITLE2 'Summary statistics using PROC MEANS'; run;

PROC univariate data=beast2;
   class sex group;
   var creat tot_bili;
title2 'Summary statistics using proc univariate'; run;
```

12.4.2 *More Descriptive Statistics: PROC FREQ and PROC TABULATE*

Frequency tables are used to show the distribution of a variable that takes on a relatively small number of values (as compared to a continuous variable like blood pressure). Suppose that in a human study, patients are randomly allocated into 4 treatment groups. After the study, an indicator of disease severity (designated DIS_SEV) is assessed. DIS_SEV takes on the values 0 = no disease, 1 = mild disease, 2 = moderate disease, 3 = marked disease, and 4 = fatal. A frequency table shows us the distribution of our sample population. This is accomplished in SAS via the procedure called PROC FREQ.

Example 12.13. The code below inputs disease severity information and uses PROC FREQ to get counts of disease severity by group. (The output is not shown.)

```
DATA SUFFER;
INPUT PATIENT $ GROUP DIS_SEV @@; *<- reads new records from same input line;
datalines;
   0001 01 0    0002 01 0    0003 01 0    0004 01 0    0011 02 1
   0012 02 2    0013 02 1    0014 02 1    0021 03 2    0022 03 2
   0023 03 3    0024 03 3    0031 04 3    0032 04 4    0033 04 4
   0091 04 4    0034 04 4    0061 01 0    0062 01 0    0063 01 0
   0064 01 0    0072 02 1    0073 02 2    0074 02 2    0075 02 2
   0081 03 3    0082 03 3    0083 03 2    0084 03 3    0092 04 1
   0093 04 2    0094 04 2
PROC SORT; *<-- no data name specified =>PROC SORT done on previous dataset;
BY GROUP; run;
PROC FREQ;
BY GROUP;
TABLES DIS_SEV;
TITLE 'FREQUENCY DISTRIBUTION OF DISEASE SEVERITY BY GROUP'; run;
```

The procedure FREQ can also be used to produce a χ^2 (chi-square) and/or a Fisher's exact test of homogeneity of proportions among two or more groups. Other tests can be performed or the output can be reduced by appropriate options in the TABLES statement. A lot more information about these options can be obtained by searching for "proc freq table" online.

Example 12.14. In the following example, the distribution of proportions of disease severities in the control group is compared to that in group 4 using a χ^2 test. Note that one must interpret the output appropriately for the particular dataset at hand.

```
PROC FREQ data=suffer;
where group = 1 or group = 4;
TABLES DIS_SEV*GROUP/chisq fisher nocolumn nopercent;
TITLE2 'COMPARISON OF CONTROL WITH GROUP 4'; run;
```

```
* ----------------------- Partial Output -----------------------;
        FREQUENCY DISTRIBUTION OF DISEASE SEVERITY BY GROUP
                  COMPARISON OF CONTROL WITH GROUP 4

                       The FREQ Procedure

                     Table of GROUP by DIS_SEV

    GROUP      DIS_SEV

    Frequency|
    Row Pct  |        0|       1|       2|       3|       4| Total
    ---------+--------+--------+--------+--------+--------+
           1 |      8 |     0  |     0  |     0  |     0  |     8
             | 100.00 |   0.00 |   0.00 |   0.00 |   0.00 |
    ---------+--------+--------+--------+--------+--------+
           4 |      0 |     1  |     2  |     1  |     4  |     8
             |   0.00 |  12.50 |  25.00 |  12.50 |  50.00 |
    ---------+--------+--------+--------+--------+--------+
    Total           8        1        2        1        4       16

              Statistics for Table of GROUP by DIS_SEV

        Statistic                      DF      Value      Prob
        -----------------------------------------------------
        Chi-Square                      4     16.0000    0.0030
        Likelihood Ratio Chi-Square     4     22.1807    0.0002
        Mantel-Haenszel Chi-Square      1     11.7391    0.0006
        Phi Coefficient                        1.0000
        Contingency Coefficient                0.7071
        Cramer's V                             1.0000

        WARNING: 100% of the cells have expected counts less
                 than 5. Chi-Square may not be a valid test.

                       Fisher's Exact Test
        -------------------------------------------
        Table Probability (P)      7.770E-05
        Pr <= P                    1.554E-04

                     Sample Size = 16
```

Example 12.15. Suppose one wishes to output the n, and the mean and median disease severities for the four groups in Example 12.14. This can be accomplished using proc tabulate as given below.

```
options ls=65 nonumber nodate;
PROC tabulate data=suffer;
var dis_sev;
class group;
table dis_sev*(n mean)*group;
title2;
run;
```

Listed below is the type of display that is outputted by proc tabulate.

```
*--------------------- Partial Output -----------------------;
        FREQUENCY DISTRIBUTION OF DISEASE SEVERITY BY GROUP

      +----------------------------------------------------
      |                     DIS_SEV                        |
      +---------------------------------------------------+
      |                       N                           |
      +---------------------------------------------------+
```

```
|                          GROUP                        |
+------------+------------+------------+----------------+
|     1      |     2      |     3      |     4          |
+------------+------------+------------+----------------+
|       8.00|        8.00|        8.00|          8.00  |
+------------+------------+------------+----------------+
```

(Continued)

```
+------------------------------------------------------+
|                       DIS_SEV                        |
+------------------------------------------------------+
|                        Mean                          |
+------------------------------------------------------+
|                       GROUP                          |
+------------+------------+------------+----------------+
|     1      |     2      |     3      |     4          |
+------------+------------+------------+----------------+
|       0.00|        1.50|        2.63|          3.00  |
+------------+------------+------------+----------------+
```

(Continued)

```
+------------------------------------------------------+
|                       DIS_SEV                        |
+------------------------------------------------------+
|                       Median                         |
+------------------------------------------------------+
|                       GROUP                          |
+------------+------------+------------+----------------+
|     1      |     2      |     3      |     4          |
+------------+------------+------------+----------------+
|       0.00|        1.50|        3.00|          3.50  |
+------------+------------+------------+----------------+
```

12.5 Diagnosing Errors in SAS programs

An irritating part of reading manuals for a programming language is that most of them show how everything works *under the best of circumstances*, that is, when the commands are all correct, but many don't elaborate about situations where programs are incorrect. This shortcoming is understandable as there are an *infinite number of ways to make mistakes*, whereas, in the case of programming, there are usually only a very few ways or sometimes even only one way to correctly implement an algorithm. As any working investigator knows, the life of a scientist is full of false starts and making annoying errors in the implementation of a larger project. When programming in SAS or any other language, one encounters situations where (s)he *thinks* that (s)he has produced a correct program but (1) no output is produced; (2) partial output is produced; or (3) the output produced doesn't appear to be correct.

Example 12.16. Consider the program given in Example 12.1. In that program, the data

```
proc ttest data=c
 class group;
 var hr;
title 'Example of the use of PROC TTEST'; run;
```

```
NOTE: Copyright (c) 2002-2008 by SAS Institute Inc., Cary, NC, USA.
NOTE: SAS (r) Proprietary Software 9.2 (TS2M2)
      Licensed to UNIVERSITY OF PITTSBURGH-T&R, Site 70082163.
NOTE: This session is executing on the X64_VSPRO  platform.


NOTE: SAS initialization used:
      real time             0.82 seconds
      cpu time              0.70 seconds

1     options nonumber nodate;*<- pages not numbered or dated;
2     data c;
3       input group $ rat hr;
4       datalines;

NOTE: The data set WORK.C has 8 observations and 3 variables.
NOTE: DATA statement used (Total process time):
      real time             0.01 seconds
      cpu time              0.01 seconds


13    ;
14    proc ttest data=c
15      class gruop;
        -----
        22
        202
ERROR 22-322: Syntax error, expecting one of the following: ;, (, ALPHA, BYVAR, CI, COCHRAN,
              DATA, DIST, H0, HO, NOBYVAR, ORDER, PLOTS, SIDES, TEST, TOST.
ERROR 202-322: The option or parameter is not recognized and will be ignored.
16      var hr;
17      title 'Example of the use of PROC TTEST'; run;

NOTE: The SAS System stopped processing this step because of errors.
NOTE: PROCEDURE TTEST used (Total process time):
      real time             0.01 seconds
      cpu time              0.01 seconds
```

Error messages in SAS are usually printed *after* the error in the code is introduced. In the above session, the `class` is underlined. Clearly, in the line before it, the semicolon (;) had been omitted. Thus, the error had to be in the statement beginning with `class` or at some line *before* that statement. In this case, the error was in the line below. This error could be fixed by either replacing the "`cards;`" statement with "`cards;`" (and *not* adding a semicolon after the data) <u>or</u> by retaining the "`cards;`" statement as is and adding a semicolon after the data. The next error as can be seen below is the misspelling of the variable `group`. This is diagnosed immediately by SAS and it is easily fixed. If there were more than one such misspelling of the variable, then a global substitution could be performed (as long as the misspelling was consistent). Be careful with such global substitutions, however, as they can lead to unintended consequences.

```
18    proc ttest data=c;
19      class gruop;
ERROR: Variable GRUOP not found.
20      var hr;
21      title 'Example of the use of PROC TTEST'; run;

NOTE: The SAS System stopped processing this step because of errors.
NOTE: PROCEDURE TTEST used (Total process time):
      real time             0.01 seconds
      cpu time              0.01 seconds
```

Lastly, the log given below is that of a program that ran properly. Of course, if one was not appropriately using a *t*-test in this case, then a "properly working" program would still not provide an appropriate analysis of the data.

```
22    proc ttest data=c;
23      class group;
24      var hr;
25      title 'Example of the use of PROC TTEST'; run;

NOTE: PROCEDURE TTEST used (Total process time):
      real time             0.04 seconds
      cpu time              0.04 seconds
```

CHAPTER 13

Appendix B: Using R

13.1 Introduction

R is a free statistical package that can be downloaded from the Internet. It is extremely useful for graphics and computing and is also useful for implementation of many types of statistical analyses.

13.2 Getting Started

To download R one can type "R" in any search engine on the web. One can then click on the entry "The **R** Project for Statistical Computing" at which point, the R website should appear. The root R webpage displays R graphics along with several general categories under which links are available: About R; Download, Packages; R Project; Documentation; and Misc. The links under sections About R, R Project, Documentation, and Misc all give useful background information about R in general, its origin, documentation, and how to contribute to its development. The download link is entitled "CRAN," which is located under the section called Download, Packages. Downloading R is usually straightforward.

Once you've downloaded R, you should be able to click onto the blue R icon to begin your R session. When you've entered the R session, you will be what is called "R console." It is at this console that you can type interactive commands. From the console, you can perform a lot of functions such as sourcing in R code from an external file, installing packages from various R, loading packages, creating and loading in "scripts."

13.3 Input/Output

13.3.1 Inputting Data and Commands from the R Console

One way to input data or commands in R is to do so directly in a session from the R console. Another way to do it is by entering data and commands into an R "script" file and then running it using the `source` command. One other way is to first capture the text of interest in the script file with your mouse or by hitting either `ctrl + c` simultaneously or `ctrl + a` simultaneously. One then runs the program by hitting `ctrl + r` simultaneously. Another way to run the program is to right click on the mouse and choose the `Run line or selection` option.

A short *R* session demonstrating some direct input is given below. Comments in *R* code are made by putting a "#" in front of the statements that serve as comments. This is similar to the "`* ;`" used for commenting in SAS.

Example 13.1. Listed below are several examples of data input into *R*.

```
> sqrt(10)
[1] 3.162278
> 10^(1/2)
[1] 3.162278
> exp(1)
[1] 2.718282
> pi
[1] 3.141593
> sin(pi)
[1] 1.224606e-16
> c(433*13,44/10,16-11,27+2)
[1] 5629.0    4.4    5.0   29.0
> 1:10 #<-- display integers 1 through 10
 [1]  1  2  3  4  5  6  7  8  9 10
> 10:1 #<-- display integers 1 through 10 BACKWARDS
 [1] 10  9  8  7  6  5  4  3  2  1
> c(rep(1:4,2),rep(1:4,each=2)) #<-- "rep" means repeat
 [1] 1 2 3 4 1 2 3 4 1 1 2 2 3 3 4 4
> c(1:6/2,1:6*2) #<- combining 2 vectors of each of length 6
 [1]  0.5  1.0  1.5  2.0  2.5  3.0  2.0  4.0  6.0  8.0 10.0 12.0
 > Z<-rnorm(7,0,1) #<- 7 standard normal deviates [could shorten to rnorm(7)]
> Z
[1] -0.1609757  0.2960181 -0.6890313 -0.6756928 -1.2690794  1.8517325 -1.2882867
> x<-seq(-10,100,10) #<- sequence of 12 numbers from -10 to 100 by 10
> x
 [1] -10   0  10  20  30  40  50  60  70  80  90 100
> length(x)
[1] 12
> AA=matrix(c(1,2,3,4),nrow=2) #<- One can also use "=" instead of "<-"
> AA
     [,1] [,2]
[1,]    1    3
[2,]    2    4
> A=matrix(c(1,2,3,4),nrow=2,byrow=T) #<- Compare A to AA
> A
     [,1] [,2]
[1,]    1    2
[2,]    3    4
> A %*% AA   #<- Matrix multiplication
     [,1] [,2]
[1,]    5   11
[2,]   11   25
> solve(A) #<-- "Solve" here is the same as matrix inversion
      [,1] [,2]
[1,] -2.0  1.0
[2,]  1.5 -0.5
```

13.3.2 Inputting Data from an External File

Analysts are often given datasets from various sources that they must read into R before they can start their analyses. R has a number of useful functions that allow the user (analyst) to do this. The most popular R routines that allow one to read in from an external data source are the `read.table`, `scan`, `read.csv`, `read.delim` and `read.fwf` statements. A broad array of these statements are discussed in Kleinman and Horton [80] and Jones, Maillardet, and Robinson [73].

Example 13.2. To demonstrate how one might read data from an ASCII file external to R, we will reconsider the data presented in Examples 12.3 and 12.4. Let us suppose in example

```
beast<-read.table(file="C:\\book\\programs\\RAT.DAT",
    col.names=c("anid","sex","group","creat","tot.bili"))
beast2<-read.fwf(file="C:\\book\\programs\\RAT2.txt",widths=c(4,1,2,2,3,4),
    col.names=c("anid","sex","group","creat","bun","tot.bili"))
beast2$creat<-beast2$creat/100 #<-- Variable reference preceded by "beast2$"
beast2

fembeast<-subset(beast2,sex=='F')
fembeast

attach(beast2) #<- can now drop the "beast2$" when referring to variables
class(beast2); class(anid); class(sex); class(group); class(creat);
```

13.3.3 Transferring from and to SAS Datasets

Because of the wide variety of procedures unique to each statistical package, it is very important to be able to transfer system files from one package to another. Probably the easiest way to transfer a SAS system into R is to use `proc export` to export the data into a comma deliminated file (which usually will have a `.csv`) extension and then use `read.csv` statement to read the file into R.

Example 13.3. Consider the SAS system file in Example 12.4 called "BEAST2." The SAS and R code given below will enable one to successfully read that data into R.

SAS code

```
libname in 'C:\example' ;
data bst2; set in.BEAST2; run;

proc export data=bst2 outfile='C:\example\BEAST2.csv' dbms=csv; run;
```

Then, upon entering R one would proceed as follows:

R code

```
beast2<-read.csv(file='C:\\example\\BEAST2.csv')
names(beast2)
[1] "ANID"      "SEX"      "GROUP"     "CREAT"     "BUN"      "TOT_BILI"
```

Since the *R* language is *case sensitive*, and since most of the commands are lower case, one might wish to use the `casefold` statement to convert the variable names to lower case.

```
names(beast2)<-casefold(names(beast2),upper=F)
names(beast2)
[1] "anid"      "sex"      "group"     "creat"     "bun"      "tot_bili"
attach(beast2)
```

13.3.4 Data Types in R

When manipulating data or doing analyses in *R*, one must be aware of the what type of data is being processed. Certain commands do not make sense if executed on improper data types. For example, when one reads data in from an external file with one of the `read` statements, the default data type is called a "data frame." In Example 13.3, `beast2` is a data frame. (Check this by typing `class(beast2)` at the *R* console.) A data frame can consist of many different kinds of variables (much like a dataset in SAS). Examples of other data types in *R* are `vector`, `list`, `matrix`, `logical`, and `factor`.

13.4 Some Simple Data Analysis Procedures

13.4.1 Descriptive Statistics: Use of the apply Functions

The family of `apply` functions in *R* are very useful and efficient for obtaining a lot of summary information on any type array of data (for example, lists, matrices and data frames). Depending on the data type that one wants to summarize, one can use any of the following functions: `apply`, `mapply` (for matrices), `tapply` (summaries for variables "by" other variables), or `lapply` (for lists).

Example 13.4. Some examples of different versions of the `apply` function are given in the interactive session below.

```
> A<-matrix(c(rnorm(5),rnorm(5,1,2),rnorm(5,-1,2)),
+ nr=3,byrow=T)
> A
             [,1]        [,2]       [,3]         [,4]        [,5]
[1,]  0.56140479  2.015333 -1.187776 -0.08938559 0.28055744
[2,] -0.01923977  2.345794  2.695181  1.28453603 0.64998415
[3,]  1.68135574 -1.250893 -3.091976  2.27286926 0.06270757
> dim(A)
[1] 3 5
> apply(A,1,mean) #<-Displays row means (row is 1st dimension)
[1]  0.31602674  1.39125117 -0.06518734
> apply(A,2,mean) #<-Displays col means (col is 2nd dimension)
[1]  0.7411736  1.0367446 -0.5281902  1.1560066  0.3310830

> names(ToothGrowth) #<-- Dataset in R: Guinea pig tooth growth
[1] "len"  "supp" "dose" #<- tooth length; delivery type; dose of Vit C
> attach(ToothGrowth)
> tapply(len,supp,mean)
```

```
        OJ        VC
20.66333  16.96333
> tapply(len,list(supp,dose),mean)
       0.5      1       2
OJ 13.23 22.70 26.06
VC  7.98 16.77 26.14
```

To get more information about this useful family of functions type `?apply` at the R console.

13.4.2 Descriptive Statistics: Use of the `summary` Function

A very useful and ubiquitous function in R is the `summary` function. The `summary` function can be applied directly to a variable or set of variables, (e.g., to a vector or a matrix) or it can be applied to an *object* that is the output of a particular R function. The output associated with the `summary` function is quite different depending on the nature of the variable or object to which it is applied. For individual variables, the `summary` function gives quantiles and other summary statistics associated with the variables. However, if the `summary` function is applied to the output of an R statistical function, then the result will be summaries tailored to the particular statistical procedure being employed.

Example 13.5. Consider the R given below. In this example, 11 values of two predictor variables are created, `x1` and `x2`, which are binary and continuous, respectively. The model simulated is given by $Y_i = 1 + x_{1i} + 0.5x_{2i} + \epsilon_i$ where $\epsilon_i \sim N(0, .5), 1 = 1, \ldots, 11$. The first use of the `summary` function is applied to the vector, $\mathbf{Y}$ and the usual summary statistics are given. We then fit the model with the `lm(Y~x1+x2)` statement, which produces minimal output, that is, the estimated coefficients of the model. The `summary` function is then applied to `lm(Y~x1+x2)` and the output given is that of the usual analysis of variance (ANOVA) table associated with a regression model.

```
> x1<-c(rep(0,6),rep(1,5))
> x2<-0:10
> y=1+x1+.5*x2
> Y<-y+rnorm(11,0,.5)
> cbind(x1,x2,y,Y)
      x1 x2   y         Y
 [1,]  0  0 1.0 1.040199
 [2,]  0  1 1.5 1.545003
 [3,]  0  2 2.0 3.119482
 [4,]  0  3 2.5 2.928603
 [5,]  0  4 3.0 3.215785
 [6,]  0  5 3.5 3.909756
 [7,]  1  6 5.0 4.866241
 [8,]  1  7 5.5 6.319827
 [9,]  1  8 6.0 6.081850
[10,]  1  9 6.5 6.212906
[11,]  1 10 7.0 7.462860
> summary(Y)
   Min. 1st Qu.  Median    Mean 3rd Qu.    Max.
  1.040   3.024   3.910   4.246   6.147   7.463
> lm(Y~x1+x2)

Call:
lm(formula = Y ~ x1 + x2)
```

```
Coefficients:
(Intercept)             x1                x2
    1.2927          0.6281            0.5335

> summary(lm(Y~x1+x2))

Call:
lm(formula = Y ~ x1 + x2)

Residuals:
    Min       1Q  Median        3Q      Max
-0.5093  -0.2540 -0.1069   0.1213   0.7598

Coefficients:
            Estimate Std. Error t value Pr(>|t|)
(Intercept)  1.29275    0.27973   4.621 0.001707 **
x1           0.62808    0.53975   1.164 0.278098
x2           0.53349    0.08499   6.277 0.000239 ***
---
Signif. codes:   0 *** 0.001 ** 0.01 * 0.05 . 0.1   1

Residual standard error: 0.4457 on 8 degrees of freedom
Multiple R-squared: 0.9639,     Adjusted R-squared: 0.9549
F-statistic: 106.8 on 2 and 8 DF,   p-value: 1.698e-06
```

13.5 Using *R* for Plots

One of the useful features of *R* (or *S*-plus) is the ease of which it allows one to visualize data summaries and mathematical functions. This allows a working scientist, engineer, clinician, statistician, or mathematician to quickly visualize information for planning a study or to view results of a data analysis.

13.5.1 Single Plot

Two primary *R* functions can be used in *R* to plot curves: the `plot` function and the `curve` function. The `plot` function provides the most general way of plotting data in *R*. However, if one wishes to quickly examine the plot of one variable, say x, as a function of another variable, say $y = f(x)$ then the *curve* function is the fastest way to do so. This *R* function, by default, plots 101 equally spaced values of y as a function of x in the interval $[0, 1]$. Of course, one can easily modify the code to plot functions on different intervals and to label axes appropriately.

Example 13.6. Suppose one wishes to graph the function, $f(x) = \sin(2\pi x)$. This could be accomplished by typing

```
curve(sin(2*pi*x))# Create a curve in [0,1] with no title & default axis labels
```

(Note that I've added a comment preceded by a # which is not necessary to implement the command.) However, one might want to choose different intervals, label the axes, and create a title for the plot. This could be accomplished as follows:

```
curve(sin(2*pi*x),-2,2,lwd=2,ylab=expression("f(x)=2"*pi*"x"),
      xlab="x",main="Plot of a sine function")
```

The output of the second command is given in Figure 13.1. The plot could have also been produced with `plot` command. One would first have to create an x variable and a y (or $f(x)$) variable.

```
x<- -200:200/100
f.x<-sin(2*pi*x)
plot(x,f.x,type="l",lty=1,lwd=2,ylab=expression("f(x)=2"*pi*"x"),
     xlab="x",main="Plot of a sine function")
```

Figure 13.1 *Example of a Single Plot: Graphing Sin$(2\pi x)$ Versus x*

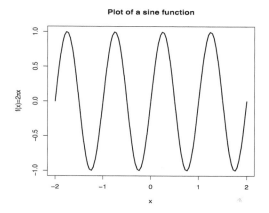

Plot of a sine function

13.5.2 Overlaying Plots

Many times, an investigator wishes to overlay plots of data on the same graph. This is especially useful when one wishes to examine the goodness of fit of regression models. This type of plot is easily implemented in *R* and *S*-plus.

Example 13.7. Suppose one wishes to compare two slightly different cubic functions, $f_1(x) = x^3 + x^2 - 2*x - 1$ and $x^3 + x^2 - x - 1$. One way to do this visually would be to display the plots of the two functions overlayed in a single graph. The code and Figure 13.2 demonstrate how this would be achieved in *R*.

```
leg<-c(expression(f(x)==x^3+x^2-2*x-2),expression(f(x)==x^3+x^2-x-2) )
curve(x^3+x^2-2*x-1,-2,2,lwd=2,ylab='f(x)',
      main="Two cubic polynomial functions overlayed")
curve(x^3+x^2-x-1,add=T,lwd=2,lty=2,col="blue")
abline(h=0,lty=4)  #<-- Creates a horizontal line, that is, f(x)=0
legend(0,6,leg,lty=1:2,lwd=c(2,2),col=c("black","blue"),cex=.8)
```

In the `plot` command, one can overlay lines by the use of either the `lines` or `points` subcommands.

Figure 13.2 *Example of How to Overlay Plots*

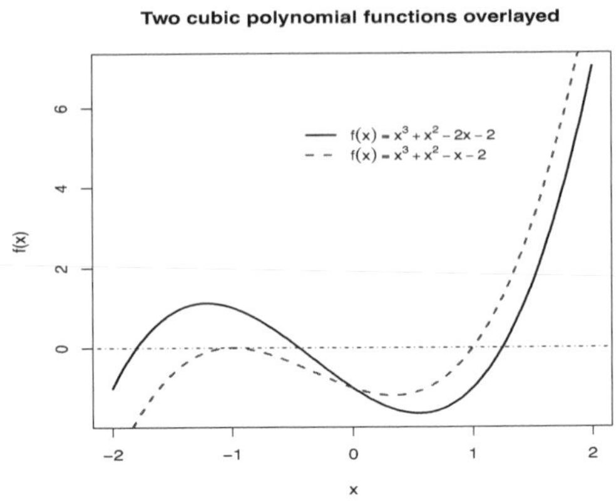

Two cubic polynomial functions overlayed

13.5.3 Multiple Plots per Page

Another powerful plotting feature in the R and S–plus packages is the ease at which one can produce many types of multiple plots displayed on a single page. Two of the most popular ways to do this is by the use of the `mfrow` option of the `par` command and the use of the `layout` command. The following example illustrates each of these features.

Example 13.8. The R (or S-plus) code given below produces two histograms on one page, as shown in Figure 13.3. The histograms are stacked on each other, which is designated by the `par(mfrow=c(2,1))` statement. If one were to plot these histograms side by side, then the statement would be `par(mfrow=c(1,2))`. This type of display could be produced for any type of plotting procedure, e.g., `plot`,

```
x<-rnorm(100)
y<-rexp(100)
par(mfrow=c(2,1))
hist(x,main="100 standard normal random deviates")
hist(y,main="100 standard exponential random deviates")
par(mfrow=c(1,1))#<-- Return to default (1 plot per frame)
```

Another interesting example of how multiple plots can be formed is via the use of the `layout` command. The code given below creates a simulated dataset that creates and fits a quadratic regression model and outputs fitted values and residuals. Three plots (raw data, fitted data overlayed by raw, and a simple residual plot) are created with the code below and displayed in Figure 13.4.

```
# The code below creates a wide plot at top & two below it
layout(matrix(c(1,1,2,3), 2, 2, byrow = TRUE))
x<-0:50/5
x2<-x^2
f.x<-1-x+x^2
y<-f.x+rnorm(length(x),0,5)
```

Figure 13.3 *Example of Two Histograms Stacked in One Frame*

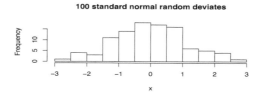

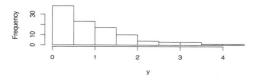

Figure 13.4 *Example of the* `layout` *Command for the Display of Different Features of a Fitted Model*

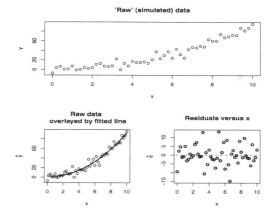

```
   fit.y<-lm(y~x+x2)
   y.hat<-fit.y$fitted.values
   e.hat<-fit.y$residuals
plot(x,y,main="'Raw' (simulated) data")
plot(x,y,main="Raw data \n overlayed by fitted line",ylab=expression(hat(y)))
   lines(x,y.hat,lwd=2)
plot(x,e.hat,lwd=2,main="Residuals versus x",ylab=expression(hat(e)))
   abline(h=0,lty=4)
def.par <- par(no.readonly = TRUE) # default
par(def.par)#- reset to default
```

To view other very powerful examples of the `layout` command, type ?layout at the *R* console.

13.6 Comparing an R-Session to a SAS Session

It is often useful to compare statistical packages by running the same analysis in both packages. The example below reproduced analyses presented in the first two examples of Chapter 12.

Example 13.9. Consider the two analyses presented in Examples 12.1 and 12.2. The R code given below parallels those two analyses.

```
> t.stat<-c(2.5,3.1,1.98)
> d.f<-c(31,44,68)
> two.sided.p<-2*(1-pt(t.stat,d.f))
> cbind(t.stat,d.f,two.sided.p)
     t.stat d.f two.sided.p
[1,]   2.50  31 0.017920227
[2,]   3.10  44 0.003370053
[3,]   1.98  68 0.051752202

> gp<-rep(c("control","treated"),each=4)
> rat<-rep(1:4,2)
> hr<-c(268,282,276,290,142,153,151,160)
> RAT<-data.frame(cbind(gp,rat,hr))
> RAT
        gp rat  hr
1 control   1 268
2 control   2 282
3 control   3 276
4 control   4 290
5 treated   1 142
6 treated   2 153
7 treated   3 151
8 treated   4 160
> var.test(hr~gp)

        F test to compare two variances

data:  hr by gp
F = 1.5758, num df = 3, denom df = 3, p-value = 0.7178
alternative hypothesis: true ratio of variances is not equal to 1
95 percent confidence interval:
  0.1020622 24.3284086
sample estimates:
ratio of variances
           1.575758

> t.test(hr~gp,var.equal=T)

        Two Sample t-test

data:  hr by gp
t = 21.4243, df = 6, p-value = 6.746e-07
alternative hypothesis: true difference in means is not equal to 0
95 percent confidence interval:
 112.9380 142.0620
sample estimates:
mean in group control mean in group treated
                279.0                 151.5
```

Notice that, in R, the test of variances is done prior to testing for the mean values. In SAS PROC TTEST, the test of variance is given automatically but the user must choose whether or not to use the usual two-sample t-test or the Satterthwaite procedure. In R, once the variance test indicated that the variances are not close to being significantly different, one must choose "var.equal=T" to implement the usual two-sample *t*-test. Without this designation, *R* uses the Welch test of equality of means. This method is very similar, in most cases, to the method suggested by Satterthwaite.

13.7 Diagnosing Problems in *R* Programs

Because *R* requires the user to input *case sensitive* commands, it is very easy to make programming errors. Also, in *R*, it is easy to apply certain functions to data types for which the particular functions are not applicable. Again, there are infinite ways of incorrectly coding a program but only a finite number of ways of correctly coding it.

Unfortunately, if the error messages are cryptic in SAS, then they are often even more so in *R*. That being said, as a rule of thumb, when one is debugging errors in *any* program (*R*, *S*-plus, SAS, Java, or otherwise), then one should go to the point where the error is generated and *scan backwards* for the offending statement(s). More subtle errors, which may allow the program to run properly, are only detected by careful analysis of the resulting output from the program.

Example 13.10. Given below are several statements that cause errors in *R*.

```
> x<-rexp(100)
> x_summary(x)  #<- This statment would work in S-plus but not in R
Error: could not find function "x_summary"
> x<-Summary(x) #<- Upper case "S" is used which is not correct in this context
Error in Summary(x) :
    function "Summary" is a group generic; do not call it directly
> y<-c("A","B","C")
> y
[1] "A" "B" "C"
> summary(y) #<- NO error message produced (only gives info about the vector)
    Length    Class    Mode
         3 character character
> class(y)
[1] "character"

> A<-matrix(c(1,2,3),nr=2) #<- Missing one element in a 2 x 2 matrix
Warning message:
In matrix(c(1, 2, 3), nr = 2) :
    data length [3] is not a sub-multiple or multiple of the number of rows [2]
```

13.8 Exercises

1. Construct the unit circle using the curve function in *R*. (Hint: Note that the equation for the unit circle is $x^2 + y^2 = 1$. Solve for the y in terms of x and note that the solution has two roots. You'll have to use the curve function twice with an add=T the second time. Also note that the range of both the x and y axes goes from -1 to 1.)

2. Redo the simulation given in Example 13.5. However, in addition to the *second* implementation of the `summary` function (to the `lm(Y~x1+x2)` function), also apply the `anova` to that function. Comment on similarities and/or differences in the outputs of the two functions.

CHAPTER 14

References

[1] Aalen, O.O. (1978), Nonparametric inference for a family of counting processes. *Ann. Statis.*, 6, 701–726.

[2] Akaike, H. (1974), A new look at the statistical model identification. *IEEE Transactions on Automatic Control*, 19(6), 716-723.

[3] Allison, P.D. (2001), *Logistic regression using the SAS System: Theory and Application*, SAS Institute, Inc. ISBN 978-1-58025-352-9 and Wiley ISBN 0-471-22175-9.

[4] Altman, D.G. (1991), *Practical Statistics for Medical Research*, Chapman and Hall, London. ISBN 0-412-38620-8.

[5] Andersen, P. K., Borgan, O., Gill, P. K., and Keiding, N. (1993), *Statistical Models based on Counting Processes*, Springer, New York.

[6] Anderson, T.W. (2003), *An Introduction to Multivariate Statistical Analysis*, 3rd ed., John Wiley & Sons, Inc., New York.

[7] Anscombe, F. J. (1973), Graphs in Statistical Analysis. *Am. Statis.*, 27(1), 17–21.

[8] Armitage, P. and Berry, G. (1988), *Statistical Methods in Medical Research*, Second Edition, Blackwell Scientific Publications. ISBN 0-632-015012-2.

[9] Atkinson, K.E. (1989), *An Introduction to Numerical Analysis*, 2nd edition, John Wiley & Sons. ISBN 0-471-62489-6.

[10] Bemjamini, Y. and Hochberg, Y. (1995), Controlling the false discovery rate: A practical and powerful approach to multiple testing. *J. R. Statist. Soc. B*, 57(1), 289–300.

[11] Box, G.E.P. (1950), Problems in the analysis of growth and wear curves. *Biometrics* 6, 362–389.

[12] Box, G.E.P., and Jenkins, G.M. (1976), *Time Series Analysis: Forecasting and Control*, Revised Edition, Holden-Day, San Francisco. ISBN 0-8162-1104-3.

[13] Breiman, L., Friedman, J.H., Olshen, R.A. and Stone, C.J. (1984), *Classification and Regression Trees*, Wadsworth, Belmont, CA. ISBN 0-5349-8054-6.

[14] Brown, H. and Prescott, R. (2006), *Applied Mixed Modles in Medicine*, 2nd ed.,

John Wiley, Ltd., West Sussex, England. ISBN 0-470-02356-2.

[15] Chambers, J.M. and Hastie, T.J., editors. (1992), *Statistical Models in S*. Wadsworth & Brooks, Pacific Grove, CA. ISBN 0-534-16765-9.

[16] Cleveland, W. S. (1979), Robust locally weighted regression and smoothing scatterplots. *J. Am. Statis. Assoc.*, 74, 829-836.

[17] Cleveland, W. S. (1981), LOWESS: A program for smoothing scatterplots by robust locally weighted regression. *Am. Statis.*, 35, 54.

[18] Cochran, W.G. (1954), Some methods for strengthening the common χ^2 tests, *Biometrics*, 10(4), 417–451.

[19] Cohen, J. (1960), A coefficient of agreement for nominal scales. *Educ. Psychol. Meas.*, 20, 37.

[20] Conover, W.J. (1974), Some reasons for not using the Yates continuity correction on 2×2 contingency tables, *J. Am. Statis. Assoc.*, 69(346), 374–376.

[21] Cook, R.D. Weisberg S. (1999), *Applied Regression Including Computing and Graphics*, John Wiley and Sons, Inc., New York. ISBN 0471-31711-X.

[22] Cox, D.R. (1972), Regression models and life tables (with discussion). *J. Royal Statis. Soc. B*, 34, 187–202.

[23] Craven, P. and Wahba, G. (1979), Smoothing noisy data with spline functions. *Numer. Math.*, 31, 377–403.

[24] Crawley, M.J. (2007), *The R book*. Wiley, New York. ISBN 978-04705-10247.

[25] Cutler, S.J. and Ederer, F. (1958), Maximum utilization of the life table method in analysing survival. *J. Chronic Dis.*, 8, 699–712.

[26] Dalgaard, P. (2002), *Introductory Statistics with R*. Springer, New York. ISBN 0-387-95475-9.

[27] Delwiche, L.D. and Slaughter, S.L. (2008), *The Little SAS Book: A Primer*, Fourth Edition. SAS Institute, Inc., Cary, NC. ISBN 978-1-59994-725-9.

[28] Deng, L.-Y. and Lin, D.K.J. (2000), Random number generation for the new century, *Am. Statis.*, 54, 145–150.

[29] Der, G. and Everitt, B.S. (2002), *A Handbook of Statistical Analyses using SAS*, Second Edition. Chapman and Hall / CRC, Boca Raton, FL. ISBN 1-58488-245-X.

[30] Draper, N.R. and Smith, H. (1981), *Applied Regression Analysis*, Second edition, John Wiley & Sons. ISBN 0-471-02995-5.

[31] Duncan, D B. (1955), Multiple range and multiple F tests. *Biometrics*, 11, 1-42.

[32] Dunnett, C.W. (1955), A multiple comparisons procedure for comparing several treatments with a control. *J. Am. Statis. Assoc.*, 50, 1096–1121.

[33] Dunnett, C.W. (1964), New tables for multiple comparisons with a control. *Biometrics*, 20, 482–491.

[34] Dunnett, C. W. (1980), Pairwise multiple comparisons in the homogeneous variance, unequal sample size case. *J. Am. Statis. Assoc.*, 75, 789-795.

[35] Efron, B. (1982), The jackknife, the bootstrap, and other resampling plans, *Soc. Indust. Appl. Math. CBMS-NSF Monographs*, 38.

[36] Efron, B., and Tibshirani, R.J. (1993), *An introduction to the bootstrap*, Chapman & Hall, New York.

[37] Einot, I. and Gabriel, K.R. (1975), A Study of the powers of several methods of multiple comparisons, *J. Am. Statis. Assoc.*, 70, 351.

[38] Everitt, B.S. (2002), *Statistical Analyses using S-plus*, Second Edition. Chapman and Hall / CRC, Boca Raton, FL. ISBN 1-58488-280-8.

[39] Everitt, B.S. and Hothorn, T. (2010), *Statistical Analyses using S-plus*, Second Edition. Chapman and Hall / CRC, Boca Raton, FL. ISBN 978-1-4200-7933-3.

[40] Feller, W.G. (1968), *An Introduction to Probability Theory and Its Applications, Volume I*, 3rd Edition. John Wiley & Sons, New York.

[41] Fieller, E.C. (1932), The distribution of the index in a bivariate Normal distribution. *Biometrika*, 24, 428-440.

[42] Fine, J. P. and Gray, R. J. (1999), A proportional hazards model for the subdistribution of a competing risk. *J. Am. Statis. Assoc.*, 94, 496-509.

[43] Finney, D.J. (1978), *Statistical Method in Biological Assay*, 3rd Ed. Griffin, London. ISBN 0-02-844640-2.

[44] Fisher, B., Costantino, J., Redmond, C., et al. (1989), A randomized clinical trial evaluating tamoxifen in the treatment of patients with node negative breast cancer who have estrogen-receptor-positive tumors. *New Engl. J. Med.*, 320, 479484.

[45] Fisher, B., Anderson, S., Tan-Chiu, E., Wolmark, N., et al. (2001), Tamoxifen and chemotherapy for axillary node negative, estrogen receptor-negative breast cancer: findings from the National Surgical Breast and Bowel Project B-23. *J. Clin. Oncol.*, 93(4), 931–942.

[46] Fisher L.D. and van Belle, G. (1993), *Biostatistics: A Methodology for the Health Sciences*, John Wiley and Sons, Inc., New York. ISBN 0-471-16609-X.

[47] Fisher, R.A. (1955), Statistical methods and scientific induction. *J. Royal Statis. Soc. B*, 17, 69–78.

[48] Fishman G.S. and Moore L.R. (1986), An exhaustive analysis of multiplicative congruential random number generators with modulus $2^{31} - 1$. *SIAM J. Sci. Statis. Comp.*, 7, 24–45.

[49] Fleiss, J. (1986), *The Design and Analysis of Clinical Experiments* Wiley & Sons, New York, 125-129. ISBN 0-471-82047-4.

[50] Freedman, L.S. (1982), Tables of the number of patients required in clinical trials using the logrank test. *Stat.Med.*, 1, 121–129.

[51] Gabriel, K.R. (1978), A simple method of multiple comparisons of means. *J. Am. Statis. Assoc.*, 73, 364.

[52] Gaynor, J.J., Feuer, E.J., Tan, C.C., Wu, D.H. et al. (1993), On the use of cause-specific failure and conditional failure probabilities: Examples from clinical oncology data. *J. Am. Statis. Assoc.* 88, 400–409.

[53] Gehan, E.A. (1965), A generalized Wilcoxon test for comparing arbitrarily singly censored samples. *Biometrika*, 52, 203–223.

[54] George, S.L. and Desu, M.M. (1974), Planning the size and duration of a clinical trial studying the time to some critical event. *J. Chronic Dis.*, 27, 15–24.

[55] Gordis, L. (1996), *Epidemiology*, W.B. Saunders Co., Philadelphia. ISBN 0-7216-5137-2.

[56] Gordon, A.D. (1999), *Classification*, 2nd ed., Chapman & Hall/CRC, Boca Raton. ISBN 1-58488-013-9

[57] Grambsch P.M. and Therneau T.M., (1994), Proportional hazards tests and diagnostics based on weighted residuals, *Biometrika*, 81, 515–526.

[58] Gray, R.J. (1988), A class of k-sample tests for comparing the cumulative incidence of a competing risk, *Ann. Statis.*, 16, 1141–1154.

[59] Gray R. (2002), *BIO 248 cd Advanced Statistical Computing Course Notes* http://biowww.dfci.harvard.edu/ gray/248-02.

[60] Grizzle, J.E. (1967), Continuity Correction in the χ^2-Test for 2×2 tables, *Am. Statis.*, 21(4), 28–33.

[61] Hand, D.J., Daly, F., Lunn, A.D., McConway, K.J. and Ostrowski, E. (editors), (1994), *Small Data Sets*, Chapman and Hall, London. ISBN 0-412-39920-2.

[62] Harrell, F. (2001), *Regression Modeling Strategies with Applications to Linear Models, Logistic Regression, and Survival Analysis*, Springer, New York. ISBN 0-387-95232-2.

[63] Harrington, D.P. and Fleming, T.R. (1982), A class of rank test procedures for censored survival data. *Biometrika*, 69, 133–143.

[64] Harvey, A.C. (1993), *Time Series Models*, 2nd Edition, MIT Press, Boston. ISBN-10: 0-262-08224-1.

[65] Harville, D. A. (1976), Extension of the Gauss-Markov theorem to include the estimation of random effects, *Ann. Statis.* 4, 384–395.

[66] Harville, D. A. (2000), *Matrix Algebra from a Statistician's Perspective*, Springer, New York. ISBN 0-387-94978-X.

[67] Hassard, T.H. (1991), *Understanding Biostatistics*, Mosby year book, St. Louis. ISBN 0-8016-2078-3.

[68] Hedeker, D. and Gibbons, R.D. *Longitudinal Data Analysis*, Wiley & Sons, 2006. ISBN 0-4714-2027-1.

[69] Hollander, R.H. and Wolfe, D.A. (1991), *Introduction to the Theory of Nonparametric Statistics*, Reprint Edition, Krieger Publishing Company, Malabar, FL (original edition 1979, John Wiley & Sons). ISBN 0-89464-543-9.

[70] Hosmer, D.W. and Lemeshow, S. (1989), *Applied Logistic Regression*, Wiley,

New York.

[71] Hotelling, H. (1931), The generalization of Student's ratio. *Ann. Math. Statis.*, 2(3), 360-378.

[72] Hsu, H. (1997), *Probability, Random Variables, & Random Processes*, Schaum's Outlines Series, McGraw-Hill, New York. ISBN 0-07-030644-3.

[73] Jones, O., Maillardet, R. and Robinson, A., (2009), *Scientific Programming and Simulation Using R*, CRC Press, Boca Raton, FL. ISBN 978-1-4200-6872-6.

[74] Kalman, R.E. (1960), A new approach to linear filtering and prediction problems. *Trans. of the ASME (J. Basic Eng.)* 82D, 35–45.

[75] Kaplan, E.L. and Meier, P. (1958), Nonparametric estimation from incomplete observations. *J. Am. Statis. Assoc.* 53, 457–481.

[76] Khattree, R. and Naik, D.N., (1999), *Applied Multivariate Statistics with SAS Software*, 2nd ed., SAS Institute, Cary NC.

[77] Kennedy, W.J. and Gentle, J.E. (1980), *Statistical Computing*, Marcel Dekker. ISBN 0-8247-6898-1.

[78] Keuls M (1952), The use of the "studentized range" in connection with an analysis of variance. *Euphytica*, 1, 112-122.

[79] Kleinbaum D.G., Kupper, L.L., Morgenstern, H. (1982), *Epidemiological Studies*, Wadsworth, Inc, Belmont, CA. ISBN 0-534-97950-5.

[80] Kleinman, K. and Horton, N.J. (2010), *SAS and R: Data Management, Statistical Analysis, and Graphics*, ISBN 978-1-4200-7057-6.

[81] Kohn, R. and Ansley, C.F. (1987), A new algorithm for spline smoothing based on smoothing a stochastic process, *SIAM J. Stat. Comput.*, 8, 33–48.

[82] Kurban, S., Mehmetoglu, I., and Yilmaz, G. (2007), Effect of diet oils on lipid levels of the brain of rats, *Indian J. Clin. Biochem.*, 22(2), 44–47.

[83] Laird, N. and Ware, J.H. (1982), Random-effects models for longitudinal data, *Biometrics*, 38, 963–974.

[84] Landis, J.R. and Koch, G.C. (1977), The measurement of observer agreement for categorical data, *Biometrics*, 33, 159.

[85] Lange, K. (1999), *Numerical Analysis for Statisticians*, Springer-Verlag. ISBN 0-387-94979-8.

[86] Lea, A.J. (1965), New observations of neoplasms of the female breast in certain European countries, *Brit. Med. J.*, 1, 488–490.

[87] Little, R.J.A. and Rubin, D.B. (2002), *Statistical Analysis with Missing Data*, 2nd ed., John Wiley & Sons, Hoboken, NJ. ISBN 0-471-18386-5.

[88] Lloyd-Jones D.M., Larson M.G., Beiser A., Levy D. (1999), Lifetime risk of developing coronary heart disease. *Lancet*, 353, 89–92.

[89] Maindonald J. and Braun J. (2003), *Data Analysis and Graphics Using R*. Cambridge Series in Statistical and Probabilistic Mathematics, Cambridge University Press. ISBN 0-521-81336-0.

[90] Manley, B.F.J. (2005), *Multivariate Statistical Methods: A Primer*, 3rd ed., Chapman & Hall/CRC, Boca Raton. ISBN 1-58488-414-2.

[91] Mann, H.B. and Whitney, D.R. (1947), On a test of whether one of two random variables is stochastically larger than the other, *Ann. Math. Statis.*, 18, 50-60.

[92] Mantel, N. (1974), Some reasons for not using the Yates continuity correction on 2×2 contingency tables: Comment and a suggestion, *J. Am. Statis. Assoc.*, 69(346), 378–380.

[93] Mantel N. and Greenhouse S.W. (1968), What is the continuity correction? *Am. Statis.*, 22(5), 27–30.

[94] McCullagh, P. and Nelder, J.A. (1989), *Generalized Linear Models*, Second Edition. Chapman & Hall/CRC, London. ISBN 0-412-317-60-5.

[95] Mardia, K.V., Kent, J.J. and Bibby, J.M. (1979), *Multivariate Analysis*, Academic Press, Inc., London. ISBN 0-12-471252-5.

[96] Marsaglia, G. and Bray T.A. (1964), A convenient method for generating normal variables, *SIAM Rev.*, 6(3), 260–264.

[97] Massey, F.J., Jr. (1952), Distribution table for the deviation between two sample cumulatives, *Ann. Math. Statist.*, 23(3), 435–441.

[98] Miettinen, O.S. (1974), Some reasons for not using the Yates continuity correction on 2×2 contingency tables: comment, *J. Am. Statis. Assoc.*, 69(346), 380–382.

[99] Miller, R.G. (1981), *Survival Analysis*, John Wiley and Sons, New York.

[100] Milton, J.S. (1992), *Statistical Methods in the Biological and Health Sciences*, Second Edition. McGraw-Hill, Inc., New York.

[101] Molenberghs G., and Kenward, M.G. (2007), *Missing Data in Clinical Studies*, John Wiley, Ltd., West Sussex, England. ISBN-1-3-978-0-470-84981-1.

[102] Morant G.M. (1923), A first study of the Tibetan skull, *Biometrika*, 14, 193–260.

[103] Muenchen, R. (2009), *R for SAS and SPSS Users (Statistics and Computing)* Springer, New York. ISBN 978-0-387-09417-5.

[104] *National Vital Statistics Report*, (2005), 54(2), September 8, 2005, 19.

[105] Nelson, W. (1969), Hazard plotting for incomplete failure data. *J. Qual. Technol.*, 1, 27–52.

[106] Nelson, W. (1972), Theory and applications of hazard plotting for censored failure data. *Technometrics*, 14, 945–965.

[107] Neter J., Kutner M.H., Nachtsheim C.J., and Wasserman W. (1996), *Applied Linear Statistical Models*, 4th edition, McGraw-Hill, Boston.

[108] Newman D (1939), The distribution of range in samples from a normal population, expressed in terms of an independent estimate of standard deviation. *Biometrika*, 31(1), 20-30.

[109] Neyman, J. (1956), Note on an article by Sir Ronald Fisher. *J. Royal Statis.*

Soc. B, 18, 288-294.

[110] Pagano, M. and Gauvreau, K. (2000), *Principles of Biostatistics*, Second Edition. Duxbury Press, Pacific Grove, CA. ISBN 0-534-22902-6.

[111] Pearson, K. and Lee, A. (1903), On the laws of inheritance in man: I. Inheritance of physical characters. *Biometrika*, 2(4), 357-462.

[112] Peto, R. and Peto, J. (1972), Asymptotically efficient rank invariant test procedures. *J. Royal Statis. Soc. A*, 135, 185–198.

[113] Piantadosi, S. (1997), *Clinical trials: A Methodologic Perspective*. John Wiley & Sons, Inc., New York. ISBN 0-471-16393-7.

[114] Pinheiro, J.C. and Bates, D.M. (2000), *Mixed-Effects Models in S and S-PLUS*, Springer, New York.

[115] Potthoff, R.F. and Roy, S.N. (1964), A generalized multivariate analysis of variance model useful especially for growth curve problems. *Biometrika*, 51, 313–326.

[116] Press, W.H., Teukolsky, S.A., Vetterling, W.T., and Flannery, B.P. (1992), *Numerical Recipes in C: The Art of Scientific Computing*. Second Edition. Cambridge University Press.

[117] *R*. (2010). *The R foundation for Statistical Computing*.

[118] Rao, C.R. (1958), Some statistical methods for comparison of growth curves. *Biometrics*, 14(1), 1-17.

[119] Rao, C.R. (1965), The theory of least squares when the parameters are stochastic and its application to the analysis of growth curves. *Biometrika*, 52, 447-458.

[120] Rao, P.V. (1998), *Statistical Research Methods in the Life Sciences*, Duxbury Press, Pacific Grove, CA. ISBN 0-534-93141-3.

[121] Rohlf, F.J. and Sokal, R.R. (1995), *Statistical Tables*, 3rd edition, W. H. Freeman and Co., New York. ISBN 0-7167-2412-X.

[122] Rosner, B. (1995), *Fundamentals of Biostatistics*, Fourth edition. Duxbury Press, Pacific Grove, CA. ISBN 0-531-20840-8.

[123] Rosner, B. (2000), *Fundamentals of Biostatistics*, Fifth edition. Duxbury Press, Pacific Grove, CA. ISBN 0-534-37068-3.

[124] Rosner, B. (2006), *Fundamentals of Biostatistics*, Sixth edition. Duxbury Press, Pacific Grove, CA. ISBN 0-534-41820-1.

[125] Ross, S.M. (2006), *Simulation*, Fourth Edition, Academic Press. ISBN 0-12-598053-1.

[126] Rothchild, A.J., Schatzberg, A.F., Rosenbaum, A.H., Stahl, J.B. and Cole, J.O. (1982), The dexamethasone suppression test as a discriminator among subtypes of psychotic patients. *Brit. J. Psychiat.*, 141, 471–474.

[127] Rubinstein, L.V., Gail, M.H., and Santner, T.J. (1981), Planning the duration of a comparative clinical trial with loss to follow-up and a period of continued observation. *J. Chronic Dis.*, 34, 469–479.

[128] *SAS Institute, Inc.*, Cary, NC.

[129] Sarkar, Deepayan (2008), *Lattice: Multivariate Data Visualization with R*, Springer. ISBN: 978-0-387-75968-5

[130] Satterthwaite, F. (1946), An approximate distribution of estimates of variance components. *Biometrics Bull.*, 2, 110–114.

[131] Schoenberg, I.J. (1964), Spline functions and the problem of graduation. *Proceed. Nat. Acad. Sci.* 52, 947–950.

[132] Scheffé, H. (1953), A method for judging all contrasts in the analysis of variance, *Biometrika*, 40, 87-104.

[133] Scheffé, H. (1959), *The Analysis of Variance*, John Wiley & Sons, New York. ISBN 0-471-75834-5.

[134] Schwarz, G.E. (1978), Estimating the dimension of a model. *Ann. Statis.*, 6(2), 461-464.

[135] Searle, S. R. (1971), *Linear Models*, John Wiley and Sons, Inc., New York. ISBN 0-471-18499-3.

[136] Seber, G.A.F. and Lee, A.J. (2003), *Linear Regression Analysis*, 2nd edition, John Wiley and Sons, Inc., New York. ISBN 0-471-41540-5.

[137] Selvin, S. (1995), *Practical Biostatistical Methods*, Duxbury Press, Belmont, CA. ISBN 0-534-23802-5.

[138] Shork, A.S. and Remington, R.D. (2000), *Statistics with Applications to the Biological and Health Sciences*, Third Edition. Prentice Hall, New Jersey. ISBN 0-13-022327-1.

[139] Shumway, R.H. and Stouffer, D.S. (2006), *Time Series Analysis and Its applications*, 3rd edition, Springer, New York. ISBN 0-387-98950-1.

[140] Snedecor, G.W. and Cochran, W.G. (1980), *Statistical Methods*, Seventh Edition, Iowa State University Press, Ames, IA. ISBN 0-813-81561-4.

[141] Sokal, R.R. and Rohlf, F.J. (1995), *Biometry*, Third Edition. W.H. Freedman and Company, New York. ISBN 0-7167-2411-1.

[142] Spearman, C. (1904), The proof and measurement of association between two things, *Amer. J. Psychol.*, 15, 72-101.

[143] *S*-plus (2008), *TIBCO Software Inc.* Copyright (c) 1988-2008 TIBCO Software Inc. ALL RIGHTS RESERVED.

[144] Sternberg, D.E., Van Kammen, D.P., and Bunney, W.E. (1982), Schizophrenia: Dopamine b–bydroxylase activity and treatment response. *Science* 216, 1423–1425.

[145] Stoline, M.R. and Ury, H.K. (1979), Tables of the studentized maximum modulus distribution and an application to multiple comparisons among means. *Technometrics*, 21(1), 87–93.

[146] Student (William S. Gosset) (1908), The probable error of a mean, *Biometrika*, 6(1), 1–25.

[147] Tarone, R.E. and Ware, J.E. (1977), On distribution-free tests for equality for survival distributions, *Biometrika*, 64, 156–160.

[148] Tausworthe, R.C. (1965), Random numbers generated by linear recurrence modulo two, *Math. Comp.*, 19, 201–209.

[149] Therneau, T.M. and Atkinson, E.J. (1997), An introduction to recursive partitioning using the RPART routines, *Mayo Foundation Technical Report*.

[150] Therneau, T.M. and Grambsch, P.M. (2000), *Modeling Survival Data: Extending the Cox Model*. Springer, New York. ISBN 978-0-387-98784-2.

[151] Thompson, J.R. (2000), *Simulation: A Modeler's Approach*, John Wiley & Sons, New York. ISBN 0-471-25184-4.

[152] Tukey, J. W. (1953), The problem of multiple comparisons. *Unpublished manuscript. In The Collected Works of John W. Tukey VIII. Multiple Comparisons: 19481983*, Chapman and Hall, New York.

[153] Venables, W.N. and Ripley, B.D. (2002), *Modern Applied Statistics with S*, Fourth Edition. Springer, New York. ISBN 0-387-95457-0.

[154] Verzani, J. (2005), *Using R for Introductory Statistics*, Chapman and Hall / CRC, Boca Raton, FL. ISBN 1-58488-4509.

[155] Wahba, G. (1978), Improper priors, spline smoothing and the problem of guarding against model errors in regression. *J. Royal Statis. Soc. B*, 40, 364–372.

[156] Wichmann, B.A. and Hill, I.D. (1982), [Algorithm AS 183] An efficient and portable pseudo-random number generator. *Appl. Statis.*, 31, 188–190. (Correction, 1984, 33, 123.)

[157] Wilcoxon, F. (1945), Individual comparisons by ranking methods. *Biometrics*, 1, 80–83.

[158] Wiklin, R. (2010), *Statistical Programming with SAS/IML Software*, SAS Press, Cary, NC. ISBN 978-1-60764-663-1.

[159] Yates, F. (1934), Contingency tables involving small numbers and the χ^2 test, *Suppl. J. Royal Statis. Soc.*, 1(2), 217–235.

[160] Ye F., Piver, W., Ando M., and Portier, C.J. (2001), Effects of temperature and air pollutants on cardiovascular and respiratory diseases for males and females older than 65 years of age in Tokyo, July and August 1980–1995. *Environ. Health Persp.*, 109(4), 355–359.

[161] Zar, J.H. (1996), *Biostatistical Analysis*, 3rd edition, Prentice and Hall. ISBN 0-13-084542-6.

[162] Zelazo, P.R., Zelazo, N.A., and Kolb, S. (1972), "Walking" in the newborn, *Science*, 176, 314–315.

Index

Printed and bound by CPI Group (UK) Ltd, Croydon, CR0 4YY

26/10/2024

01779669-0002